The GEO Handbook on Biodiversity Observation Networks

Michele Walters · Robert J. Scholes
Editors

The GEO Handbook
on Biodiversity
Observation Networks

Editors
Michele Walters
Natural Resources and Environment
Council for Scientific and Industrial
    Research (CSIR)
Pretoria
South Africa

and

Centre for Wildlife Management
University of Pretoria
Pretoria
South Africa

Robert J. Scholes
Global Change and Sustainability Research
    Institute
University of the Witwatersrand
Johannesburg
South Africa

Additional material to this book can be downloaded from http://extras.springer.com.

ISBN 978-3-319-27286-3   ISBN 978-3-319-27288-7   (eBook)
DOI 10.1007/978-3-319-27288-7

Library of Congress Control Number: 2016951648

Printed on acid-free paper

This Springer imprint is published by Springer Nature
The registered company is Springer International Publishing AG
The registered company address is: Gewerbestrasse 11, 6330 Cham, Switzerland

What gets measured, gets managed.

—Peter Drucker

Foreword

The Group on Earth Observations (GEO) is a voluntary international partnership of 102 governments and 92 participating organisations which share a vision of a future in which decisions and actions for the benefit of humankind are informed by coordinated, comprehensive and sustained Earth observations. GEO achieves its mission largely through self-organising communities focused on important Earth observation domains where decision-making will benefit from data that is shared broadly and openly. These communities form connected systems and networks, creating a Global Earth Observation System of Systems (GEOSS). During its first ten-year implementation period, 2005–2015, GEO identified biodiversity as a key 'Societal Benefit Area', resulting in the formation of the GEO Biodiversity Observation Network, GEO BON.

As GEO moves into its second, ten-year implementation period, GEO BON is recognised as one of its strongest communities. It has helped to mobilise and coordinate the data and information needed for an effective response to the global threats faced by organisms, species and ecosystems. In collaboration with international treaty bodies such as the Convention on Biological Diversity (CBD) and the Ramsar Convention on Wetlands of International Importance, GEO BON has worked with national conservation agencies and non-governmental organisations at scales from regional to global. These efforts have revealed both the benefits of working together and the challenges of such a complex, but urgent task, not least of which is filling the remaining large gaps in data and information.

The practical experience which GEO BON has accumulated through its own actions, and through the efforts of its network partners, is a valuable resource to biodiversity information systems everywhere—from those just starting out in places where there has previously been little information, to large operations holding enormous amounts of data and wishing to know how better to use it. This handbook is a powerful resource that will provide valuable guidance to those committed to protecting, sustaining and preserving biodiversity across the planet.

I congratulate GEO BON on creating this powerful mechanism and wish the GEO BON community great success in each of its future endeavours.

Geneva, Switzerland   Barbara J. Ryan
Executive Director: Group on Earth Observations

Acknowledgements

This book is based on the collective knowledge of the GEO BON network of volunteers, working towards the establishment of a global, sustainable observation system for monitoring changes in biodiversity. As such, a great number of colleagues have contributed their time and energy towards the production of this handbook and huge thanks go to all of them.

Chapter 6 acknowledges information obtained from the World Register of Marine Species (WoRMS), the European Ocean Biogeographic Information System (EurOBIS) and Marine Regions, which are supported by data and infrastructure provided by VLIZ as part of the Flemish contribution to LifeWatch, and is funded by the Hercules Foundation. Also thanks to Chap. 6 team: Jan Mees, Francisco Hernandez and Pieter Provoost (IOC) for being very helpful in accessing data from OBIS.

The authors gratefully acknowledge the 30 reviewers who provided invaluable inputs and comments for individual chapters.

The editors and authors offer their immense gratitude to Helen Matthey for her tireless assistance with coordination of the review and revision process, and the many other ways in which she contributed to the finalisation of the manuscript.

The editors thank the South African Council for Scientific and Industrial Research (CSIR) for the in-kind support provided them during the production of this work.

Finally, we are thankful for the support of the German Centre for Integrative Biodiversity Research (iDiv) Halle-Jena-Leipzig funded by the German Research Foundation (FZT 118), which generously funded the open-access publication of this book.

Contents

Chapter 1
Working in Networks to Make Biodiversity Data More Available

**Robert J. Scholes, Michael J. Gill, Mark J. Costello,
Georgios Sarantakos and Michele Walters**

Abstract It became apparent a few decades ago that biodiversity is declining worldwide at nearly unprecedented rates. This poses ethical and self-interested challenges to people, and has triggered renewed efforts to understand the status and trends of what remains. Since biodiversity does not recognise human boundaries, this requires the sharing of information between countries, agencies within countries, non-governmental bodies, citizen groups and researchers. The effective monitoring of biodiversity and sharing of the data requires convergence on methods and definitions, best achieved within a relatively loose organisational structure, called a network. The Group on Earth Observations Biodiversity Observation Network (GEO BON) is one such structure. This chapter acts as an introduction to the GEO BON biodiversity observation handbook, which documents some of the

R.J. Scholes (✉)
Global Change and Sustainability Research Institute, University of the Witwatersrand,
Private Bag 3, Wits 2050, Johannesburg, South Africa
e-mail: bob.scholes@wits.ac.za

M.J. Gill
Polar Knowledge Canada, 360 Albert Street, Suite 1710, Ottawa, ON K1R 7X7, Canada
e-mail: mike.gill@polar.gc.ca

M.J. Costello
Institute of Marine Science, University of Auckland, Auckland, New Zealand
e-mail: m.costello@auckland.ac.nz

G. Sarantakos
Swiss Competence Center for Energy Research—FURIES, Swiss Federal Institute
of Technology in Lausanne, Lausanne, Switzerland
e-mail: georgios.sarantakos@epfl.ch

M. Walters
Natural Resources and Environment, Council for Scientific and Industrial Research,
PO Box 395, Pretoria 0001, South Africa
e-mail: mwalters@csir.co.za

M. Walters
Centre for Wildlife Management, University of Pretoria, Pretoria 0002, South Africa

© The Author(s) 2017
M. Walters and R.J. Scholes (eds.), *The GEO Handbook on Biodiversity
Observation Networks*, DOI 10.1007/978-3-319-27288-7_1

co-learning achieved in its first years of operation. It also addresses the basic
questions of how to set up a biodiversity observation network, usually consisting of
a number of pre-existing elements.

Keywords Network · Management · Biodiversity · Observations · Indicators ·
EBV · Organisation

1.1 Observing Biodiversity

People have observed biodiversity—the variety of life on Earth, in all its forms and
levels (Fig. 1.1; based on Noss 1990)—throughout history. Indeed, having a deep
understanding of biodiversity was an essential element for survival for most of the
human past. The description of new species and mapping of their distribution was
an important activity in post-enlightenment science (Costello et al. 2013a). Today
there are hundreds of millions of observations of biodiversity in museums, herbaria,
databases, field notebooks and learned publications (Wheeler et al. 2012). Despite
this abundance, the fraction of the information which is available and accessible

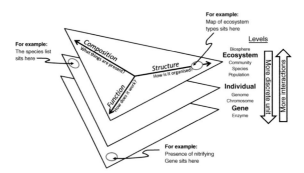

Fig. 1.1 The contemporary definition of biodiversity embraces three aspects of variation
(differences in composition, structure and function) and several levels of biological organisation
(from the enzyme, to the biosphere). There is not a 'right' level to observe biodiversity, nor a
'right' aspect to observe: ideally you should be capturing elements of all aspects and all levels, and
be able to move seamlessly between them. In practice, in any particular situation there will
inevitably be stronger emphases on some levels or aspects. Historically, many people considered
'biodiversity' to consist only of composition, at the species level. Be guided primarily by what the
users of the information need, secondly by what is observable using the available technology, and
only then by what happened to have been collected in the past. As you shift downward from the
ecosystem towards the organism and ultimately the gene, the entities with which you are dealing
become more focussed and precise, but the price you pay is a loss of information about interactions
between them and the emergent properties which arise from those interactions (*Source* based on
Noss 1990)

remains inadequate to address the emerging challenges to biodiversity, human development and planetary management (Costello et al. 2013d).

It is well known that biodiversity is in world-wide decline (Butchart et al. 2010 summarises recent evidence). This impoverishment takes the form of local and global extinctions, but also more pervasive and subtle simplification, hollowing-out and dominance by a few species of formerly complex, abundant and equitable ecosystems (e.g., see Pereira et al. 2012). The resources of the Earth—land, oceans, water, primary productivity and nutrients—are increasingly appropriated by humans and their client species (Haberl et al. 2007). The process of human domination has been underway for nearly ten thousand years, ever since the domestication of crops and livestock, but has accelerated over the past century or two. It has reached such proportions that we have entered the 'Anthropocene'—the era when human actions are the dominant Earth-shaping force (Crutzen 2002). There is little doubt that the current and projected rate of biodiversity loss exceeds its rate of generation. As a result, the world is getting poorer in terms of the biological variation it supports.

The loss of biodiversity has well-established immediate causes: the loss, degradation and fragmentation of habitat needed for the completion of life histories; over-harvesting of organisms which have commercial value (and the collateral damage to other organisms and ecosystems in the process); pollution of the environment by biocides and the waste products of human activity; and competition, predation or infection by invasive alien species deliberately or inadvertently introduced from other parts of the world are the leading causes (SCBD 2010). Climate change during the 21st century is projected to be high up on this list of the causes for biodiversity loss.

The contemporary decline in diversity is not entirely without precedent. On at least five previous occasions in the approximately five billion year history of this living planet, biodiversity has undergone relatively abrupt decreases (Leakey and Lewin 1995). In some cases, this has been the result of the rise to dominance of a new group of organisms, such as the evolution of oxygen-generating algae three billion years ago, which confined the previously dominant anaerobic bacteria to low-oxygen niches. In other cases, it is attributed to cataclysmic events such as the impact of an asteroid. Although previous episodes of biodiversity loss have left a lasting imprint on the biota of the world, biodiversity overall has always recovered, often in different forms. Disruption of the old order may even have been the stimulus for biological innovation. For instance, the end of domination by dinosaurs may have allowed a relatively obscure group of proto-mammals to evolve, ultimately, into our own species. Why then are we concerned about the current loss of diversity?

First, the current loss of biodiversity is just one element of an interconnected syndrome known as 'Global Change'. Another element is climate change, mostly driven by human activities, including the burning of fossil fuels and release of other waste gases. A key driver of both climate change and biodiversity loss is the ongoing transformation of the surface of the planet due to human activities, including agriculture, deforestation, settlements, transport infrastructure, fishing and mining. Underpinning these changes have been transformations in how people organise themselves economically, politically, socially and technologically—the accelerating processes of development, globalisation and modernisation. The fact that biodiversity loss is intimately connected to these other momentous reorganisations makes it both an indicator of change—a canary in the mine, warning of potentially life-threatening dangers—and a key part of that change itself. It also makes halting biodiversity loss difficult, because it requires addressing the development expectations of billions of people.

Second, although past extinctions appear sudden (and perhaps some of them were), the fossil record from which we derive much of our knowledge of them tends to distort our view of their actual rate. Previous episodes of species loss may have extended over many millions of years. The current loss of biodiversity is, by contrast, extremely rapid. Furthermore, although biodiversity in the abstract sense recovered from past crises, whole groups of affected species did not. From the particular perspective of our species, we run the risk of being in the latter group.

Third, despite amazing advances in biotechnology, the loss of biodiversity in its ultimate form (the global extinction of unique genetic lineages) remains effectively irreversible. It represents the loss of millions of years of evolutionary experimentation through mutation, adaptation and natural selection. With this loss, we lose options for the future, and knowledge of the past and present.

Finally, there is emerging evidence that diversity itself (variety, as opposed to the presence of one or more particular species) is important for maintaining the productivity and stability of ecosystems, from the local to global scale (Díaz et al. 2005; Hooper et al. 2005). As humanity enters what promises to be a critical phase of its development—the transition from a 'weedy species' to one in some form of equilibrium with its environment—ensuring the resilience of the biosphere is of crucial importance. Maintaining diversity is one element of a strategy for an adaptive Earth.

Three broad reasons have been invoked as to why humans have a responsibility to conserve biodiversity. The first is essentially aesthetic: the diversity of organisms is a thing of beauty and wonder, and that is a sufficient reason to preserve them. The second class of reasons are ethical: the desire to ensure that future generations of humans are able to enjoy and use their natural heritage; or increasingly, a view that organisms have unalienable rights to existence, just as humans have. The third category is utilitarian: humans depend for their present and future well-being on the presence and functional health of other organisms, and on the fact that those organisms are diverse in composition, structure and function.

Whatever the combination of motivations, the desire to know biodiversity and protect it from further loss is now widespread. It is expressed in many cultures, and at scales from the local to the global. It takes many forms: the biodiversity-aware actions of ordinary people, resource custodians, managers and consumers; the rise of biodiversity-oriented organisations, especially in urban societies; the promulgation of laws and regulations to protect biodiversity at all levels of government, including the proclamation of protected areas and the establishment of conservation agencies; and the emergence of international treaties and organisations dedicated to biodiversity conservation. All these initiatives share a need for information to assist them to fulfil their mandates effectively and efficiently: 'what gets measured, gets managed'.

Several assessments have concluded that the current state of knowledge about biodiversity is far from adequate for the purpose of conserving it and managing it sustainably (Walpole et al. 2009; GEO BON 2011). Many existing biodiversity monitoring programs lack the power needed to detect and attribute trends in biodiversity (Legg and Nagy 2006). Even the most fundamental step, knowing what species exist on Earth, may be at best two-thirds complete and will only be achieved before a significant fraction goes extinct with coordinated international efforts (Costello et al. 2013b, c, e). This book is a contribution to fixing that problem. Better biodiversity information is essential to slow biodiversity loss and achieve a sustainable planet. To this end, several hundred countries and organisations pooled their skills and knowledge to form the Group on Earth Observations (GEO). One of its areas of concern is biodiversity, and the 'community of practice' that arose to help implement global data sharing on this topic is called the Biodiversity Observation Network (BON), or GEO BON. This handbook represents the pooled wisdom of that network.

1.2 Working Together Makes Sense

It has never been possible for any individual to know more than a tiny fraction of the biological diversity on Earth. Therefore, the investigation of biodiversity has always been a collaborative effort. Even Linnaeus, originator of the scientific system for classifying biological diversity, personally knew only a few thousand varieties and relied on a network of colleagues' observations. We now estimate that the total number of species on Earth runs into millions and at least hundreds of thousands remain to be described (Costello et al. 2013b).

The species that exist within one defined area may be different from those in another area (Gaston 2000). Thus, local experts may misapply the name of a similar species from another region to a local endemic, or describe a local species as new to science without realising it has been described from another region. The biological world is spatially organised in a way that bears little relationship to how humans have chosen to divide up the world. Considerations of political jurisdiction, culture, language and human history are ignored by biodiversity, but often form an impediment to the sharing of information about it. Improved communication, online species checklists, and greater access to species descriptions should minimise such problems and increase taxonomic efficiency (Wheeler et al. 2012; Costello et al. 2013b).

Contemporary global environmental consciousness began to emerge in the late twentieth century. It led, in 1992, to the 'Rio Conventions' on climate change, biodiversity and desertification. Each of these international treaties contains language about the need to share information relating to the topic between countries.

For example, the Convention on Biological Diversity (UNCBD) states, in article 17.1 '*The Contracting Parties shall facilitate the exchange of information, from all publicly available sources, relevant to the conservation and sustainable use of biological diversity...*'. On the tenth anniversary of the Rio meeting, one of the outcomes of the World Summit on Sustainable Development was the realisation that the management of globally pervasive issues required the global sharing of pertinent data and information. This led to the formation of the voluntary association of countries and member organisations known as GEO, dedicated to data sharing on a range of topics deemed to be of 'societal benefit', including those of biodiversity and ecosystems (GEO 2005).

The principle benefits of cooperation in the collection, sharing and coordinated analysis of biodiversity information are self-evident, but bear repeating.

Whatever biodiversity level is under consideration—for instance gene, species or ecosystem—often either has an extent of occurrence which goes beyond the jurisdiction of a single organisation, or a set of influences (acting on it, or from it) which does. Furthermore, many biodiversity elements are highly variable in space and time, thus requiring significant effort to establish baselines and detect trends. Therefore, even the largest and best-resourced institutions depend on information collected and curated elsewhere.

A full accounting, which is seldom done, of the costs of biodiversity observation and data curation would show that it represents a large historical and ongoing expense. The benefits that flow from this outlay result from the use of the information, not its collection. The benefits to society multiply synergistically as the information is made available in such a way that it can be combined with other sources of information. Even the benefits to the host organisation usually outweigh the additional costs of making such information available: having many eyes scan it and many minds interrogate it is better than a few.

Efficiencies in observation, storage, analysis and application can be achieved by learning from others. The benefits of harmonisation of methods become progressively greater as the degree to which information needs to be 'interoperable'—i.e., visible and exchangeable between systems—increases.

1.3   Networks as an Organisational Structure

The network—defined as a relatively loose affiliation of organisations that agree to create value by collaborating towards a common purpose while retaining their individual mandates, resources and management—has risen to prominence as a way of organising many activities in the modern era. A cynic might say this is because the world has lost the appetite for creating and funding new institutions or that networking is a way to suggest that some action has been initiated without actually taking responsibility for ensuring that it gets done (Provan and Milward 2001). However, if a global-scale source of biodiversity data is the desired goal, it would be hard to achieve except via the mechanism of a network, simply because sampling and species identification is more cost-effective and situation-appropriate if conducted using local and regional expertise.

A more positive view is that networks are the appropriate structure for addressing certain categories of problems, which happen to be pervasive in the modern era. These include complex and interconnected issues (like biodiversity loss) in which there are many affected parties, none of whom can solve the issue by working alone (Kickert et al. 1997). Networks are intrinsically adaptive, arguably more so than top-down structures, despite the *apparent* power and responsiveness of traditional command-and-control approaches. This paradox is explained by the fact that centrally-directed action is only effective if the goal is clearly defined, relatively unchanging and shared by all parties. Polymorphous, emerging and shifting objectives are better served by a more devolved approach. Anyone who has been part of a large, hierarchical organisation will know they have inherent inertia.

Notable examples of biodiversity networks are the Global Biodiversity Information Facility (GBIF), Species 2000 (Roskov et al. 2013), and World Register of Marine Species (WoRMS; Boxshall et al. 2014; Costello et al. 2014). GBIF is a network of countries and affiliated NGOs. Species 2000's members publish species databases through its website, and WoRMS is a network of over 200 individual taxonomists who edit parts of a common online database. Other forms of partnerships also exist, such as consortium agreements (e.g., FishBase) (reviewed by Costello et al. 2014), but the most enduring initiatives are international.

GEO BON is a 'network of networks'. Its parent body (GEO) was formed to catalyse a 'coordinated, comprehensive and sustained Earth Observation' system in support of informed decision-making worldwide'. Like its parent body, GEO BON is a voluntary 'community of practice' that serves to translate user needs in the broad arena of biodiversity (but especially at national to global scales, where the needs are often related to international treaties), into observational products and services, through collaboration between the many existing biodiversity information sources and other Earth observation systems.

Biodiversity observation, while intrinsically a collaborative activity, has not always been achieved through networks. Even in the present time, much of the primary work is done within centrally-managed organisations. As the scope of the activity increases and as larger scale drivers of biodiversity change increase in prominence, those organisations are increasingly dependent on the activities of other organisations to effectively detect and attribute biodiversity change. It is possible to imagine a global unitary organisation focussed on biodiversity observations, but it would almost certainly be unachievable in the foreseeable future given issues of national sovereignty and the sheer scale of the task. To address the urgent current needs for increased and shared biodiversity observations, some form of collaborative network seems inevitable.

While networks are often presented as a 'low-cost' option involving little more than existing efforts, they come with additional transactional costs which can be large enough to overwhelm the benefits flowing from collaboration (Costello et al. 2014). Apparently-simple guidelines can avoid this outcome: don't work through a network unless it is the most effective and achievable option for reaching the objective; include key partners; keep the network structure simple and efficient; ensure continuity through high-level commitment; be mindful of ensuring value-addition exceeds incremental costs for both network members and network funders; have well-defined roles and responsibilities; and pay close attention to minimising the transactional costs and budgeting for them—especially the hidden ones. The key transactional costs include the high level of communication required in networks and the additional costs of data management across multiple platforms. The product of the network must also be sufficiently unique, of appropriate size, quality assured, and thus prestigious, that host institutions, individual scientists and funding agencies will commit to its long-term support (Costello et al. 2014).

1.4   Managing Networks

Every bookstore has shelves overflowing with management texts, but few offer useful advice on the management of networks, which is surprising given how pervasive networks are. There are some exceptions, such as Ford et al. (2011) and, in the context of biodiversity databases, Costello et al. (2014). The principal difference between networks and more conventional, centrally-controlled organisational forms (often referred to as 'hierarchical' or 'top-down') is the degree of direct control which the manager has over human and financial resources. A useful way for network managers to think of their environment is as consisting of three concentric spheres; a visualisation attributed to Covey (1989). The central sphere contains the things over which they have direct, almost assured control. The next larger one contains those things over which they can exert some influence—by persuasion, relationship management and co-allocation of resources. The outside sphere contains those things that are out of their control, but nevertheless have an impact on the attainment of their objectives. The manager must be aware of trends and events in this outer sphere, and adapt to them, without being able to change them. Traditional management takes place almost entirely in the central sphere. Network management occurs mostly in the middle sphere. The currency of network management is influence and information rather than authority or power. No single person or organisation really fully 'owns' or 'controls' a network, even if it is centrally managed. The network looks subtly different when viewed from the differing perspective of its various partners (Ford et al. 2011). Similarly, the outcomes of a network cannot be legitimately claimed by any single participant. There is usually a trade-off in organisational structures between efficiency—which comes with centralisation—and innovation, which benefits from more distributed approaches such as networking.

The distinction between 'standardisation' and 'harmonisation' of data collection, storage and exchange follows from this understanding of what is under direct control, and what can be influenced (and can influence you), but not controlled. Within networks, 'harmonisation' is often achievable where rigid 'standardisation' is not. Fortunately, for most purposes harmonisation is sufficient. Within a unitary organisation, it is usually possible and preferable to insist on a single method ('standard'), but precisely because of this legacy, it is generally unreasonable to expect other organisations to abandon their standards in favour of yours. The solution is to permit network partners to continue, as far as possible, to apply their own approaches, but to (1) ensure those methods are explicit and visible; (2) work out how the various combinations of standards within the network relate to one another, in order to allow inter-calibrations, conversions and sorting of data; and (3) sometimes to run several approaches in parallel. This is called 'harmonisation'. It may not seem efficient (though in the long run it is more efficient than being locked into a single, increasingly inappropriate standard), but it is achievable.

Two broad aspects of network management are equally important. The first relates to the content of the network—what information is passed between partners, in what form and through what channels, and who is responsible for its collection, quality control, storage and analysis. The second relates to 'soft systems', the management of the behaviours and social relationships that hold networks together. Both aspects need active management. GEO, and GEO BON, manage the former through collectively developing, documenting and disseminating protocols for data exchange. GEO BON manages the latter by a mixture of periodic 'face-to-face' meetings, interspersed with electronic exchanges.

While an argument can be made that the societal value addition achieved by networks is large, the incremental costs of networking are usually borne by individual organisations. This is a fatal problem for networks if institutional budget decisions are based on narrowly defined, short-term cost-benefit analysis. This highlights the need for networks to show, rapidly and convincingly, the value-addition of integrating efforts to these individual organisations. Fortunately, 'social capital' often provides the bridge that permits the realisation of larger, longer-term outcomes despite near-term deficits in 'financial capital'. Successful networks are inevitably driven by people who enjoy working together and have a strong sense of the collective and individual benefits of doing so. This element of human behaviour should not be left to chance in networks. It has to be nurtured through providing opportunities and incentives for people to get to know one another, to have fun, and to develop a shared vision and purpose.

1.5 Guiding the Enterprise

'Governance' is a topic that typically bores the action-oriented denizens of the biodiversity observation world. Nonetheless, an effective but minimal set of rules and structures is essential to guide collaborative activities, especially if they are built up of many organisations with independent and possibly divergent mandates and potential conflicts of interest. Informal arrangements are effective when the number of participants is small and the level of social trust is high. The need for formal organisational design and rules of procedure rapidly emerges as the scale increases and stakes are higher. In the field of scientific assessments, also often conducted in network-like structures, three key factors for success have been identified: legitimacy (which relates especially to having transparent governance, including traceability to an 'authorising environment' that establishes the mandate); salience, which means a focus on addressing the needs of the user group; and credibility, which in this context means due attention to scientific quality (Cash et al. 2002).

The simplest governance approach, which can work if the number of stakeholders (including users) is small, is to include representatives of all of the stakeholder groups in a single steering committee, which meets on a regular basis. Once procedures and trust have been established, many of the meetings can be 'virtual', making use of telecommunication technology to minimise time and travel costs; but there is currently no satisfactory substitute for physical meetings, at least initially, that allow the development of the interpersonal relationships ('social capital') alluded to above. It is these interpersonal relationships that lead to a sense of commitment and obligation from each member to advance the work of the network.

For larger and more complex problems, such as biodiversity monitoring, a single, all-encompassing governing body may not work. A minimally more complex model that has been effective in similar contexts is to create two bodies, with clearly differentiated roles and responsibilities. One consists of representatives of intended beneficiaries, users and funders. It acts as the proxy for the authorising and receiving environment. This 'direction-setting body' addresses the questions of what to observe, and whether the result is fit for its intended purpose, as defined by this representative body. The second body consists of technical experts from all the essential implementation elements of the network, and addresses the question of 'how' to make and share the observations. Another way to think of the distinction between the two is that the first asks 'is this network observing the right things?' while the second asks 'is the network observing things the right way?' The direction-setting body defines the scope of the observation system, establishes an authorising environment, nominates the technical experts, and facilitates access to the resources needed to implement the network. The technical body then responds by developing a detailed implementation plan and a periodically updated description of activities, timelines, budget, and progress in terms of the plan. The direction-setting body approves these (or asks for revision if they are deemed inadequate to meet the goals) and resolves any conflicts that may arise between the implementation partners, for instance over roles or resources. Finally, the direction-setting body monitors and evaluates progress and acts as the final quality-control step: are the objectives being achieved? Each body may, if necessary, create sub-committees in order to address particular topics more efficiently. Financial and content-related accountability resides with both bodies, but sequentially. The direction-setting body has the final responsibility.

GEO BON, as a network of networks, is governed by an implementation committee, composed of working group leaders, regional and thematic Biodiversity Observation Network coordinators, and representatives of key projects and activities. GEO BON also has an advisory board, which provides guidance to the implementation committee, and is composed of representatives of organisations, governments, and experts, in a geographically balanced manner. Members of the advisory board serve 3 years, renewable once, and often combine, in one person, expertise in many parts of the observation-analysis-use chain—for instance, data collection in a particular biodiversity domain, scientific research, and use of data for policy purposes. The Chair and Vice-Chair of GEO BON are elected unpaid positions. The GEO BON committees reconstitute themselves in a staggered fashion, striving to keep a disciplinary, regional and other balance while adapting to emerging challenges. GEO BON working groups are established around specific tasks or themes and are open to membership by any expert or practitioner. Working groups are not permanent features, but last as long as they need to achieve a given objective, or for as long as that objective is a priority, and for as long as they are deemed effective.

Biodiversity Observation Networks (BONs) contribute to the collection and analysis of harmonised biodiversity observations, develop interoperable biodiversity monitoring programs, and help make biodiversity data and data products available. BONs can cover a political unit such as a country (National BON), a region (Regional BON), or a specific theme (Thematic BON) such as a taxonomic group, ecosystem type, or even monitoring approach. Working groups and BONs report to the implementation committee, but are given a great deal of individual freedom—and minimal resourcing—with respect to how they constitute themselves and achieve their objectives. GEO BON is supported by a small secretariat of employed officers, typically funded by a host organisation. GEO BON reports to GEO on its activities and responds to GEO initiatives as appropriate. Its activities are funded primarily by participating organisations through proposals, often endorsed or coordinated by GEO BON, to donor agencies.

1.6 Working Backwards to Move Forwards

The majority of current observing and data systems, such as GBIF and the Ocean Biogeographic Information System (OBIS), originated with the data collectors rather than the data users. This is fine where collectors and users are within the same or closely connected organisations—but increasingly they are not. As a result, what is provided by the observation system may deviate from what is needed (Sheil 2001), thus diminishing the viability of the observation system. An alternate approach is to start with the demands and work backwards to define what observations must be collected to satisfy them, including how often and where the observations must be made (Durant 2013). In defining needs, it is critical that they be clearly described, measurable and achievable in order to ensure successful outcomes. There may be several steps between primary observations and final products; each of these steps needs equal attention.

In practice, defining what to observe and how to process it so that it is of maximum utility is a two-way process: a negotiation (or conversation, if you prefer less adversarial metaphors) which in the best cases converges on a solution that is both useful and feasible. The design is said to be co-determined or co-produced, and is neither 'user-driven' nor 'supply-driven', but both. This approach helps to remove a sense that one group is in charge, and the others are subservient. That situation is detrimental to accountability, creativity and the sense of partnership that makes networks work. While it is customary to talk of 'data providers' and 'data users' as non-overlapping sets (with 'data brokers' sometimes interposed between them), in reality individual partners often play multiple roles simultaneously—they are providers of some observations, but users of others.

GEO BON is a meeting place for both 'providers' and 'users', and does not make a mutually exclusive distinction between them. They are all part of a continuum of stakeholders. It helps to refine user needs by organising periodic topically-focussed user workshops, where both users and potential suppliers are present. The outcome is thus 'co-generated', and takes the form of a discussion rather than a unilateral instruction in one direction or the other. If the needs cannot be currently met, the outcome is a set of specifications for future Earth observation activities.

A second key way of identifying needs is to be closely engaged with bodies that have a mandate to define such needs collectively. In the case of GEO BON, this includes for instance the Convention on Biological Diversity (CBD), whose agreed 'Aichi Targets' for national reporting towards global objectives include many explicit observational needs.

1.7 The Purpose, Structure and Content of This Volume

This handbook captures the collective learning, at the time of writing, of the organisations involved in the GEO Biodiversity Observation Network. We do not believe that it is the last word on the topic of biodiversity observations, since this is a rapidly evolving field. It is already clear, however, that a degree of convergence in biodiversity observation and information storage methods is highly beneficial to all parties, and easier to achieve if implemented early rather than late. There is a surge of biodiversity observation network activity at present, driven by the urgent need to address biodiversity loss effectively and efficiently and specific actions such as the CBD Aichi targets for the year 2020. As new networks start up and existing networks expand and reconfigure, some guidance can help them to avoid problems that have been encountered and solved elsewhere, and get going more quickly along a path that allows for better networks in the future.

A number of chapters in this handbook is structured around the Essential Biodiversity Variable (EBV) framework, which GEO BON started developing in 2012 with the purpose of representing a minimal set of fundamental observations needed to support multi-purpose, long-term biodiversity information needs at various scales (see Pereira et al. 2013).

By combining EBV observations with other information, such as on the attributes of biodiversity, or drivers and pressures of biodiversity change, indicators can be developed which are directly useful for policy support. EBVs can thus have multiple uses. For instance, an observation system that collects data on species abundance for several taxa at multiple locations on our planet, can support the derivation of the Living Planet Index (Collen et al. 2009), the Wild Bird Index (Butchart et al. 2010), the Community Temperature Index (Devictor et al. 2012), measures of species range shifts (Parmesan 2006), and a number of other high-level indicators on the CBD's indicative list of indicators for the strategic plan for biodiversity 2011–2020 (CBD 2015; Fig. 1.2).

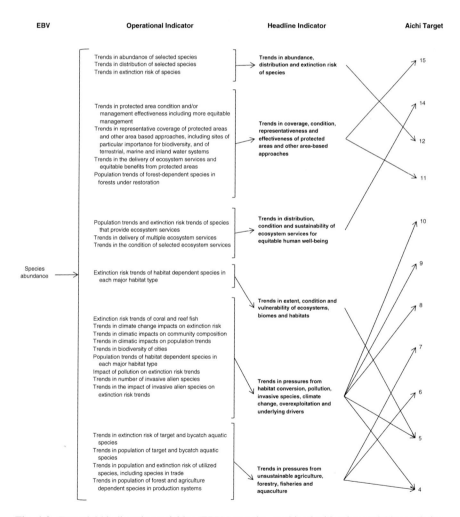

Fig. 1.2 Essential biodiversity variables (EBVs) may be combined with other variables to derive multiple high-level indicators used to measure progress against multiple targets. In this example the EBV 'species abundance' feeds into 24 possible indicators that may be used to derive the headline indicators for monitoring progress towards 11 of the Aichi biodiversity targets

Essential Biodiversity Variables may fall within six classes: genetic composition; species populations and ranges; species traits; community composition; ecosystem structure; and ecosystem function. Whilst the EBVs are currently still under development, a number of candidates have been suggested by the broader GEO BON community. The subsequent chapters of this handbook touch on some of these and provide details of how to measure EBVs in many different environments—on land, in freshwater ecosystems such as lakes and rivers, on the coast and in oceans; and for different types of organisms and at various scales.

Chapter 2 of this handbook addresses biodiversity observations at the ecosystem scale—the scale at which many policy, management and societal needs are focussed. It covers terrestrial ecosystems and leaves the practical special considerations for biodiversity observations in marine and freshwater environments to Chaps. 6 and 7, respectively.

An increasing number of countries are including ecosystem services and natural capital accounting in their national accounts, to better inform decision-making. Chapter 3 addresses the data requirements and the toolkits and models available for assessing and monitoring ecosystem services.

The observations needed for detecting changes in the abundance of individuals in populations of particular species are addressed in Chap. 4, which includes identification of the question to be addressed, the choice of variables, taxa and spatial sampling scheme.

Chapter 5 introduces the fast-growing field of gene-level observations, including the current state-of-the-art in genetic monitoring, with an emphasis on new molecular tools and the richness of data they provide to supplement existing approaches.

Chapter 6 expands on marine and coastal systems and the special approaches that are required when observing biodiversity in a three-dimensional, fluid environment that is often remote, unexplored and not owned by any particular country.

Chapter 7 deals with observing biodiversity in freshwater systems, and highlights special considerations for freshwater biodiversity and methods and tools available for monitoring these systems.

Chapter 8 discusses the use of remote sensing for observing biodiversity and provides a baseline set of information about using remote sensing for conservation applications in three realms: terrestrial, marine, and freshwater.

Biodiversity has long had a tradition of citizen observers, which is the topic of Chap. 9. How can ordinary people be organised and incentivised using modern technology, and how can the quality of the observations be assured?

The old distinction between observations and models is rapidly breaking down. Chapter 10 addresses the question of how models can help to fill gaps in space and time, and how one can use in situ and remotely sensed observations to detect changes in biodiversity.

Modern observation networks cannot function without paying attention to cyber-infrastructure (Chap. 11). How is data captured, stored, made discoverable and interoperable?

Chapter 12 explores the use of biodiversity data in decision-making processes, as well as the realities of indicator development and use. It reflects on what data might be used for, how it is packaged, and what the challenges are.

Finally, Chap. 13 reflects, through the presentation of several case studies, on various approaches for capacity building in the field of biodiversity monitoring.

References

Boxshall, G., Mees, J., Costello, M. J., Hernandez, F., Gofas, S., Hoeksema, B. W., et al. (2014). *World register of marine species.* Available via VLIZ. http://www.marinespecies.org.Cited. December 7, 2014.

Butchart, S. H. M., Walpole, M., Collen, B., van Strien, A., Scharlemann, J. P. W., Almond, R. E. E., et al. (2010). Global biodiversity: Indicators of recent declines. *Science 328,* 1164–1168.

Cash, D., Clark, W., Alcock, F., Dickson, N., Eckley, N. & Jäger, J. (2002). *Salience, credibility, legitimacy and boundaries: linking research, assessment and decision making.* Research Working Papers Series, Cambridge, USA: Faculty John F. Kennedy School of Government, Harvard University.

CBD (2015). UNEP/CBD/ID/AHTEG/2015/1/3 and UNEP/CBD/SBSTTA/19/INF/5 Convention on Biological Diversity, Montreal, Canada. Available from https://www.cbd.int/doc/strategic-plan/strategic-plan-indicators-en.pdf. September, 2016.

Collen, B., Loh, J., Whitmee, S., McRae, L., Amin, R. & Baillie, J. E. (2009). Monitoring change in vertebrate abundance: The living planet index. *Conservation Biology, 23*(2), 317–327.

Costello, M. J., Bouchet, P., Boxshall, G., Fauchald, K., Gordon, D. P, Hoeksema, B. W., et al. (2013a). Global coordination and standardisation in marine biodiversity through the World Register of Marine Species (WoRMS) and related databases. *PLoS ONE, 8,* e51629.

Costello, M. J., Appeltans, W., Bailly, N., Berendsohn, W. G., de Jong, Y., Edwards, M., et al. (2014). Strategies for the sustainability of online open-access biodiversity databases. *Biological Conservation, 173,* 155–165.

Costello, M. J., May, R. M. & Stork, N. E. (2013b). Can we name Earth's species before they go extinct? *Science, 339,* 413–416.

Costello, M. J., May, R. M. & Stork, N. E. (2013c). Response to comments on "Can we name Earth's species before they go extinct?" *Science, 341,* 237.

Costello, M. J., Michener, W. K., Gahegan, M., Zhang, Z-Q. & Bourne, P. (2013d). Data should be published, cited and peer-reviewed. *Trends in Ecology and Evolution, 28,* 454–461.

Costello, M. J., Wilson, S. P. & Houlding, B. (2013e). More taxonomists but a declining catch of species discovered per unit effort. *Systematic Biology, 62,* 616–624.

Covey, S. (1989). *The 7 habits of highly effective people.* New York, USA: Fireside.

Crutzen, P. J. (2002). Geology of mankind. *Nature, 415*, 23.

Devictor, V., Van Swaay, C., Brereton, T., Chamberlain, D., Heliölä, J., Herrando, S., et al. (2012). Differences in the climatic debts of birds and butterflies at a continental scale. *Nature Climate Change, 2*(2), 121–124.

Díaz, S., Tilman, D., Fargione, J., Chapin III, F. S., Dirzo, R., Kitzberger, T., et al. (2005). Biodiversity regulation of ecosystem services. In: R. Hassan, R. J. Scholes, & N. Ash (Eds.), *Ecosystems and human wellbeing: Current state and trends: Findings of the condition and trends working group* (pp. 297–329). Washington D.C., USA: Island Press.

Durant, S. M. (2013). Building sustainable national monitoring networks. In B. Collen, N. Pettorelli, J. E. M. Baillie, & S. M. Durant (Eds.), *Biodiversity monitoring and conservation: Bridging the gap between global commitment and local action* (pp. 311–334). Oxford, UK: Wiley-Blackwell.

Ford, D., Gadde, L.-E., Hakansson, H., & Snehota, I. (2011). *Managing business relationships.* Hoboken, USA: Wiley.

Gaston, K. J. (2000). Global patterns in biodiversity. *Nature, 405*, 220–227.

GEO (2005). *Global Earth Observation System of Systems (GEOSS): 10-Year implementation plan reference document.* Noordwijk, The Netherlands: ESA Publications Division.

GEO BON (2011). *Adequacy of biodiversity observation systems to support the 2020 CBD targets.* Pretoria, South Africa: GEO BON Secretariat. http://www.earthobservations.org/documents/ cop/bi_geobon/2011_cbd_adequacy_report.pdf. Cited December 17, 2014.

Haberl, H., Erb, K.-H., Krausmann, F., Gaube, V., Bondeau, A., Plutzar, C., et al. (2007). Quantifying and mapping the human appropriation of net primary production in earth's terrestrial ecosystems. *Proceedings of the National Academy of Sciences of the USA, 104*, 12942–12947.

Hooper, D. U., Chapin, F. S., III, Ewel, J. J., Hector, A., Inchausti, P., Lavorel, S., et al. (2005). Effects of biodiversity on ecosystem functioning: A consensus of current knowledge and needs for future research. *Ecological Monographs, 75*, 3–35.

Kickert, W. J. M., Klijn, E-H. & Koppenjan, J. F. M. (1997). Introduction: A management perspective on policy networks. In W. J. M. Kickert, E-H. Klijn, & J. F. M. Koppenjan (Eds.), *Managing complex networks* (pp. 1–13). London, UK: Sage Publications.

Leakey, R. E., & Lewin, R. (1995). *The sixth extinction: Patterns of life and the future of humankind.* New York, USA: Doubleday.

Legg, C. J., & Nagy, L. (2006). Why most conservation monitoring is, but need not be, a waste of time. *Journal of Environmental Management, 78*, 194–199.

Noss, R. F. (1990). Indicators for monitoring biodiversity: A hierarchical approach. *Conservation Biology, 4*, 355–364.

Parmesan, C. (2006). Ecological and evolutionary responses to recent climate change. *Annual Review of Ecology, Evolution, and Systematics*, 637–669.

Pereira, H. M., Navarro, L. M., & Martins, I. S. (2012). Global biodiversity change: The bad, the good, and the unknown. *Annual Review of Environment and Resources, 37*, 25–50.

Pereira, H. M., Ferrier, S., Walters, M., Geller, G. N., Jongman, R. H. G., Scholes, et al. (2013). Essential biodiversity variables. *Science, 339*(6117), 277–278.

Provan, K. G., & Milward, H. B. (2001). Do networks really work? A framework for evaluating public-sector organizational networks. *Public Administration Review, 61*, 414–423.

Roskov, Y., Kunze, T., Paglinawan, L., Abucay, L., Orrell, T., Nicolson, D., et al. (Eds.). (2013). *Species 2000 & ITIS catalogue of Life*, November 19, 2013. Leiden, The Netherlands: Species 2000: Naturalis. www.catalogueoflife.org/col

SCBD (Secretariat of the Convention on Biological Diversity). (2010). *Global biodiversity outlook 3.* Montréal, Canada: Secretariat of the Convention on Biological Diversity.

Sheil, D. (2001). Conservation and biodiversity monitoring in the tropics: Realities, priorities, and distractions. *Conservation Biology, 15*, 1179–1182.

Walpole, M., Almond, R. E. A., Besancon, C., Butchart, S. H. M., Campbell-Lendrum, D., Carr, G. M., et al. (2009). Tracking progress toward the 2010 biodiversity target and beyond. *Science, 325*, 1503–1504.

Wheeler, Q. D., Knapp, S., Stevenson, D. W., Stevenson, J., Blum, S. D., Boom, B. M., et al. (2012). Mapping the biosphere: Exploring species to understand the origin, organization and sustainability of biodiversity. *Systematics and Biodiversity, 10*, 1–20.

Chapter 2
Global Terrestrial Ecosystem Observations: Why, Where, What and How?

Rob H.G. Jongman, Andrew K. Skidmore, C.A. (Sander) Mücher, Robert G.H. Bunce and Marc J. Metzger

Abstract This chapter covers the questions of ecosystem definition and the organisation of a monitoring system. It treats where and how ecosystems should be measured and the integration between in situ and RS observations. Ecosystems are characterised by composition, function and structure. The ecosystem level is an essential link in biodiversity surveillance and monitoring between species and populations on the one hand and land use and landscapes on the other. Ecosystem monitoring requires a clear conceptual model that incorporates key factors influencing ecosystem dynamics to base the variables on that have to be monitored as well as data collection methods and statistics. Choices have to be made on the scale at which monitoring should be carried out and eco-regionalisation or eco-logical stratification are approaches for identification of the units to be sampled. This can be done on expert judgement but nowadays also on stratifications derived from multivariate statistical clustering. Data should also be included from individual research sites over the entire world and from organically grown networks covering many countries. An important added value in the available monitoring

R.H.G. Jongman (✉) · C.A. (Sander) Mücher
Alterra Wageningen UR, Wageningen, The Netherlands
e-mail: rob.jongman@xs4all.nl

C.A. (Sander) Mücher
e-mail: sander.mucher@wur.nl

R.H.G. Jongman
JongmanEcology, Wageningen, The Netherlands

A.K. Skidmore
Faculty of Geo-Information Science and Earth Observation, University of Twente, Enschede, The Netherlands
e-mail: a.k.skidmore@utwente.nl

R.G.H. Bunce
Estonian University of Life Sciences (EMU), Tartu, Estonia
e-mail: bob.bunce@emu.ee

M.J. Metzger
School of GeoSciences, University of Edinburgh, Edinburgh, Scotland, UK
e-mail: marc.metzger@ed.ac.uk

© The Author(s) 2017
M. Walters and R.J. Scholes (eds.), *The GEO Handbook on Biodiversity Observation Networks*, DOI 10.1007/978-3-319-27288-7_2

technologies is the integration of in situ and RS observations, as various RS technologies are coming into reach of ecosystem research. For global applications this development is essential. We can employ an array of instruments to monitor ecosystem characteristics, from fixed sensors and in situ measurements to drones, planes and satellite sensors. They allow to measure biogeochemical components that determine much of the chemistry of the environment and the geochemical regulation of ecosystems. Important global databases on sensor data are being developed and frequent high resolution RS scenes are becoming available. RS observations can complement field observations as they deliver a synoptic view and the opportunity to provide consistent information in time and space especially for widely distributed habitats. RS has a high potential for developing distribution maps, change detection and habitat quality and composition change at various scales. Hyperspectral sensors have greatly enhanced the possibilities of distinguishing related habitat types at very fine scales. The end-users can use such maps for estimating range and area of habitats, but they could also serve to define and update the sampling frame (the statistical 'population') of habitats for which field sample surveys are in place. Present technologies and data availability allow us to measure fragmentation through several metrics that can be calculated from RS data. In situ data have been collected in several countries over a longer term and these are fit for statistical analysis, producing statistics on species composition change, habitat richness and habitat structure. It is now possible to relate protocols for RS and in situ observations based on plant life forms, translate them and provide direct links between in situ and RS data.

Keywords Ecosystem monitoring · Habitat · Hyperspectral sensor · In situ observation · Plant life form · Stratification · Sensor networks

2.1   Introduction

In the last decades it has been emphasised that we still lack empirical baseline data on local patterns of biodiversity and their dynamics and interactions within communities and habitats (Scholes et al. 2008). The lack of empirical biodiversity observation data is obvious at various levels of complexity; even basic inventories of current local-to-global biodiversity are missing. There are several reasons for this. Firstly, global cooperation in biodiversity research and monitoring is a relatively recent phenomenon. We lack standards, we do not yet share protocols, we do not consider strategic sampling and there is limited exchange of data at and between spatial scales. Noss (1990) flagged this problem and developed a general concept for a hierarchical approach to monitoring biodiversity. Ecosystem monitoring is needed to track the impacts of various drivers such as land use change and climate change.

   In this chapter we deal with dryland terrestrial ecosystems (marine and freshwater species and ecosystems are dealt with in Chaps. 7 and 8, respectively).

Several long term ecosystem monitoring networks, based on coordinated long-term observation systems, do exist. Examples include the networks of the International Long-Term Ecological Research Network (ILTER/LTER; global, national scale), the Important Bird and Biodiversity Areas program (IBA; global scale), the Biodiversity Monitoring Transect Analysis in Africa project (BIOTA; Africa, São Paulo State), the Global Observation Research Initiative in Alpine Environments program (GLORIA; mountain summits at a global scale), the South African Environmental Observation Network (SAEON; South Africa), the Federal System of Protected Areas (SiFAP; Argentina), the Terrestrial Ecosystem Research Network (TERN; Australia), the National Ecological Observatory Network (NEON; USA), and the Amazon Forest Inventory Network (RainFor; Amazon).

This Chapter Covers Four Main Issues and Comprises Four Sections:

- what is an ecosystem?
- where to measure ecosystems,
- what to measure and how to measure it, and
- how to link the various approaches and protocols.

All four issues require choices by decision-makers concerning effort, budget, human resources and infrastructural capacities.

2.2 Ecosystems and Ecosystem Variables

Ecosystems are universally understood as systems of biotic communities interacting with themselves and with their abiotic environment. Ecosystems can be conceptualised as the integration of living and non-living components in nature. They are characterised by their composition, function and structure which depends on the local environment, as well as management approaches. Each of these three dimensions should be included in ecosystem monitoring.

In biodiversity surveillance and monitoring, ecosystems are an essential link between species and populations on the one side, and land use and landscapes on the other (e.g., Noss 1990; also see Fig. 1.1 in Chap. 1). What could be measured in ecosystems potentially touches on all the major dimensions of biodiversity. Therefore strategic choices have to be made about what should be measured, and how and where to measure it.

Ecosystems in the most general sense are conceptual rather than physical entities, and are therefore dimensionless. Their spatial or structural aspects do have physical manifestations, with units, and can be defined as ecotopes or habitats. Definitions of the term 'habitat' range from how species are associated with landscape-scale units to very detailed descriptions of the physical environment used by species (Hall et al. 1997). They also include aspects such as snow cover, openness and patchiness. Bunce et al. (2008) gave a practical definition of habitats and rules for assignment of a given patch to a habitat class. They define habitat as '*an element of the land surface that can be consistently defined spatially in the field in order to define the principal environments in which organisms live*. Functional aspects of ecosystems can be defined as the cycling of matter and energy expressed in biomass, seasonal changes, succession and soil development, growth, energy storage and regulation processes. Compositional aspects of ecosystems are species richness, diversity of species and guilds, and presence of certain species assemblages.

In many cases ecosystems and habitats are, in practice, defined based on their vegetation compositional and/or structural aspects. Classical phytosociology was designed for description, rather than long-term monitoring and change detection, but individual plots that have been studied in the past can be resampled, if the sites are re-locatable. Vegetation structure and biomass are more important for animal populations than vegetation composition and some widely recognised habitats may not be directly linked to vegetation composition. The TERN project (www.tern.au) stipulates that a monitoring design needs to pay careful attention to:

- the question(s) of interest;
- statistical principles;
- a conceptual model that incorporates the key factors influencing ecosystem dynamics;
- the type of entities that need to be monitored;
- the data collection methods that will be effective; and
- the scale of the required monitoring program.

It is important to realise that errors are inevitable and that in some cases absence of a feature (a dry lake with no water) or taxon (no birds in a forest) is as important as its presence. Measuring a non-stable variable that may be associated with a particular error to boot, can lead to a poor level of understanding. In other words, too many constraints in a monitoring scheme may reduce the likelihood of a monitoring system being successful. Therefore an appropriate and sound statistical design that, for instance, can deal with variability and the presence of null records (zeros) is essential in the set-up of long-term monitoring schemes.

Because we are interested in detecting trends, long-term quantitative approaches in measurement are important. There are many different variables that could be measured, so choices have to be made. Land cover forms a valuable basis for practical applications like forest and rangeland monitoring, but also for monitoring climate change, biodiversity and desertification (Jansen and Di Gregorio 2002). Climate and agricultural variables are measured under the umbrella of the World Meteorological Organisation (WMO) and the Food and Agriculture Organization of the United Nations (FAO) respectively. The key variables to be measured for biodiversity are variables related to ecosystem status and trends.

After the Nagoya Conference of the Parties of the Convention on Biological Diversity (CBD), the Group on Earth Observations Biodiversity Observation Network (GEO BON) organised a series of workshops to assess the possibility of collecting data relevant to reporting on progress in reaching the targets of the convention. In the process GEO BON developed the concept of Essential Biodiversity Variables (EBVs; Table 2.1; Pereira et al. 2013).

Table 2.1 Some candidate ecosystem related Essential Biodiversity Variables (EBVs)

Class	Candidate EBV	Example parameters	Definition
Ecosystem process/function	Photosynthetic activity	Forest tree diameter, biomass production, sphagnum growth, ocean plankton	Fixation of carbon and production of O_2 by the biosphere
	Aerobic respiration	Autotrophic and heterotrophic respiration from soil and vegetation	Uptake of oxygen and production of CO_2 by the biosphere
	Phosphorus tolerance/limitation		Phosphorus load in an ecosystem in relation to a reference condition
	Secondary production	Fish population, krill density, wildebeest population, migratory bird population	Sum of production by herbivores and higher trophic levels
	Nutrient retention	Nitrogen retention	Capacity of ecosystem to store, fix or retain a given nutrient or chemical
	Disturbance regime	Wind-blow, hurricanes, el Niño, forest fires, flooding	Type, timing and intensity frequency of external influences on ecosystems
Ecosystem structure (mean and seasonality)	Cover, amount per volume, height clumping	Fractional cover, vegetation height	Organisation of habitats in a three dimensional space
	Ecosystem extent and distribution	Land cover	Total occupied area by type
	Ecosystem connectivity and fragmentation	Landscape heterogeneity	Connectivity is defined by patch size, shape, pattern and the ability to move between patches
Ecosystem composition (in space and time)	Community or functional composition	Species or functional profile based on relative abundance	Species or functional profile based on relative abundance
	Phylogenetic diversity (PD) of community		Sequence data based on phylogeny is used to assess genetic sequence based diversity in communities
Ecosystem level responses	Degree of protection		Level of ecosystem extent protected (through legal status, actual management and ownership)
Ancillary attributes of ecosystems	Richness and endemism		Sum of known species within an area (or ecosystem) and of species endemic to that area (or ecosystem)
	Uniqueness		Percentage of the global extent of an ecosystem found in a certain area

2.3 Where to Measure Ecosystem Variables

The question of where to measure ecosystems and ecosystem variables for an analysis at a particular scale calls for a 'sampling frame' that is strategically located across the globe, continent, country or region. The use of remotely-sensed land cover maps provides the first part of the picture of habitat change. It will therefore be an important tool for reporting change.

In addition to the overview of structural ecosystem change provided by repeated habitat maps there is a need for statistics on change and a need for monitoring of ecosystem processes. Here the question of where to measure becomes critical. For many purposes, such as consistent input to climate impact models, or reporting towards the Aichi targets, standardised frameworks and methods are required among different studies or countries to enable integration of data and reporting. The development and adoption of harmonised methods is a complex and difficult process, because ecological data collection tends to be coordinated at the regional or national level, following country specific methods, classifications and priorities. It is made more difficult by the long-term nature of the data: it may not be possible to harmonise data from old studies, and those responsible for the collection and curation of long-term records are typically reluctant to change their methods in substantive ways.

Ecosystems can be as extensive as the entire arctic tundra, or as small as a particle of soil. They are thus understood to exist at multiple scales. This means that choices have to be made on the scale at which monitoring should be carried out. Mapping ecologically homogenous regions across the planet to select monitoring sites has been accomplished through a process of eco-regionalisation as in the WWF global ecoregions map. However, this and most other approaches rely heavily on expert judgement for interpreting class divisions. This makes it difficult to ensure reliability across the world and limits their use in scientific analysis. The Global Environmental Stratification (GEnS) is the first high-resolution global bio-climate stratification derived from multivariate statistical clustering (Fig. 2.1). The GEnS also provides sufficient detail to support the design of regional monitoring programmes that can be nested within the global network.

A cost-efficient and data-effective selection of sites for data collection should be based on a stratified random selection procedure for the whole land surface of the target area. The GEnS (Fig. 2.1) is a way to provide a common global framework for positioning fixed monitoring stations, the development of LTER sites as well as for stratified random sampling and global statistics (Metzger et al. 2013a). The GEnS consists of 125 strata, which have been aggregated into 18 global environmental zones. The stratification has a 30 Arcsec resolution (equivalent to 0.86 km^2 at the equator). One of the recent applications of the GEnS is the ecological monitoring project in the Kailash Sacred Landscape (KSLCI). This is the first

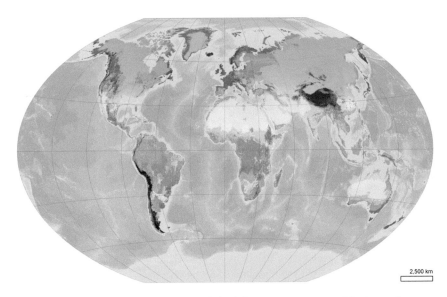

Fig. 2.1 Global environmental zones map derived from temperature, precipitation, and seasonality data and with a grid of 30 Arcsec squares. The stratification exists of 125 strata in 18 global zones. *Source* Metzger et al. (2013a)

cooperation of its kind among China, India, and Nepal seeking to conserve the area through application of transboundary ecosystem management and enhanced regional cooperation (Metzger et al. 2013b). A comparable ecoregion based approach has been used in the USA to identify the NEON monitoring sites. The outcome of the geographical analysis resulted in twenty domains in which the observatories have been placed.

Data are collected at individual research sites or by national monitoring systems, all over the world. This process is currently not globally coordinated. The Long Term Ecological Research sites network (LTER) in Europe is an example of an organically grown network that covers many countries. There are at present approximately 1000 facilities with LTER activities, ranging in extent from less than 10 ha to several thousand hectares. They differ in monitoring objectives, methods of measurements, and spatial extent. However, as Metzger et al. (2010) showed, their distribution is not even (Fig. 2.2).

One may of course question whether one site per region can adequately address the eco-climatic variability in a large, diverse areas. In the NEON design this problem has been tackled by including both permanent core sites and relocatable auxiliary sites that should allow for covering the variation within a region. Remote sensing observations can allow generalisation of point samples over larger areas.

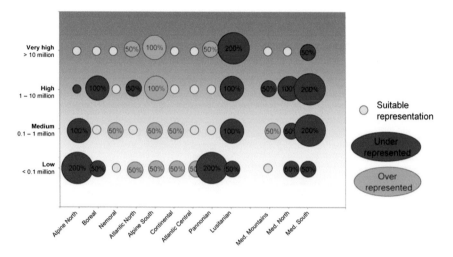

Fig. 2.2 Representation of LTER facilities per socio-ecological region based on the Environmental Stratification of Europe. The strata in the X-axis are European Environmental Zones; the Y-axis indicates population density. *Source* Metzger et al. (2010)

The BIOTA observatories in Africa (Morocco, West Africa and South Africa) are situated on transects and each consists of a series of 1 km² squares where species and ecosystem variables are measured regularly (Jürgens et al. 2011). They also provide ground-truthing for remote-sensing observations. In this example, several 'auxiliary observatories' have also been established at a variety of scales, for process and pattern observations.

In global and continental stratifications climate plays a dominant role. This changes when stratifications are made at national and regional scales, especially in smaller countries and mountainous areas. Then the stratification should be broken down in a hierarchical flexible structure. In Fig. 2.3 such an approach is shown for the Alpine region in Europe. In Fig. 2.3a the Alpine region is shown in an aggregated way, and consists of large climate zones. This level is appropriate for reporting at the European level. Figure 2.3b shows the Alpine zone at the more detailed level of environmental strata (ALS1, ALS3 and ALS5) based on mainly climate variables. At this level, summits, valley sides and valley floors are still included in the same stratum, because of the smoothing effect of the climate data. The ecosystems and taxa in these different topographic locations will be very different. Therefore a subdivision based on altitude is made (Fig. 2.3c). This demonstrates the full complexity of the Alpine zone and will enable any sample of 1 km² plots to be dispersed efficiently through the landscape, i.e., on valley floors, valley sides and summits. At an even lower level, not only geomorphology, but also other information such as soil types and hydrology can be used for further refinements.

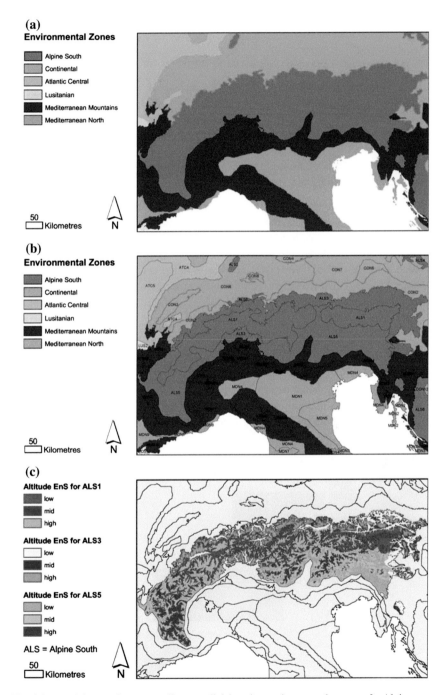

Fig. 2.3 a Alpine regions according to division in environmental zones; **b** Alpine zone subdivided in environmental strata (ALS1, ALS3 and ALS5) within Alpine zone; **c** Alpine zone with environmental strata subdivided according to altitudinal bands. *Source* Jongman et al. (2006)

2.4   How to Measure Ecosystem Variables

There are generally three ways to measure ecosystem variables.

1. Most of the functional processes can be measured as fluxes, using in situ sensors.
2. Precise monitoring of composition, abundance, extent and change is commonly done by in situ monitoring through habitat surveillance combined with vegetation plots.
3. Structural change is monitored using in situ habitat surveillance in combination with remote sensing from space or aircraft.

There are advantages and differences between the methodologies and one solution does not satisfy all data questions. Remote sensing technologies are increasingly becoming integrated with in situ measurements as various new technologies become available for ecosystem research. For global applications this development is essential. Nowadays we can employ an array of instruments to monitor ecosystem characteristics, from fixed sensors and in situ measurements, to drones, planes and satellite sensors (Fig. 2.4).

2.4.1   Sensor Networks

Biogeochemical components determine much of the chemistry of the environment (air, water, and soil) and the geochemical regulation of ecosystems. Key measurements, among others include the greenhouse gases CO_2, CH_4 and NO_x, which

Fig. 2.4 An overview of the available array of sensors to measure ecosystem variables and metrics, varying from in situ sensors and surveillance to drones, airplanes and satellites

determine the climate change process and are important drivers of change in bio-diversity. These and other chemicals such as NH_4 also can cause acidification and eutrophication and in this way lead to ecosystem degradation, involving a sustained loss of ecosystem services and/or biodiversity. The water, carbon and nitrogen cycles have a direct influence on ecosystems globally and are measured using sensor networks in many countries in the world. Long-term, patch-scale measure-ments using eddy covariance (EC) are, for example, employed to estimate ecosystem carbon budgets. This is mainly done in research sites or dedicated monitoring sites. A global database of soil respiration data has been developed by the US Oak Ridge National Laboratory (http://daac.ornl.gov). It can be used as a reference database, because the number of sites is small, but it covers the globally important terrestrial ecosystems.

The extent to which pollutants are detrimental to ecosystem function and bio-diversity is not always known, but clear effects have been reported for nitrogen, phosphorus, sulphur, pesticides, herbicides, aerosols and ozone. For an indication of excess pollutant exposure, it is important to know the difference between natural versus anthropogenic exposure levels. For this purpose emission, dispersion and deposition model calculations are generally used. Measurements of pollutants are made in many countries, but mostly at irregular intervals and patchily over space. Global coordination and harmonisation are lacking, but there are attempts to improve this, for instance in the way nitrate is measured in networks in Europe (EMEP), North-America (NADP), Canada (CapMon), and East Asia (EANET).

2.4.2   In Situ Mapping

Common approaches for in situ monitoring of ecosystem extent require definitions that are harmonised nationally, continentally and globally, which is not the situation at present. Forest definitions differ between international organisations such as FAO, CBD and UNFCCC and between European countries.

Surveillance involves recording of features at a specific location at one moment, i.e., taking stock. Monitoring involves repeated observation, to create a time series which enables the detection of change. This requires that the location of monitoring is known, and preferably kept constant over time. Moreover, in most cases the field assessment of biodiversity or habitats is based on samples. Sampling procedures must not be compromised by spatial heterogeneity or complexity. As sampling effort (i.e., the time taken to record information) is usually fixed, a choice has to be made between recording basic information in many sample units, or more detailed information in fewer units; similarly there is a trade-off between many small and few large units (Bunce et al. 2008). This has consequences for the statistical inference which can be made using the data. Often the optimal solution is neither one nor the other, nor an intermediate state, but a clever combination which has many simple sites for extrapolation purposes and a few comprehensively monitored sites to understand the details and processes.

For recognising trends and sudden changes in ecosystem composition and diversity it is important to produce statistics based on direct measurements. These can be used to derive indicators such as pattern and changes in species richness, patchiness and linear features. This has been done in the Great Britain Countryside Survey since 1978, producing statistics on species composition change, habitat richness and habitat structure to support policy (www.countrysidesurvey.org.uk/). The configuration and fragmentation of structural biodiversity, species composition, age of systems and their components as well as biomass, ecological relations and extinction rates are important aspects related to ecosystem health and integrity.

For statistically-robust trend detection it is essential to return periodically to the same sites to record changes. National and regional in situ networks exist for monitoring ecosystems and biodiversity change. They employ various size units from 16 km^2 down to 0.25 km^2. Some, such as the META project in Hungary, use hexagonal units of 35 ha, because a hexagon has six neighbouring cells with all more or less the same distance from the centre (Molnár et al. 2007). The most common emerging scale for the field recording of habitats is 1 km^2, making a compromise between detail and generality.

In the EU-FP7 EBONE project a habitat and vegetation recording procedure was elaborated and made generally available (www.wageningenur.nl/ebone). It includes a manual and a database with a digital field form that helps to support consistent mapping. The protocols have adopted plant life forms as the basis of a system of General Habitat Categories (GHCs). The GHC system includes some classes such as mud flats and scree slopes which do not have vegetation, in order to cover the terrestrial world from forests and grasslands to deserts. At a continental level, ecosystems can best be defined in terms of the physiognomy and life forms of the dominant species, because individual species are too limited to encompass widely dispersed geographical locations. Moreover, life forms can provide direct links between in situ and remotely-sensed data and dynamic global vegetation models. GHCs have been tested successfully throughout Europe, Israel, South Africa and Western Australia. The GHC framework also made it possible to harmonise different national habitat mapping systems so that they could be used to produce consistent indicator information across borders. It is therefore a good candidate to be tested globally.

2.4.3 Remote Sensing

Traditionally, ecologists map biodiversity and ecosystems based on in situ observations, perhaps generalised using aerial photography. However, existing Remote Sensing (RS) tools can be used to measure and map a number of ecosystem variables and metrics directly, much more effectively than can be done using field measurements. RS is recognised as a powerful tool to acquire synoptic data on habitats, but to date, its use for operational monitoring and reporting of biodiversity is still limited. One reason for this appears to be the knowledge gap between the agencies and individuals responsible for biodiversity monitoring and the remote sensing community. To overcome this gap requires mutual awareness, willingness to collaborate and technology transfer.

RS observations can complement field observations as they deliver a synoptic view and offer the opportunity to provide consistent information in time and space (Vanden Borre et al. 2011). It must be determined in each case what variable can be measured best by using RS, alone or in a hybrid scheme with an optimally-distributed set of in situ measurements. Recognition of habitat types on images is easier for widely distributed habitats than for rare habitat types. In general, rare ecosystems have to be specially targeted and small habitat elements (smaller than the minimum resolution of space-based sensors, which is in the region of 1–5 m for non-military instruments, and down to 0.3 m using airborne sensors) can only be monitored by in situ observations. Habitat distribution maps, change detection and even habitat quality and composition change at various scales can be cost-effectively monitored with these types of sensors (Turner et al. 2003). Although these techniques are promising, they still fall short in several aspects (Mücher et al. 2013): (i) airborne hyperspectral data or airborne Lidar are suitable, but coverage is still limited; (ii) existing methods have not fully addressed the issue of habitat structure and functioning, which is a key factor for assessing habitat quality; and (iii) most existing remote sensing methodologies have not been tested rigorously for operational purposes.

Monitoring of habitat quality information in enough detail remains challenging as this requires sensors and methods which can deal with complex transitional gradients in natural vegetation. Hyperspectral sensors offer finer spectral measurements than multispectral instruments, with often hundreds of spectral bands of narrow width being recorded, allowing a near continuous spectrum to be reconstructed for each pixel. This presents opportunities for more precise identification of biochemical and biophysical properties of the vegetation compared to when broadband multispectral sensors are used. The downside is the substantial increase in data volume and complexity.

Direct approaches to assess biodiversity using RS are based on analysis of dominant species over larger areas (Turner et al. 2003). These methods map the composition, abundance and distribution of individual species or assemblages and can be used to directly quantify habitats. Indirect approaches use remotely sensed data to measure environmental variables or indicators that are known or understood through biological principles to capture aspects of biodiversity (Duro et al. 2007). These include measures of: (i) the physical environment itself, such as climate and topography; (ii) vegetation production, productivity or function; (iii) habitat characteristics such as spatial arrangement and structure; and (iv) metrics of disturbance which can provide indirect measures of changes in biodiversity.

A wide range of in situ and remote sensing products [e.g., vegetation indices such as the Normalized Difference Vegetation Index (NDVI) and Foliage Projected Cover (FPC)] are beginning to be used for ecological monitoring in a variety of research projects and operational programs. Several satellite sensors [e.g., those on board of Landsat, Indian Remote Sensing Satellite (IRS) and SPOT satellites] have been providing repeated global coverages for several decades. However, significant new opportunities are being presented with the increased availability of very high resolution images, hyperspectral data, Synthetic Aperture Radar, and LiDAR data. Their application has yet to be developed into routine and operational use in surveillance and monitoring of ecosystems, but soon will be.

2.4.3.1 Ecosystem Extent and Distribution

Trends in ecosystem extent and distribution are highly dependent on the scale of the evaluation being undertaken. For example, at a given scale, coastal wetlands may appear to be uninterrupted and uniform. However, at a more resolved scale, edges, patches, corridors associated with tidal creeks, and discontinuous distributions of species become evident. Forested and tree rich landscapes have a high connectivity for forest birds, but that may not be the case for carabid beetles and butterflies. Defining systems in terms of local organisation or dominant species facilitates discussion and analysis, but may also obscure the important linkages between systems across landscapes. It is therefore important to define the systems under consideration and the appropriate scale and resolution at which to observe and analyse them, before discussing trends in their extent and distribution.

Trends in the extent and distribution of ecological systems depend on the temporal and spatial scale of the assessment. Temporal changes occur naturally over long time scales, such as those associated with geological and climatological forces (e.g., glaciation). Change can also occur more quickly as a result of direct shifts in land use such as deforestation and urbanisation or the drainage of wetlands. Thus, trends can be the result of natural forces but may be accelerated by human pressure or exclusively due to human activities.

RS products have a high potential for mapping habitat extent and distribution maps at various scales. Hyperspatial (very high resolution) and hyperspectral sensors have greatly enhanced the possibilities of distinguishing related habitat types at very fine scales. The end-users can use such maps for estimating range and area of habitats, but they could also serve to define and update the sampling frame (the statistical 'population') of habitats for which field sample surveys are in place.

2.4.3.2 Phenology

Phenology is defined as the change in the life cycles of ecosystems and species through the seasons, for example the emergence of leaves or flowers. Phenology can be measured and analysed at different time scales, for example in hours to monitor water stress in crops and irrigation, days to manage plant stress from pests, quarters to monitor seasons, or years to understand seasonality and climate change. A convenient measure of plant phenology is the Normalised Difference Vegetation Index (NDVI)—an index which is available as a consistent data set for the entire Earth every 10 days at a resolution of 250 m (MODIS) and since 1982 for 8 km imagery (NOAA AVHRR; see http://phenology.cr.usgs.gov/ndvi_avhrr.php). Other vegetation indices, such as the Enhanced Vegetation Index (EVI) avoid some the problems associated with NDVI (such as interferences caused by certain soils) and are possible to calculate using data from satellites launched after about 1995. Even better are direct measures of ecosystem function, such as the Fraction Absorbed Photosynthetic Radiation (FAPAR), which relates directly to Gross Primary Production, and is also a standard product of many modern Earth observation satellites.

Seasonal variations in any of the vegetation indices mentioned above can be used to track changes in vegetation phenology (Beck et al. 2007). 'Hypertemporal' imagery (i.e., observed every few days) can be parameterised using unsupervised classifiers and then used to map species distribution, such as a recent demonstration of mapping the extent of *Boswellia papyrifera* in Ethiopia. Such maps of species and biodiversity demonstrate a key advantage of long time series, an advantage of NDVI. Increasingly, landscapes are considered as gradients of particular traits, attributes and species rather than as discrete land cover classes. Treating the landscapes as gradients allows higher map accuracies to be achieved.

Vegetation indices have a spatial and a temporal dimension and so analysis and display of phenological processes can be challenging. For example, hypertemporal NDVI shows how vegetation greenness changes in time and with altitude. Remote sensing technology is being increasingly applied to studies of vegetation and ungulate habitats. For example, superimposing the movement data of radio-tracked giant pandas facilitates the visualisation of correlations between vegetation phenology and seasonal animal movement.

2.4.3.3 Connectivity and Fragmentation

Fragmentation is the process of breaking apart of previously uninterrupted patches of habitat and can have either negative or positive impacts on particular communities. Land and water development, land use and land use change are strongly fragmenting many landscapes and ecosystems e.g., by building highways through forests or damming rivers for hydro-electric power. The latter limits fish migration and separates essential parts of river ecosystems. Dams also reduce the populations of some species groups living in these ecosystems e.g., those that depend on running water, but increases habitat of others e.g., those that need still water. Fragmentation and the increasing length of edge habitat may force migrating species to find new ecological corridors, but may also allow new species (e.g., competitors, pathogens, weeds) to enter new areas. Regardless of specific impacts, fragmentation will in general result in smaller and more vulnerable ecosystems and in shifting the distribution of species.

Fragmentation can be measured through several metrics that can be calculated from RS data. The most simple is the Habitat Patch Density (HPD) that is defined as the total number of areal elements within an area, for instance per km^2. It is related to landscape grain and the composition of the landscape because the higher number of patches that are present a given area the higher is the landscape grain. The increase in HPD indicates an increase of the number of discrete elements in the landscapes and could lead to patch isolation when considering patches of the same habitat. According to meta-population theories, the increase in fragmentation and isolation may cause reductions in the flows of individuals and genes between habitat patches and can therefore threaten the viability of populations (Hanski 1998). The interpretation of HPD should be associated with the type of habitat, since the sensitivity to fragmentation and changes in connectivity associated with isolation, are dependent on constituent habitats and species.

Fragmentation can also be measured through Habitat Patch Size (HPS) that is defined as the average size of a patch in a given area. The HPS is linked to the number of patches within a given area. Although the link between HPD and HPS is not simple, in general if the number of patches within a given area increases there is a reduction in the average patch area. HPS is an indicator related to fragmentation since when a decrease in HPS is related to habitat shrinkage and could results in loss of core habitat, favour edges and decrease connectivity between patches. It has a negative impact on the abundance of habitat specialist species, particularly in forests. It would be interesting to differentiate the HPS by habitat types in order to follow time trends and comparisons between regions. Some animal species, including birds, mammals and reptiles prefers large habitat patches that provide sufficient area to provide them with all the resources needed. A decrease in HPS will often result in a reduction of biodiversity. At the landscape level the effect could however be counterbalanced by habitat diversity and connectivity especially for insects and other small mobile species.

2.5 Relating RS and in Situ Observations: LCCS and GHC

In recent years work has been done to enable harmonisation between RS land cover and in situ habitat data. The monitoring of changes in land cover is important for the monitoring of changes in structural biodiversity. In many cases land use can be inferred from the land cover through virtue of its spatial configuration and context, e.g., a field of maize. Habitat maps can be derived from land cover maps based on RS data along with ancillary geographic information (e.g., soil maps) and other data derived from remote sensing data, e.g., Digital Elevation Models (Mücher 2011).

Where more than one system is used, the relationships between the components of these systems need to be made explicit (Scholes et al. 2012). Additionally, the harmonisation of land cover maps and habitat maps is very important, as habitats have strong associations with floristic and faunal taxa and are therefore considered significant as indicators of biodiversity (Bunce et al. 2013). It is a challenge to combine RS and in situ biodiversity observation systems to monitor changes in land cover and habitat reliably and to better understand the implications on habitat quality and the flora and fauna that it contains. Various initiatives have produced an increasing number of datasets with different classification schemes and mapping integrated yet.

To harmonise global ecosystems (or habitats as their spatial expression), use can be made of Plant Life Forms as first developed Raunkiaer (1934), elaborated by Küchler and Zonneveld (1988) and recently elaborated in the FAO-Land Cover Classification System (LCCS) for land cover interpretation of RS images (Jansen and Di Gregorio 2002), and in the GHCs (Bunce et al. 2008). Plant Life Forms are correlated with the main environmental gradient from the equator to the arctic and therefore can be used in both land cover and habitat mapping. Although LCCS and GHCs both use plant life forms as a basis, they were independently developed and therefore have small differences. Habitat classes are invariably related to land cover classes, but have more ecosystem-based characteristics. A translation system between GHCs and LCCS is important because this links land use as a driver of change and habitats as the spatially explicit representation of biodiversity.

LCCS has been used and proved valuable in land cover interpretation in Africa and Europe. The GHCs represent an important level of information on the status of biodiversity and habitats of good quality can be considered as a proxy for species occurrence. For instance, birds such as the bittern (*Botaurus stellaris*) can only be found in reed marshes and the European large blue butterfly (*Phengaris arion*) only in calcareous grasslands. Vegetation structure is central to both LCCS and the GHC classification and it therefore facilitates interaction between the GHC and LCCS taxonomies (Kosmidou et al. 2014). The main height categories of life forms are comparable between the two approaches with minor differences as shown in Table 2.2. As GHCs have in some cases a more detailed system, the translation between the two approaches requires in some cases ancillary data (Fig. 2.5).

Table 2.2 Vegetation height definitions in the LCCS and GHC taxonomies

Height (m)	LCCS		GHC
>40	A12.A3.B5 Trees		Giga Phanerophytes, GPH
14–40			Forest Phanerophytes, FPH
7–14	A12.A3.B6 Trees		
5–7	A12.A3.B7 Trees		
3–5	A12.A3.B7 Trees	A12.A4.B8 Shrub	Tall Phanerophytes, TPH
2–3	A12.A4.B9 Shrub		
0.6–2			Mid Phanerophytes, MPH
0.3–0.6	A12.A4.B10 Shrub		Low Phanerophytes, LPH
0.05–0.3			Shrubby Chamaephytes, SCH
<0.05			Dwarf Chamaephytes, DCH

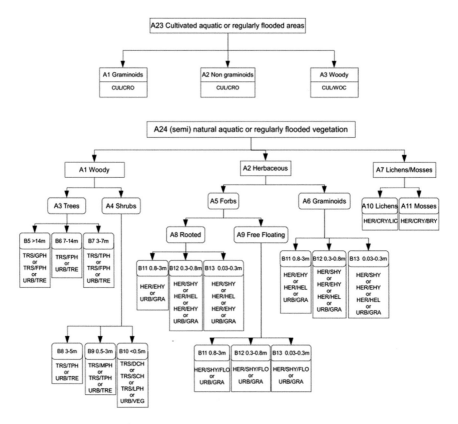

Fig. 2.5 Relation table between the A23 and A24 LCCS categories and the corresponding GHC classes. *Source* Kosmidou et al. (2014)

References

Beck, P. S. A., Jönsson, P., Høgda, K.-A., Karlsen, S. R., Eklundh, L., & Skidmore, A. K. (2007). A ground-validated NDVI dataset for monitoring vegetation dynamics and mapping phenology in Fennoscandia and the Kola Peninsula. *International Journal of Remote Sensing, 28*, 4311–4330.

Bunce, R. G. H., Bogers, M. M. B., Evans, D., Halada, L., Jongman, R. H. G., Mücher, C. A., et al. (2013). The significance of habitats as indicators of biodiversity and their links to species. *Ecological Indicators, 33*, 19–25.

Bunce, R. G. H., Metzger, M. J., Jongman, R. H. G., Brandt, J., de Blust, G., Elena Rossello, R., et al. (2008). A standardized procedure for surveillance and monitoring european habitats and provision of spatial data. *Landscape Ecology, 23*, 11–25.

Duro, D. C., Coops, N. C., Wulder, M. A., & Han, T. (2007). Development of a large area biodiversity monitoring system driven by remote sensing. *Progress in Physical Geography, 31*, 235–260.

Hall, L. S., Krausman, P. R., & Morrison, M. L. (1997). The habitat concept and a plea for standard terminology. *Wildlife Society Bulletin, 25*, 173–182.

Hanski, I. (1998). Metapopulation dynamics. *Nature, 396*, 41–49.

Jansen, L. J. M., & Di Gregorio, A. (2002). Parametric land cover and land-use classifications as tools for environmental change detection. *Agriculture, Ecosystems & Environment, 91*, 89–100.

Jongman, R. H. G., Bunce, R. G. H., Metzger, M. J., Mücher, C. A., Howard, D. C., & Mateus, V. L. (2006). Objectives and applications of a statistical environmental stratification of Europe. *Landscape Ecology, 21*, 409–419.

Jürgens, N., Schmiedel, U., Haarmeyer, D. H., Dengler, J., Finckh, M., Goetze, D., et al. (2011). The BIOTA biodiversity observatories in Africa—A standardized framework for large-scale environmental monitoring. *Environmental Monitoring and Assessment, 184*, 655–678.

Kosmidou, V., Petrou, Z., Bunce, R. G. H., Mücher, C. A., Jongman, R. H. G., Bogers, M. M. B., et al. (2014). Harmonization of the Land Cover Classification System (LCCS) with the General Habitat Categories (GHC) classification system. *Ecological Indicators, 36*, 290–300.

Küchler, A.W., & Zonneveld, I.S. (1988). *Handbook of vegetation science*. Dordrecht, The Netherlands: Kluwer Academic Publishers.

Metzger, M. J., Brus, D. J., Bunce, R. G. H., Carey, P. D., Gonçalves, J., Honrado, J. P., et al. (2013a). Environmental stratifications as the basis for national, European and global ecological monitoring. *Ecological Indicators, 33*, 26–35.

Metzger, M. J., Bunce, R. G. H., Jongman, R. H. G., Sayre, R., Trabucco, A., & Zomer, R. (2013b). A high resolution bioclimate map of the world: A unifying framework for global biodiversity research. *Global Ecology and Biogeography, 22*, 630–638.

Metzger, M. J., Bunce, R. G. H., van Eupen, M., & Mirtl, M. (2010). An assessment of long term ecosystem research activities across European socio-ecological gradients. *Journal of Environmental Management, 91*, 1357–1365.

Molnár, Z., Bartha, S., Seregélyes, T., Illyés, E., Botta-Duká, Z., Tímár, G., et al. (2007). A grid-based, satellite-image supported, multi-attributed vegetation mapping method (MÉTA). *Folia Geobotanica, 42*, 225–247.

Mücher, C. A. (2011). Land use, climate change and biodiversity modeling: Perspectives and applications. In Y. Trisurat, R. P. Shrestha, & R. Alkemade (Eds.), *Land use, climate change and biodiversity modeling: Perspectives and applications* (pp. 78–102). Hershey, PA, USA: IGI Global.

Mücher, C. A., Kooistra, L., Vermeulen, M., Vandenborre, J., Haest, B., & Haveman, R. (2013). Quantifying the structure of Natura 2000 heathland habitats using spectral mixture analysis and segmentation techniques on hyperspectral imagery. *Ecological Indicators, 33*, 71–81.

Noss, R. F. (1990). Indicators for monitoring biodiversity: a hierarchical approach. *Conservation Biology, 4*, 355–364.

Pereira, H. M., Ferrier, S., Walters, M., Geller, G. N., Jongman, R. H. G., Scholes, R. J., et al. (2013). Essential biodiversity variables. *Science, 339*, 277–278.

Raunkiaer, C. (1934). *The life forms of plants and statistical plant geography, being the collected papers of C Raunkiaer*. Oxford, UK: Clarendon.

Scholes, R. J., Mace, G. M., Turner, W., Geller, G. N., Jürgens, N., Larigauderie, A., et al. (2008). Toward a global biodiversity observing system. *Science, 321*, 1044–1045.

Scholes, R. J., Walters, M., Turak, E., Saarenmaa, H., Heip, C. H. R., Ó Tuama, É, Faith, D. P., et al. (2012). Building a global observing system for biodiversity. *Current Opinion in Environmental Sustainability* 4, 139–146.

Turner, W., Spector, S., Gardiner, N., Fladeland, M., Sterling, E., & Steininger, M. (2003). Remote sensing for biodiversity science and conservation. *Trends in Ecology & Evolution, 18,* 306–314.

Vanden Borre, J., Paelinckx, D., Mücher, C. A., Kooistra, L., Haest, B., De Blust, G., et al. (2011). Integrating remote sensing in Natura 2000 habitat monitoring: Prospects on the way forward. *Journal for Nature Conservation, 19,* 116–125.

Chapter 3
Ecosystem Services

Patricia Balvanera, Sandra Quijas, Daniel S. Karp, Neville Ash,
Elena M. Bennett, Roel Boumans, Claire Brown, Kai M.A. Chan,
Rebecca Chaplin-Kramer, Benjamin S. Halpern, Jordi Honey-Rosés,
Choong-Ki Kim, Wolfgang Cramer, Maria José Martínez-Harms,
Harold Mooney, Tuyeni Mwampamba, Jeanne Nel, Stephen Polasky,
Belinda Reyers, Joe Roman, Woody Turner, Robert J. Scholes,
Heather Tallis, Kirsten Thonicke, Ferdinando Villa, Matt Walpole
and Ariane Walz

Abstract Ecosystem services are increasingly incorporated into explicit policy targets and can be an effective tool for informing decisions about the use and management of the planet's resources, especially when trade-offs and synergies need to be taken into account. The challenge is to find meaningful and robust indicators to quantify ecosystem services, measure changes in demand and supply and predict future direction. This chapter addresses the basic requirements for

P. Balvanera (✉) · S. Quijas · T. Mwampamba
Instituto de Investigaciones en Ecosistemas y Sustentabilidad, Universidad Nacional
Autónoma de México, AP 27-3, Mexico City, Michoacán, Mexico
e-mail: pbalvanera@cieco.unam.mx

S. Quijas
e-mail: squijas@gmail.com

T. Mwampamba
e-mail: tuyeni@cieco.unam.mx

S. Quijas
Centro Universitario de la Costa, Universidad de Guadalajara, Guadalajara, Jalisco, Mexico

D.S. Karp
Institute for Resources, Environment, and Sustainability, University of British Columbia,
Vancouver, BC, Canada
e-mail: dkarp@ucdavis.edu

N. Ash
Director, UNEP World Conservation Monitoring Centre, 219 Huntingdon Road,
Cambridge CB3 0DL, UK
e-mail: neville.ash@unep-wcmc.org

E.M. Bennett
EWR Steacie Fellow, Associate Professor, Natural Resource Sciences and McGill School
of Environment, Montreal, Quebec, Canada
e-mail: elena.bennett@mcgill.ca

© The Author(s) 2017
M. Walters and R.J. Scholes (eds.), *The GEO Handbook on Biodiversity
Observation Networks*, DOI 10.1007/978-3-319-27288-7_3

collecting such observations and data on ecosystem services. Biodiversity regulates the ability of the ecosystem to supply ecosystem services, can be directly harvested to meet people's material needs, and are valued by societies for its non-tangible contributions to well-being. Societies are deeply embedded within ecosystems, depending on and influencing the ecosystem services they produce. The different types of ecosystem services (provisioning, regulating, and cultural), and their different components (supply, delivery, contribution to well-being, and value) can be monitored at global to local scales. Different data sources are best suited to account for different components of ecosystem services and spatial scales and include:

R. Boumans · J. Roman
Afordable Futures, Charlotte, VT, USA
e-mail: rboumans@afordablefutures.com

J. Roman
e-mail: jroman@uvm.edu

C. Brown · M. Walpole
Director, UNEP World Conservation Monitoring Centre, 219 Huntingdon Road,
Cambridge CB3 0DL, UK
e-mail: Claire.Brown@unep-wcmc.org

M. Walpole
e-mail: matt.walpole@unep-wcmc.org

K.M.A. Chan
Institute for Resources, Environment and Sustainability, University of British Columbia,
Vancouver, British Columbia, Canada
e-mail: kaichan@ires.ubc.ca

R. Chaplin-Kramer
Natural Capital Project, Stanford University, Stanford, CA, USA
e-mail: bchaplin@stanford.edu

B.S. Halpern
National Center for Ecological Analysis and Synthesis, University of California Santa
Barbara, Santa Barbara, CA, USA
e-mail: halpern@bren.ucsb.edu

B.S. Halpern
Bren School of Environmental Science and Management, UCSB, Santa Barbara, CA, USA

B.S. Halpern
Silwood Park Campus, Imperial College London, Ascot, UK

J. Honey-Rosés
School of Community and Regional Planning, University of British Columbia,
Vancouver, British Columbia, Canada
e-mail: jhoney@mail.ubc.ca

C.-K. Kim
Environmental Policy Research Group, Korea Environment Institute, Sejong,
Republic of Korea
e-mail: ckkim@kei.re.kr

census data at national scales, remote sensing, field-based estimations, community monitoring, and models. Data availability, advantages and limitations of each are discussed. Progress towards monitoring different types of services and gaps are explored. Ways of exploring synergies and trade-offs among services and stakeholders, using scenarios to predict future ecosystem services, and including stakeholders in monitoring ecosystem services are discussed. The need of a network for monitoring ecosystem services to synergise efforts is stressed. Monitoring ecosystem services is vital for informing policy (or decision making) to protect human well-being and the natural systems upon which it relies at different scales. Using this information in decision making across all scales will be central to our endeavours to transform to more sustainable and equitable futures.

W. Cramer
Mediterranean Institute for Biodiversity and Ecology, Aix Marseille University,
CNRS, IRD, Avignon University, Aix-en-Provence, France
e-mail: wolfgang.cramer@imbe.fr

M.J. Martínez-Harms
Centre for Biodiversity and Conservation Science, School of Biological Sciences,
University of Queensland, Brisbane, Queensland, Australia
e-mail: m.martinezharms@uq.edu.au

H. Mooney
Department of Biology, Stanford University, Stanford, CA, USA
e-mail: hmooney@stanford.edu

J. Nel · B. Reyers
Director, GRAID programme, Stockholm Resilience Centre, Stockholm University,
Kräftriket 2B, George, Stockholm, Sweden
e-mail: JNel4water@gmail.com

B. Reyers
e-mail: belinda.reyers@su.se

S. Polasky
Departments of Ecology, Evolution, and Behaviour, University of Minnesota,
St. Paul, MN, USA
e-mail: polasky@umn.edu

S. Polasky
Department of Applied Economics, University of Minnesota, St. Paul, MN, USA

W. Turner
Earth Science Division, NASA Headquarters, Los Angeles, Washington, DC, USA
e-mail: Woody.Turner@nasa.gov

R.J. Scholes
Global Change and Sustainability Research Institute, University of the Witwatersrand,
Johannesburg, South Africa
e-mail: bob.scholes@wits.ac.za

H. Tallis
The Nature Conservancy, San Francisco, CA, USA
e-mail: htallis@tnc.org

Keywords Biodiversity · Policy targets · Trade-offs · National statistics · Remote sensing · Field estimations · Community monitoring · Models

3.1   Introduction

Ecosystem services are the benefits people obtain from ecosystems and are co-produced by the interactions between ecosystems and societies. Since the Millennium Ecosystem Assessment (MA 2005) governments have embedded ecosystem services and natural capital in explicit policy targets. Globally, for example, the Parties to the Convention on Biological Diversity (CBD; www.cbd.int) have committed to 'enhancing the benefits to all from biodiversity and ecosystem services'. The CBD Aichi Target 14 is of particular relevance to ecosystem services: '*By 2020, ecosystems that provide essential services, including services related to water, and contribute to health, livelihoods and well-being, are restored and safeguarded, taking into account the needs of women, indigenous and local communities, and the poor and vulnerable*'. Beyond the conservation sector, interest in ecosystem services is increasingly aimed at the development of policies at national and global scales (Griggs et al. 2013). Regionally, the European Union Biodiversity Strategy to 2020, for example, aimed to halt the degradation of ecosystem services, and to map and assess the state of ecosystems and their services in their national territories by 2014 (Maes et al. 2016). This study also aimed to assess the economic value of such services, and promote the integration of these values into accounting and reporting systems at EU and national levels by 2020. Non-EU governments of nations such as Australia, Canada and Mexico are also incorporating ecosystem services and natural capital into national accounts.

At a national and sub-national scale, ecosystem services can be an effective tool for informing decisions about the use and management of the planet's resources, especially when trade-offs and synergies need to be taken into account. Without this information, decisions that determine the fate of terrestrial, coastal, and marine systems and the benefits they provide, are made in the dark, with little understanding of the ecosystem services outcomes (benefits and costs) of any given

K. Thonicke
Potsdam Institute for Climate Impact Research, Potsdam, Germany
e-mail: Kirsten.Thonicke@pik-potsdam.de

F. Villa
Basque Centre for Climate Change (BC3); IKERBASQUE, Basque foundation for Science, Burlington, Bilbao, Spain
e-mail: ferdinando.villa@bc3research.org

A. Walz
Institute of Earth and Environmental Science, University of Potsdam, Potsdam, Germany
e-mail: ariane.walz@pik-potsdam.de

decision or its consequences for the different stakeholders depending on these services.

While many observations and datasets are available to measure progress towards global, regional, and national goals for ecosystem services, and to ensure effective decision-making for sustainable human use of the planet's resources (Egoh et al. 2012), their coverage is patchy, incomplete and inconsistent. The challenge is to find meaningful and robust indicators to quantify ecosystem services, measure changes in demand and supply and predict future scenarios. At present, most governments are not effectively measuring or monitoring ecosystem services. This chapter addresses the basic requirements for collecting information on ecosystem services.

3.2 Biodiversity and Ecosystem Services

Biodiversity is related to ecosystem services through a variety of mechanisms operating at different spatial scales (Fig. 3.1) (Mace et al. 2012). Biodiversity regulates the state, the rates and in many cases the stability of ecosystem processes fundamental to most ecosystem services (Cardinale et al. 2012). Components of biodiversity are also directly harvested to meet people's material needs, and are also valued by societies for their non-tangible contributions to well-being, for example to psychological health, people's identity and the asset it can be for future generations. Fundamentally, biodiversity provides the evolutionary building blocks of

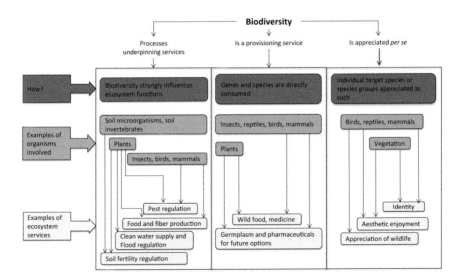

Fig. 3.1 Biodiversity is linked to ecosystem services in three different ways: (*i*) as a regulator of the ecosystem functions that lead to the supply of provisioning, regulating or supporting services, (*ii*) as a provisioning service, (*iii*) as something that is appreciated in itself rather than for the benefits obtained from it. Selected examples are used to illustrate these linkages. *Source* Modified from Mace et al. (2012), Reyers et al. (2012)

life on Earth and therefore provides important adaptive capacity through its continued ability to support desired ecosystem services and processes in the face of often rapidly changing selective pressures (Mace et al. 2014).

Due to the complexity of the links between biodiversity and ecosystem services, as well as the important role played by other non-biophysical inputs into the goods and benefits we obtain from ecosystems (Díaz et al. 2015), monitoring biodiversity alone is not sufficient to understand the status and trends of the services it provides. In fact, monitoring annual changes in the state of ecosystems and determining trends in ecosystem services, can contribute to our understanding of changes in biodiversity and inform on the underlying dynamics of the complex interactions between societies and ecosystems.

3.3 Key Ecosystem Service Concepts

Societies are embedded within ecosystems, depending on and influencing the ecosystem services they produce. The characteristics of ecosystems, such as species composition, tree cover or growth conditions, modulate the type and magnitude of ecosystem services that can flow to societies. Management regimes, technologies, as well as tenure and access arrangements modulate the ways by which ecosystem services are produced and benefit societies. In other words, ecosystem services result from the interactions between ecosystems and societies, which together form a social-ecological system.

Four types of ecosystem services can be distinguished (MA 2005), though we focus only on three of them in this chapter. *Provisioning services* are the goods that can be extracted and consumed from ecosystems and are often valued in markets: for example, water, food, wood and biofuels. *Regulating services* are the benefits derived from ecosystem processes that modulate the conditions which we experience: such as the regulation of climate, soil fertility or floods. They seldom have markets, and must be valued indirectly. *Cultural services* are the real but not physical ('intangible') benefits that emerge from interactions between humans and ecosystems (Chan et al. 2012), for instance employment, sense of identity, spiritual value, aesthetic value and cognitive development. Some cultural services, such as recreation, do have markets, while others do not. The fourth category, which we do not elaborate on, is *supporting services*, the fundamental ecosystem processes such as photosynthesis, nutrient cycling and evolution, which permit the delivery of the first three categories, and thus find societal benefit through them.

In order to fully understand ecosystem services, we need to measure and monitor four different components: supply, delivery, contribution to well-being, and value (Tallis et al. 2012). Table 3.1 provides a detailed examination of each of these components across different categories of ecosystem services. The table includes a definition and some popular metrics or indicators used in the quantification and assessment of services. This list is not exhaustive since it does not cover all services or potential indicators, but rather presents a range of different types of services that have been found to be very relevant to societies.

Table 3.1 Examples of provisioning, regulating and cultural ecosystem services, including descriptions, drivers and potential indicators for each of the four components of the ecosystem service (supply, delivery, contribution to well-being and value)

Service	Description	Drivers	Ecosystem service component			Economic value
			Supply	Delivery	Contributions to well-being	
Provisioning						
Crops	All cultivated edible plant products (e.g. maize, wheat, olive, apple)	Biophysical (e.g. climatic, soil), crop choice (e.g. species, genetic characteristics, amount of seed/plant), management (e.g. fertilizers, irrigation, labour and machinery) and societal (e.g. agricultural policies, crop market price, land tenure, agricultural institutional arrangements)	Potential amount of important crops	Total production of all commercial crops (t), caloric or micronutrient content of all commercial crops (g)	% caloric or micronutrient intake contributed by crops, % income or number of jobs contributed by crops	Market value of all commercial crops (US$)
Fodder	All vegetable tissue and grains grown in rangelands and pastures as well as in agricultural fields to feed livestock	Similar to that of crops	Amount of biomass available for fodder (pasture or forage) (t)	Total production of fodder (t), amount of protein, number of animals grazed	% contribution of fodder (to support cattle) to protein consumed	Market value of fodder (US$)

(continued)

P. Balvanera et al.

Service	Description	Drivers	Ecosystem service component			Economic value
			Supply	Delivery	Contributions to well-being	
Livestock	Includes beef, pork, goat and other species grown to obtain meat, milk and skin	Biophysical (climatic, feed, fodder), livestock choice (e.g. species, genetic characteristics), management (e.g. number of animals, target age, type of enclosure) and societal (e.g. demand, livestock policies, livestock grower culture, land tenure of feeding grounds)		Total production of meat, milk and other livestock products (t, m^3), protein content of all livestock products	% protein consumption contributed by livestock, % population reliant on livestock for income or food, % income or jobs contributed by livestock	Market value of all livestock products (US$)
Aquaculture	All fish and invertebrate species cultivated in continental, coastal or marine water bodies	Biophysical (e.g. water temperature), species choice (e.g. taxonomic identity), management (e.g. target harvest weight, feed, use of antibiotics), and societal (e.g. legislation, market price, access to suitable areas)		Total harvest (t), total protein content of landings (kg), total subsistence aquaculture production (t)	% protein consumption contributed by aquaculture, % population reliant on aquaculture for food or income, % income contributed by fishing, % jobs contributed by fishing	Market value of all aquaculture products (US$)

Table 3.1 (continued)

Service	Description	Drivers	Ecosystem service component		Contributions to well-being	Economic value
			Supply	Delivery		
Fisheries	Includes aquatic invertebrate and fish species harvested from continental, coastal or marine water bodies	Biophysical (e.g. biomass and abundance of target species), management (e.g. vessel type, fishing devices, fishing intensity), and societal (e.g. demand, institutional arrangements)	Biomass or abundance of (commercially) important species	Volume of weight of landings-harvest (t), protein content of landings (kg), total subsistence fish catch (t)	% protein consumption contributed by fisheries, % population reliant on fisheries for food or income, % income contributed by fishing, % jobs contributed by fishing	Market value of all fisheries products (US$)
Wood	Includes tree trunks (normally with diameter at breast height larger than 30 cm) harvested from natural forests, plantations, or some agro-ecosystems	Biophysical (e.g. biomass and abundance of target species, land cover), management (e.g. machinery, target size, forest management), and societal (e.g. demand, market price, institutional arrangements for forest management, legal access to forests)	Amount of woody biomass generated per year (m^3/y)	Volume of harvested wood (m^3)	% income contributed by wood harvesting, % jobs contributed by wood harvesting, % house constructed with wood	Market value of harvested wood (US$)

(continued)

Table 3.1 (continued)

Service	Description	Drivers	Ecosystem service component			Economic value
			Supply	Delivery	Contributions to well-being	
Biofuels	Refers to fuels in which energy is derived from photosynthesis including woody materials, plant carbohydrates, vegetable oils and crop seeds	Biophysical conditions (e.g. climate, soil), biofuel choice (e.g. species), management (e.g. harvested from wild or cultivated), and societal factors (e.g. biofuel policies, demand, market price)		Weight or volume of biofuel, fuelwood or charcoal produced (kg, m^3), amount of energy produced (kJ)	Amount of fossil fuel use avoided (t), total GHG emissions avoided (t), % income or jobs contributed by biofuel production, % energy consumptions contributed by biofuels	Market value of biofuels (US$)
Harvested wild goods (including game meat, construction or weaving materials, medicinal plants)	All goods harvested from ecosystems, including fisheries, wild vertebrates consumed for food, wood, poles and other uses (e.g. honey, medicinal plants)	Biophysical (e.g. biomass of abundance of target species), biofuel choice (e.g. species), management (e.g. hunting intensity, technology), and societal (e.g. legal access to game meat, demand, cultural preferences)	Biomass or abundance of all potentially harvestable wild food products	Amount of wild products harvested (t). Amount of game meat harvested (t), protein content of game meat (kg)	% population reliant on harvested wild products for food, income and other uses, % protein consumption contributed by game meat, % population reliant on fisheries for food or income	Market value of harvested wild products (US$), market value of game meat (US$)

(continued)

Table 3.1 (continued)

Service	Description	Drivers	Ecosystem service component		Contributions to well-being	Economic value
			Supply	Delivery		
Water	Volume of surface water flow and the amount of water stored in groundwater for domestic, industrial and agricultural use	Biophysical (e.g. climatic, land cover) and societal (e.g. location of user, demand per type of user)	Volume of superficial or ground-water available (m^3)	Volume of superficial or ground-water withdrawn per user (agricultural production, domestic and industrial)	% of population or water user or economic sector with available water above water needs	Market value of water to agriculture, tourism, industry, etc. (US$), marginal contribution of irrigation to crop market value
Hydropower generation	Energy produced in dams derived from water produced by the watershed	Biophysical (water yield, timing of water release) and societal (e.g. water consumption, dam location, energy production per dam, energy policies)	Potential energy produced by hydropower (kW)	Hydropower energy produced (kW)	% energy needs contributed by wind hydropower, % GHG emissions reduced by production of hydropower	Market value of hydropower (US$), avoided water replacement costs (US$)
Regulating						
Climate regulation (Carbon stocks and uptake)	Mediated by carbon stored over the long-term in vegetation that is not released and carbon taken from the atmosphere via photosynthesis	Biophysical (climatic, soil, land cover)	Amount of emissions avoided by maintaining carbon stocks (t of C), amount of carbon taken by vegetation from the atmosphere (t of C)	Amount of emissions avoided by maintaining carbon stocks (t of C), amount of carbon taken by vegetation from the atmosphere (t of C)	Reduced negative impacts (from floods, wind, drought) on society (% population, % more vulnerable area or people) from climate regulation	Market value of maintained carbon stocks and carbon uptake (US$), avoided costs from climate change (US$)

(continued)

Table 3.1 (continued)

Service	Description	Drivers	Ecosystem service component		Contributions to well-being	Economic value
			Supply	Delivery		
Regulation of marine and fresh water quality	Can be impaired by nutrients (phosphorus and nitrogen), sediment, dissolved organic carbon content, temperature, pH, and concentrations of pathogens or toxic compounds. The abiotic and biotic components of ecosystems can contribute to mitigate such contaminants	Biophysical (e.g. land and sea bottom cover, aquatic biodiversity), management (e.g. fertilization or sewage upstream, water treatment) and societal (e.g. sanitation regulations, water quality standards per type of use)	Mass of nutrients, organic matter, sediments, or toxic organisms or compounds removed (kg), changes in temperature, pH	Water conditions in relation to standards for different water users at or above withdrawal point	Avoided disease by water treatment	Avoided water treatment costs (US$); cost of wetland construction for nutrient removal (US$)
Regulation of soil fertility	Refers to the physical, chemical, and biological characteristics of soils that underpin the amount of nutrients available for agriculture, fodder, forestry and biofuel production	Biophysical (e.g. geologic, topographic, soil, land cover) and management (e.g. rotation cycles, soil preparation, fertilizer, irrigation)	Soil nutrient availability (mg)	Marginal contribution of soils to agricultural, forestry and biofuel production	Marginal contribution of soils to food, wood or biofuel consumption	Marginal contribution of soils to economic value of agricultural, forestry and biofuel production (US$)

(continued)

Table 3.1 (continued)

Service	Description	Drivers	Ecosystem service component		Contributions to well-being	Economic value
			Supply	Delivery		
Regulation of soil erosion	Mediated by the characteristics of a landscape, land cover and soils that regulate the amount of soil loss driven by rain and reduce the amount of sediments accumulated in hydraulic infrastructure	Biophysical (e.g. rain intensity, topographic, soil, land cover), management (e.g. soil preparation, fertilizer, irrigation) and societal (e.g. characteristics of dams and human made water canals)	Mass of retained soils (kg)	Mass of soils retained to support productive activities or to avoid dams, reservoirs and water canals; Mass of soils retained to prevent soil sedimentation in residential or industrial areas (kg)	Reduced negative impacts of soil loss and of sediment flows to different stakeholders	Marginal contribution of soils retained to productive activities and avoided costs of dredging (US$)
Flood regulation	A function of the vegetation and soils that increase infiltration rates and thus reduce the amount of surface water flow that contributes to floods	Biophysical (e.g. climate, soil, aquatic vegetation), management (e.g. hydraulic infrastructure) and societal (e.g. people's location, infrastructure characteristics)	Flood volume regulated by vegetation and soils (m^3)	Area of avoided flood damage due to regulation by vegetation and soils (ha)	Number of people protected from flood by regulation from vegetation and soils	Avoided economic loss by flood regulation from vegetation and soils (US$)

(continued)

Table 3.1 (continued)

Service	Description	Drivers	Ecosystem service component			
			Supply	Delivery	Contributions to well-being	Economic value
Coastal protection	Refers to the idea that coastal habitats can serve as natural shields against waves, storms and wind that may lead to infrastructure loss, flooding and erosion	Biophysical (e.g. climatic, wave, land and sea bottom cover) and societal (e.g. people's location, infrastructure characteristics)		Area of avoided infrastructure loss, flood and erosion (ha)	Number of people protected from infrastructure loss, flooding and erosion from coastal protection	Avoided economic loss by coastal protection (US$)
Regulation of commercially important marine species populations	Mangroves, coral reefs and sea grass provide nursery grounds and refuge for many recreationally and commercially valuable marine species	Biophysical (e.g. land and sea bottom cover, dependence of target species on these habitats) conditions that contribute to marginal increased fisheries yield		Marginal contributions of coastal habitats (e.g. mangroves) to fisheries production		Marginal contribution of coastal habitats (e.g. mangroves) to market value of fisheries production (US$)
Pollination	Bees, bats, birds and other animals pollinate fruit and seed crops, contributing to increased yield, quality, and stability	Biophysical (e.g. pollinator identities and abundances) and management (e.g. types and density of crops, land use and land cover type around agricultural fields)	Pollinator abundances and pollination rates	Marginal contribution of pollinators to crop production	Marginal contribution of pollinators to food or biofuel production	Marginal contribution of pollination to crop market value (US$)

(continued)

Table 3.1 (continued)

Service	Description	Drivers	Ecosystem service component			
			Supply	Delivery	Contributions to well-being	Economic value
Pest control	Insects, bats and birds regulate the abundances of agricultural pests	Biophysical (e.g. pest identity and abundance, trophic interactions among insects, birds and bats) and management (e.g. crop type, landscape configuration)	Abundances of pests and their natural enemies	Regulation of pests by their natural enemies	Marginal contribution of pest control to food or biofuel production	Marginal contribution of pest control to crop market value (US$)
Cultural						
All non-tangible benefits	Includes a large array of non-tangible benefits from ecosystems that include heritage (cultural or religious), inspiration (spiritual or artistic), sense of place, identity, social relations, and education, among others	A suite of biophysical (e.g. biodiversity, topography), management (e.g. dominant management activities) and societal (e.g. culture) conditions	Non-material benefits from ecosystems and the interactions among them			

(continued)

Table 3.1 (continued)

Service	Description	Drivers	Ecosystem service component			Economic value
			Supply	Delivery	Contributions to well-being	
Aesthetic views	Refers to various landscape features that convey aesthetic characteristics that are appreciated and enjoyed	Biophysical (e.g. topography), management (e.g. land use and land cover type), and societal (e.g. access roads or boating areas, number of visitors, cultural preferences)	Area that provides aesthetic views	Area that is enjoyed by visitors or local inhabitants for its aesthetic views, number of visitors	Marginal contributions to income or well-being of visitors and to local inhabitants derived from aesthetic views	Economic revenues derived from visits to aesthetic areas or marginal contribution to real estate prices by aesthetic characteristics (US$)
Nature-based tourism	A function of multiple characteristics of landscapes, water bodies and biodiversity that determine whether areas are attractive to tourists	Biophysical (e.g. bird richness, characteristics of water bodies), management (e.g. land use and land cover type), and societal (e.g. protection status, facilities to support visits, distance from cities)	Area that is suitable for nature-based tourisms	Area where nature-based tourism occurs, number of visitors	Marginal contributions to income or well-being of visitors and local inhabitants derived from nature-based tourism	Economic revenues derived from or costs associated with undertaking nature-based tourism (US$)

(continued)

Table 3.1 (continued)

Service	Description	Drivers	Ecosystem service component		Contributions to well-being	Economic value
			Supply	Delivery		
Recreation	Includes hiking, angling, cycling, birding, swimming, diving, and others	A suite of biophysical (e.g. biodiversity, topography), management (e.g. land use and land cover type) and societal (e.g. access roads or boating areas, number of visitors, facilities to support visits, distance from cities)	Areas that are suitable for recreation-based tourisms	Area where recreation-based tourism occurs, number of visitors	Marginal contributions to income or well-being of visitors and local inhabitants derived from recreation-based tourism	Economic revenues derived from or costs associated with undertaking nature-based recreation (US$)

Supply refers to the potential of a social-ecological system to generate a service, typically quantified as a flow (i.e., an amount per unit time). Ecosystem condition (e.g., intact or degraded, stressed or unstressed) and processes (e.g., primary productivity), as well as the way ecosystems are managed, are taken into account when determining supply. This is the component of ecosystem services that has been most commonly measured.

Delivery accounts for how much of the service is actually extracted (e.g., amount of timber harvested), used (e.g., area of avoided flood damage, area that is enjoyed by visitors), and delivered to societies (e.g., spatial location of those benefiting from flood regulation), and how societies have access to these services (e.g., laws rules, norms and restrictions that limit access to a service). Delivery thus depends on the links between ecosystem services supply and people's location, activities and societal factors determining access to services.

Contribution to well-being accounts for the change in people's well-being, which results from consuming, using, or having access to the service. Changes in living standards, nutrition status, mortality rates, social conflicts, security in the face of extreme environmental conditions, or happiness partially depend on the delivery of ecosystem services. This component of ecosystem services is the least understood and seldom quantified. One of the issues is that well-being typically has many components and many causes, so it hard to isolate the contributions of a particular service.

Value refers here to the relative importance society attributes to the service. The value of ecosystem services is often accounted in monetary terms, but other ways of establishing the socio-cultural value are potentially equally valid, and may be more appropriate than monetary valuation for some services. For instance, contributions to longevity or perceived quality of life need not be expressed in monetary terms. The monetary value of most provisioning services (e.g., timber) is provided by markets. Where freely-traded markets do not exists (for instance, this is frequently the case for water service), the value can be estimated through a variety of methods, such as the cost of delivering a substitute, or the marginal value addition of the service to other services which do have markets. Valuation approaches, based on willingness to pay, damage costs avoided, travel costs, or hedonic values, have been used to attribute economic value to many regulating and cultural services. Socio-cultural values of ecosystem services to an individual can be assessed through various valuation methods, such as through preference surveys, paired comparisons, and narrative or participatory methods. What is frequently reported is the aggregate societal value resulting from some combination of individual valuations.

These components of ecosystem services feed back into the way social-ecological systems are managed and governed. Supply allows for delivery which allows for contributions to well-being which, in turn, influences value. Ecosystem service contributions to well-being, shape the status of and vision for the well-being of individuals and societies, which directly influences the way formal and informal institutions are designed to modulate interactions with the environment. Value determines which services are fostered, and shape institutions and management interventions, aimed at modifying social-ecological conditions to promote the supply of the desired services at the cost of other services (Díaz et al. 2015).

3.4 Monitoring Ecosystem Services

Ecosystem services can be monitored at multiple spatial scales. For global observation systems, emphasizing the nation state as the focal unit allows for better tracking of progress towards national targets for ecosystem services. In addition, many key global policies, such as the Convention on Biological Diversity (CBD; www.cbd.int), the Sustainable Development Goals (https://sustainabledevelopment.un.org/), and the Commission on Climate Change and Development (www.ccdcommission.org) are governed by mutual agreement of participating nations, requiring monitoring of progress toward global targets. Monitoring, however, can also take place at the local scale, and data can then be aggregated up to the national and global scales, but this is not always a straightforward procedure (Scholes 2009). A multiple scale approach makes it possible for information from one spatial scale to be tested or refined using data produced at other scales. Such comprehensive monitoring at different spatial scales can include national statistics and remote sensing to cover national to global scales, as well as remote sensing and field-based assessments to cover local scales. Models can be developed at all spatial scales.

Different data sources are best suited to account for different components and spatial scales of ecosystem services (see Table 3.2). Supply is best characterised by data sources that consider the condition of social-ecological systems, for example, from remote sensing and models. Delivery is often based on societal characteristics and can be accounted for from national statistics, field-assessment and models. Contributions to well-being are documented in different ways (mostly field assessments, national statistics and census) and have seldom been explicitly incorporated into models. Economic value can be derived from markets, national statistics or from economic models. Sociocultural value can be obtained from field assessments of preferences, or from the analysis of cultural norms. Different types of value have been incorporated into models.

3.5 National Statistics

Census data at national scales are readily available for several ecosystem services. In most cases the census has been conducted at a much more resolved scale (the census district, which may be as small as a neighbourhood). Sometimes such data is available for local analysis, subject to special procedures designed to protect the privacy of individual respondents. The United Nation's Food and Agriculture Organisation publishes a global database (http://faostat.fao.org/) of the amount produced or extracted (delivery), traded, and the monetary value (value) of several ecosystem services, for example, total production of all commercial crops for countries or regions, export or import quantity of trade crops and their economic value per unit. Other databases, such as that of the World Bank (http://data.worldbank.org) report water withdrawals and water availability to people. Some of

58

Table 3.2 Comparison between different ecosystem service data sources

	National statistics	Remote sensing		Field estimations		Models				
	FAOSTAT WORLD BANK	High resolution	Low resolution	TESSA	Natura	InVEST	LPJmL	ARIES	ESTA	MIMES
Ecosystem service component										
Supply		✓	✓			✓	✓	✓	✓	✓
Delivery	✓	✓		✓	✓	✓		✓		✓
Contribution to well-being				✓	✓					✓
Economic value	✓			✓	✓	✓		✓	✓	✓
Spatial scale										
Local/landscape		✓		✓	✓	✓		✓	✓	✓
National	✓	✓	✓			✓	✓	✓		✓
Global			✓			✓	✓			✓

FAOSTAT: The Statistics Division of the Food and Agriculture Organization of the United Nations (FAO 2012), TESSA: Toolkit for Ecosystem Service Assessments (Peh et al. 2014), Natura: Assessing Socioeconomic Benefits (Kettunen et al. 2009), InVEST: Integrated Valuation of Environmental Services and Tradeoffs (Tallis et al. 2013), LPJmL: Lund-Potsdam-Jena managed Land Dynamic Global Vegetation and Water Balance Model (Bondeau et al. 2007), ARIES: ARtificial Intelligence for Ecosystem Services (Bagstad et al. 2013a), ESTA: Ecosystem Service Tradeoff Analysis (White et al. 2012), and MIMES: Multiscale Integrated Models of Ecosystem Services (Altman et al. 2014) the services are monitored in most countries and updated annually (e.g., crops), while others are only available for a small subset of nation states and updated infrequently (~ 5 years; e.g., water withdrawal). While these statistics provide very relevant information for assessing provisioning ecosystem services, they imperfectly reflect their delivery and economic value. They cannot, for instance, inform on the supply of the services. They further inform only partially on the delivery of the services, as they can only account for the fraction of the food production that enters markets and national statistics. The stronger biases are for economic values, which are the product of markets and incentives, and do not necessarily account for the marginal contribution of ecosystems to food production through primary productivity, water for irrigation, soil fertility, pollination, or pest regulation, relative to those contributed by society. Also, these values do not include the negative impacts of agricultural intensification and expansion, nor that of industrial fisheries, on biodiversity conservation and the degradation of supporting and regulating ecosystem services. The societal costs of intensive agriculture or fisheries are not accounted for either.

Data accuracy in national statistics is quite variable and is dependent on national monitoring infrastructure (human and technical capacity), relative importance of informal activities (e.g., subsistence production or unreported extraction cannot be accounted for), and governmental policies on transparent reporting. Temporal data gaps are common for many countries and are often filled using a variety of techniques, including interpolation, models or expert judgement, which all have well-documented biases. In all cases, uncertainty analyses are needed to quantify and help improve reliability of existing data.

3.6 Remote Sensing

Remote sensing (see Chap. 8) consists of data collection 'at a distance': from sensors on the ground, in the water, on aircraft, or in space. Remote sensing of ecosystem services relies on hybrid methods, that use models to combine in situ information (collected either by humans or machines) with that collected at coarser spatial scales (e.g., climate, landform, social or economic variables).

Remote sensing has not been used directly to measure ecosystem services, yet in combination with other data sources it can contribute to the assessment of many ecosystem services (e.g., water quantity and quality, erosion prevention, moderation of extreme events; Horning et al. 2010). These data sources can either contribute to assessing the potential supply of ecosystem services or to assess the social-ecological drivers that influence the supply, delivery, contribution to well-being, and value of ecosystem services (Andrew et al. 2015).

Products from multiple frequencies within the range of visible and near-infrared bands contribute to vegetation indices, such as greenness measures like the Normalized Difference Vegetation Index (NDVI) that indicates plant vigour. Such information can be used as one of several data sources to assess crop delivery (through potential productivity of known plant/crop species), carbon stocks and carbon uptake, fisheries (through ocean productivity), water quality (through changes in water colour), and land use change (a driver). High-resolution data can inform on small-scale ecological features, such as individual trees. Information on roads, fields and habitat patches can be used to provide information on drivers of many ecosystem services. Products from radar devices provide high-resolution information for topography, vegetation and water cover, and potentially on the aboveground biomass. These can contribute to assessing land use change, crops, or water cover (superficial water bodies) over a targeted region. Products based on Laser Imaging Detection and Ranging (LIDAR) devices provide high resolution information on above-ground carbon stocks, water (water surface elevation, and in combination with bathymetry, the volume of freshwater bodies), and ecosystem structure, that can be used to model a range of provisioning, regulating and cultural services. High resolution images (with individual pixels of around 1 m^2) are increasingly available from commercial satellites and can be used to refine information for particular locations. The cost is currently high, but may still be cost-effective if compared with manual mapping on the ground, and is being driven down by the advent of unmanned autonomous vehicles or 'drones' (e.g., see www.conservationdrones.org) equipped with cameras.

3.7   Field-Based Estimations

Field-based estimations contribute to local or site-based monitoring and assessment, as well as to validation of models and remotely sensed data products. Ultimately, field-based estimations are a principal source of new data on the supply, delivery, contributions to well-being and value for all services. Some services, such as the flow of water in rivers, are routinely monitored by in-field devices, and new technologies such as eddy covariance are extending the range of in situ observations of services such as carbon sequestration.

Conducting primary data collection can be costly, time consuming and technically specialised, and the methods and information from different data sources need to be standardized. Toolkits are emerging to deal with these issues, and promote standardized rapid assessments at the site scale. Such toolkits provide guidance on the steps to be followed, the kind of data to be gathered and the methods suggested to gather or model quantitative data at this scale that can then be used in an assessment under a range of contexts. Assessments incorporate local knowledge, basic local data collection and other data sources to create fine scale, locally-relevant assessments of multiple ecosystem services.

Two of these toolkits have been particularly useful (Table 3.3). The Toolkit for Ecosystem Service Site-based Assessments (TESSA; Peh et al. 2014) was developed to assist site-scale users with limited capacity and resources, to develop simple estimates of ecosystem services. The Natura toolkit was developed for assessing the

Table 3.3 Examples of toolkits available to assess ecosystem services and their advantages and disadvantages

Model (website)	Basic principles	Advantages	Disadvantages
TESSA: Toolkit for Ecosystem Service Assessments www.birdlife.org/datazone/info/estoolkit	Field-based estimations to develop and deploy a rapid assessment tool to understand how far conserving sites for their biodiversity importance also helps to conserve different ecosystem services relative to a converted state	Aimed at local decision-makers. Easy to use. Allows for the assessment of multiple components of ecosystem services. Can be applied to a range of conditions. Emphasizes alternative states and the identification of stakeholders that win or lose from these states	Applicable only at local scales. Not scalable from local to regional as its use is highly context dependent
Natura: Assessing Socioeconomic Benefits www.natura.org/	Practical guide for practitioners (e.g. site managers, landowners and other land users) involved in the management of sites in Europe. Toolkit will help these practitioners in exploring the different values and socio-economic 'potential' of their sites, e.g. possible socio-economic benefits gained by managing sites and land in a sustainable manner	Aimed at local decision-makers. Easy to use. Applicable at local to regional scales. Emphasizes what benefits are obtained by which stakeholders	Mainly focused on conservation projects and thus current and potential protected areas. Emphasizes only economic and social and cultural benefits obtained from ecosystem services

socio-economic benefits associated with the ecosystem services of 200 conserved or protected sites in Europe (Kettunen et al. 2009).

3.8   Community Monitoring of Ecosystem Services

Ecosystem services that are locally relevant can be monitored by local stakeholders, such as land owners and consumers (see Chap. 9 on Citizen Science). Several studies have shown that local communities without conventional scientific training have successfully collected accurate data on a wide range of ecosystem services such as forest carbon storage and sequestration, water quantity and quality, and their links to well-being (Hein et al. 2006; Dinerstein et al. 2013).

Involving communities in data generation enables year-round, low cost generation of local data (plot to landscape level) and wide spatial coverage. It provides information for local-level decision-making for ecosystem service management, and it can also generate employment, enthusiasm, and personal investment in ecosystem service based initiatives. Additionally, it can better incorporate traditional ecological knowledge and help maintain cultural heritage, identity, and values. Community involvement in monitoring can increase local interest and investment in the maintenance of ecosystems and the services they provide.

Information generated by locally-based monitoring systems, however, can be influenced by power struggles and incentives surrounding the monitored resource and validation mechanisms need to be implemented.

Numerous data collection and management tools have been developed in the last 5–10 years to facilitate gathering, storage, and sharing of data by communities.

3.9 Models

Numerical models, understood here as practical tools that predict how ecosystem services change through time and space, are increasingly being used to support decision-making. These models are often developed when data availability is scarce, when spatially explicit information is needed, and in order to assess trade-offs among services under alternative future management scenarios.

A wide variety of approaches have been used for building and applying such models. Five of the more commonly used modelling platforms are described here (Table 3.4).

- The Integrated Valuation of Environmental Services and Tradeoffs (InVEST) suite is a free and open-source software tool to help inform and improve natural resource management and investment decisions (Tallis et al. 2013).
- The Lund-Potsdam-Jena managed Land Dynamic Global Vegetation and Water Balance Model (LPJmL; www.pik-potsdam.de/research/climate-impacts-and-vulnerabilities/models/lpjml) is a tool that was not specifically designed for ecosystem service assessment, but still allows deducing a number of ecosystem services consistently from the same process based model (Bondeau et al. 2007).
- The ARtificial Intelligence for Ecosystem Services (ARIES; www.ariesonline. org) can be used to model supply, demand (delivery), flow (the link between the areas of supply and those of delivery), depletion (the balance between supply and delivery), and values (differential preferences among stakeholders) of ecosystem services (Bagstad et al. 2013b). A range of tools (www.ariesonline. org/resources/toolkit.html) and models for a range of case studies (www. ariesonline.org/resources.html) is available.

Table 3.4 Comparative table of the ecosystem services models described in this chapter and their individual advantages and disadvantages

Model (website)	Basic principles	Advantages	Disadvantages
InVEST (www.naturalcapitalproject.org/InVEST)	Set of spatially-explicit process-based models. Predict services from social-ecological conditions. User-defined future scenarios. Biophysical and monetary assessments of ecosystem services. Emphasis on relationships among multiple services	Broadly applicable across a variety of social-ecological contexts. Models use the minimum data required allowing application in many data-scarce regions. Moderate time consuming models and not technically specialized allowing its broad use. Modules of either biophysical modelling and economic valuation	Models do not simultaneously feedback on one another. Simple models, assuming that the provision of ecosystem services change linearly with land use change. High uncertainty when models are applied with coarse secondary data and no validation
LPJmL (www.pik-potsdam.de/research/climate-impacts-and-vulnerabilities/models/lpjml)	Simulates vegetation dynamics and their impacts on hydrological processes up to global scale; sensitive to land use and climatic change. 35 land cover classes including potential natural vegetation, 9 plant functional types and 13 crop types (irrigated or not)	Useful for modelling mid- to long-term change in ecosystem services provision under alternative climate change and land-use scenarios. Variability estimates over time	Models require high resolution climate data that is only available in few countries. Time consuming models requiring technically specialized skills. Low resolution of final outputs (50 km²) for most countries
ARIES (www.ariesonline.org)	Models built from Bayesian belief networks informed by user data. Uncertainty associated with its estimates quantified. Generic models adapted to specific applications at different spatial scales and for particular social-ecological contexts	Useful to quantify flows of the services to beneficiaries. Models incorporate an uncertainty measure in its estimates done through Bayesian networks and Monte Carlo simulation	Time consuming models requiring technically specialized skills. Models have a low level of generalization (specific application at particular social-ecological contexts)

(continued)

Table 3.4 (continued)

Model (website)	Basic principles	Advantages	Disadvantages
ESTA	Coupled, dynamic bio-economic models to simulate the production and value of multiple ecosystem services. Focus on understanding the trade-offs that emerge when management has multi-service objectives	Models developed using best available data for a region. Include direct and indirect interactions among services. Any number of services can be assessed simultaneously	Time consuming models requiring technically specialized skills and data-rich contexts
MIMES (www.ebmtools.org/ mimes)	Models simulate changes in biophysical conditions and economic activities over time and through space. Developed in collaboration with stakeholders. Functional and dynamic models over space and time developed from multiple data sources	Integrated dynamics and interactions among services. Values emerge from trade-offs and impacts on human well-being. Incorporate an uncertainty measure in its estimates	Time consuming models requiring technically specialized skills. Models have a low level of generalization
Co\$ting Nature (www. policysupport.org/ costingnature)	Web-accessible tool to map ecosystem services and conservation priority areas. Also analyses the benefits provided by the natural environment, the beneficiaries of those ecosystem services, and assesses the impacts of possible human interventions on the continued provision of these benefits	Rapid analysis of indexed, bundled services based on global data, along with conservation priority maps. Models have a high level of generalization	Models require high resolution biophysical and socio-economic data that is only available in few countries
WaterWorld (www. policysupport.org/waterworld)	Details process-based modelling of selected provisioning and regulating hydrological services. It incorporates high resolution spatial datasets for the entire world, spatial models for biophysical and socio-economic processes along with scenarios for climate, land use and economic change	Rapid analysis of bundled services based on global data. The biophysical and socio-economic consequences of alternative interventions (policy options) can be modelled Models have a high level of generalization	The high resolution datasets needed are only available for a few countries

- The Ecosystem Service Trade-off Analysis (ESTA) was initially developed to inform and evaluate the trade-off between biodiversity and fisheries objectives, and has been applied to an increasing number of case studies with a range of ecosystem services, including offshore wind and wave energy, aquaculture, and ecotourism (White et al. 2012).
- The Multi-scale Integrated Models of Ecosystem Services (MIMES; www. ebmtools.org/mimes.html) platform is designed to address the magnitude, dynamics, and spatial patterns of ecosystem service values (Altman et al. 2014).
- Co$ting Nature (www.policysupport.org/costingnature) is a web-based tool for natural capital accounting and analysing the ecosystem services provided by natural environments (i.e., nature's benefits), identifying the beneficiaries of these services and assessing the impacts of human interventions (Mulligan 2015a).
- WaterWorld (www.policysupport.org/waterworld) is a web-based tool can be used to understand the hydrological and water resources baseline and water risk factors associated with specific activities under current conditions and under scenarios for land use, land management and climate change (Mulligan 2015b).

3.10   Current Tools to Monitor Ecosystem Services

Ecosystem services can be monitored and assessed at different spatial scales using readily available data sources (Table 3.5). However clear gaps exist, especially when one considers all four components requiring data per ecosystem service (see Table 3.6). We explore progress and gaps per ecosystem service category below.

Mismatches can occur between data sources and data needs. Some data sources, such as LPJmL models or the older remote sensing data, are only available at low spatial resolution (50 km^2 grid cells in the case of LPJmL) and might not be suitable for assessments at landscapes scales. Similarly, assessments of changes in services within very short time frames are incompatible with some data sources that are only available on a yearly basis, as is the case of national statistics, or those that are modelled from data for which data sources are not updated regularly, as is the case of governmental land use and land cover maps in Mexico. The converse situation can also be true: changes in soil carbon or soil fertility within the same land cover type through time could be estimated from repeated remote sensed data, but changes would not be observed given the long time frame over which the processes that regulate them operate.

The data needed for ecosystem service estimation is often the flow of service rather than the particular conditions of the service in one point in time. This is the case of water flowing from a river, or the amount of carbon being taken up by vegetation. The most commonly found approach is for rates to be estimated from differences in the magnitude of the stock which provides or receives the service between two selected dates, as is the case of carbon uptake, most commonly

Table 3.5 Data sources for ecosystem services

Service	Global and National statistics	Remote sensing	Field estimations	Models	Additional data sources and comments
Provisioning					
Crop	FAOSTAT	✓	TESSA	ARIES, LPJmL, MIMES	(www.teebweb.org/agriculture-and-food/) for further discussion on limitations to FAOSTAT data
Fodder		✓		MIMES	
Livestock	FAOSTAT			MIMES	
Aquaculture	FAOSTAT	✓		InVEST, ESTA	
Fisheries	FAOSTAT	✓		ARIES, ESTA, MIMES	Only subsistence fisheries from ARIES
Wood	FAOSTAT	✓		InVEST, LPJmL, MIMES	
Biofuels	FAOSTAT	✓		MIMES	IEA, CDM, ISO14040/44
Game meat	FAOSTAT	✓		MIMES	
Harvested wild goods		✓ i	Natura	ARIES, MIMES	
Water	FAOSTAT, WORLD BANK	✓ i	TESSA	InVEST, LJPmL, ARIES, MIMES, Co$ting Nature, WaterWorld	
Hydropower energy		✓ i		InVEST, ESTA, MIMES	
Regulating					
Climate regulation (Carbon stocks and uptake)	WDCGG	✓	TESSA	InVEST, LJPmL, ARIES, MIMES, Co$ting Nature	IPCC, National statistics available for selected countries. Carbon uptake needs monitoring through time
Regulation of marine and freshwater quality		✓	Natura	InVEST, ESTA, MIMES, Co $ting Nature, WaterWorld	Only nutrients-freshwater for Natura. Highly patchy data availability. Quality defined with respect to users
Regulation of soil fertility				MIMES	Multiple local survey methods

(continued)

Table 3.5 (continued)

Service	Global and National statistics	Remote sensing	Field estimations	Models	Additional data sources and comments
Regulation of soil erosion		✓ i	Natura	InVEST, ARIES, MIMES, WaterWorld	Marine/coastal and terrestrial erosion models from InVEST
Flood regulation		✓ i		ARIES, MIMES, Co$ting Nature	
Coastal protection		✓ i		InVEST, ESTA, MIMES, Co$ting Nature	
Contribution of coastal habitat to fisheries		✓ i		InVEST, ESTA, MIMES	
Pollination			Natura	InVEST	
Pest control			Natura, IPM		
Cultural					
All non-tangible benefits				MIMES	Growing literature available on this topic
Aesthetic views		✓ i		InVEST, ARIES	
Nature-based tourism		✓ i	Natura, TESSA	InVEST, ESTA, Co$ting Nature	
Recreation		✓ i	TESSA	ARIES	

This list of data sources is not exhaustive but rather refers to the data sources reviewed in this chapter. Additional sources: IEA: International Energy Agency (www.iea.org/stats/prodresult.asp?PRODUCT=Renewables), provides information on land cover by biofuel crops. CDM: Methodologies developed by the Clean Development Mechanism (CDM; http://cdm.unfccc.int/methodologies/index.html), ISO14040/44: Standard methodologies for full life cycle assessments of biofuels (Finkbeiner et al. 2006), TEEBAgFood: The Economics of Ecosystem and Biodiversity for Agriculture and Food (http://www.teebweb.org/agriculture-and-food/), WDCGG: World Data Centre for Green House Gases (WDCGG; http://ds.data.jma.go.jp/gmd/wdcgg), IPCC: Standards for measuring carbon stocks and uptakes developed by the Intergovernmental Panel on Climate Change (www.ipcc.ch/ipccreports/sres/land_use/index.php?idp=7), IPM: Integrated Pest Management protocols for field surveys developed by University of California, Davis (www.ipm.ucdavis.edu). i: Contribution of remote sensing as one of the information layers

Table 3.6 Ecosystem service data sources for different ecosystem services components

	National statistics	Remote sensing		Field estimations		Models						
	FAOSTAT	High resolution	Low resolution	TESSA	Natura	InVEST	LPJmL	ARIES	ESTA	MIMES	CoSting nature	WaterWorld
Ecosystem service component												
Supply		✓	✓			✓	✓	✓	✓	✓	✓	✓
Delivery	✓			✓	✓	✓		✓		✓	✓	✓
Contribution to well-being				✓	✓					✓	✓	
Value	✓			✓	✓	✓		✓	✓	✓		
Spatial scale												
Local/landscape		✓		✓	✓	✓		✓	✓	✓	✓	✓
National	✓		✓			✓	✓	✓		✓	✓	✓
Global			✓			✓				✓	✓	✓

National statistics: FAOSTAT, The Statistics Division of the Food and Agriculture Organization of the United Nations (FAO 2012). TESSA: Toolkit for Ecosystem Service Assessments (Peh et al. 2014), Natura: Assessing Socioeconomic Benefits (Kettunen et al. 2009), InVEST: Integrated Valuation of Environmental Services and Tradeoffs (Tallis et al. 2013), LPJmL: Lund-Potsdam-Jena managed Land Dynamic Global Vegetation and Water Balance Model (Bondeau et al. 2007), ARIES: ARtificial Intelligence for Ecosystem Services (Bagstad et al. 2013b), ESTA: Ecosystem Service Tradeoff Analysis (White et al. 2012), MIMES: Multi-scale Integrated Models of Ecosystem Services (Altman et al. 2014) estimated from changes in carbon stocks. Actual flows of ecosystem services, such as in the case of water, can be assessed by some of the models such as ARIES, or by in situ flow measuring devices.

3.11 Provisioning Services

Most provisioning services are already observed at national and local scales in most parts of the world using one or more of the data sources above. National statistics are available (at least partially) for many provisioning services, but are typically blind to subsistence ('informal', family consumption, not traded in monitored markets) or illegal operations that can contribute to large proportions of delivery in some countries. Remote sensing data are available for services related to vegetation primary productivity, biomass harvest and water quantity. Field estimations are available for provisioning services (from e.g., TESSA and Natura). Models are available for most provisioning services, from at least one of the four platforms described above.

Observations of supply, that largely depend on biophysical conditions are only available for a few provisioning services. Instead, delivery data sources are commonly reported for services associated with commonly used goods, although only those that are accounted for in statistics. As many provisioning services are commercialised in markets, economic (especially monetary) values are also readily available, but such values do not reflect all the contributions of the ecosystem to these services, nor the consequences. Data on the contributions to well-being are largely missing or in development for most services.

Information on the balance between the demand of the services and the supply, or other estimators of the long-term ability of the ecosystem to sustain the supply of these services are not currently available for most provisioning services.

3.12 Regulating Services

Data on regulating services is increasingly available from national statistics or from remote sensing in conjunction with models, particularly for carbon stocks and uptake (climate regulation). The emphasis has been put on carbon stocks and carbon uptake through primary productivity, which is relatively easily measured and quite relevant to climate change mitigation, while the links to actual carbon dynamics and climate processes is largely absent. Models of regulating services associated with hydrological processes (water quality, erosion regulation), those on the impacts of extreme meteorological events (flood and coastal regulation), as well as those for pest regulation and pollination are increasingly available. Today models are available for most regulating services and most of these models have been developed at landscape and regional scales, but seldom at national scales. Field estimations are available for services (most of which are available from TESSA or Natura, and from a plethora of approaches).

Both supply and delivery of regulating services are accounted for in most models. Data and models for contributions to well-being are absent or in development. Economic values are largely related to avoided costs or marginal contributions to economic activities from regulating services.

Given that regulating services depend on multiple social-ecological processes operating at several spatial and temporal scales, data, models and field estimations of regulating services are necessarily a simplification and, in some cases, they may be an oversimplification which is more misleading than useful.

> **Box 3.1. The Demand for Ecosystem Services at Drinking Water Treatment Facilities in Barcelona**
>
> Engagement with drinking water managers in Barcelona, Spain allowed for the identification of ecosystem services relevant for decision-makers. Discussions revealed that treatment costs were particularly sensitive to three water quality parameters: stream temperature, ammonium and conductivity. In particular, high stream temperature increased water treatment costs because of the water treatment technology used and the high concentration of sterilisation products during warm summer months (Valero and Arbós 2010). Understanding the demand for reduced stream temperatures by water treatment managers allowed for the development of a targeted research program focusing on ecosystem structures that would reduce thermal heating in the Llobregat River. It was found that the restoration of riparian forests upstream would be able to recover ecosystem processes, reduce stream temperature in the summer and therefore reduce water treatment costs. After modelling multiple restoration scenarios, nearly half of the investment in riparian river restoration was estimated to be recovered in a 20 year period through a reduction in water treatment costs (Honey-Rosés et al. 2013). Understanding the demand for reduced stream temperatures by water treatment managers allowed for the development of a targeted research program focusing on ecosystem structures that would reduce thermal heating in the Llobregat River.

3.13 Cultural Services

Cultural services present a challenge when it comes to observation and assessment because some of them are not easily disentangled from other ecosystem services, such as provisioning services. For instance many important cultural services are co-produced by the same ecosystem components and human activities that produce material objects for consumption (Chan et al. 2012), such as agricultural landscapes or harvested forests. The different cultural services are highly intertwined, and unlike with provisioning or regulating services, it is not possible to clearly delineate the different components of the services. Cultural services are highly context dependent and thus information on these is often only available and relevant at local scales. This is not true for all cultural services: some are well-defined, discrete and routinely monitored, such as the use of national parks, or the income from nature-based tourism and recreation.

Readily available sources of information on cultural services are very wide ranging. These include local assessments of cultural preferences (for aesthetic views; Bagstad et al. 2013c) (can be obtained from the above toolkits), and databases on use of particular areas or ecosystems for ecotourism at national scales (governmental database). Further sources of information on cultural services are embedded into local artistic expression (e.g., poetry, music) or in social norms that articulate a value or impact of nature on the human condition.

3.14 Observing Multiple Ecosystem Services

Historically, ecosystem management has often focused on delivery of a single service from that ecosystem (often a provisioning service, such as timber or grazing) without recognition that the same ecosystem produces multiple, often interacting services which are also affected by management interventions. This often leads to trade-offs (where one service decreases while the other increases), but can also lead to synergies (where increasing the supply of one services also increases the supply of another). Moving observation systems beyond single services to the full bundle of services (a set of services that tend to co-occur in space or time), to quantify and reflect the synergies (positive interactions) and trade-offs (negative interactions) is a major challenge for current research efforts. Also, an understanding of the interactions among stakeholders that have differential preferences for the traded-off services is needed.

The identification of bundles of services that arise under particular biophysical, management, and societal conditions is particularly relevant. Data needed for these assessments is hindered by the reduced replicability of the same measurements across different social-ecological conditions. It is seldom that they supply exactly the same sets of provisioning, regulating and cultural services, at the same spatial and temporal scale, and measuring the same components (e.g., supply or value). While still patchy, such datasets have been increasingly available in the past few years. Comparisons across studies are nevertheless faced with the lack of interoperability among them.

Additional observations of biodiversity (see other chapters) and multiple ecosystem services at different spatial scales will contribute to a better understanding of their inter-linkages, patterns of interactions across scales and time, and common trade-offs and synergies.

3.15   Using Scenarios in Modelling to Predict Future Ecosystem Services

Scenarios are stories about plausible futures, with the power to capture public attention and inform more sustainable decisions (Henrichs et al. 2010). They can help communicate the outcomes of different choices for societies and ecosystems while at the same time involving stakeholders in a powerful learning process. It is important to consider the explicit goals for the use of scenarios in determining which type of scenario will best address those goals and reach their intended audience. Three main uses of scenarios include: (1) assessing the impact of decisions under consideration, (2) exploring hypothetical but plausible futures, and (3) building consensus around a shared vision for the future (e.g., see IPBES 2016).

Certain characteristics can make scenarios more effective. Scenarios that are relevant to the decision context or stakeholder interests will align with the problems and questions of interest to stakeholders. To be legitimate, the scenario development process should include diverse stakeholder views and beliefs. To be credible, scenario storylines should be developed using scientifically robust methods. To be plausible, scenarios should tell coherent stories that could conceivably happen. Finally, to tell a compelling story, scenarios should be distinct enough from one another that they show contrasting ecosystem service impacts. Iteration of scenarios can greatly enhance many of these characteristics, as they are refined over time to incorporate stakeholder feedback, as well as emerging knowledge, trends and issues.

Translating scenarios to decision-support tools requires that storylines be made spatially-explicit, with each scenario corresponding to a map of land cover, or coastal or marine habitats and uses that feed into the biophysical and/or economic models underlying ecosystem service assessment. Converting scenario storylines into maps can be accomplished by asking stakeholders to simply draw maps for each scenario; more analytical methods of forecasting where change is most likely to occur on the landscape or seascape are based on past trends; rule-based approaches define which areas are likely to be most suitable for particular uses or activities. Models of future supply, delivery, value and benefit of ecosystem services into alternative scenarios are increasingly being developed.

All the modelling platforms described above may be used to predict ecosystem services under different future scenarios for land/sea use and management patterns. Different models have been built to be differentially sensitive to alternative future issues. For instance, the LPJmL, is highly sensitive to climate change, which is particularly helpful when looking for mid- to long-term effects.

3.16 Linking Ecosystem Service Observations to Decision-Making

Monitoring for ecosystem services to support decision-making is greatly enhanced with early involvement of the actual stakeholders involved in the decisions. One key advantage to examining ecosystem services with a stakeholder driven agenda includes the easy identification of key services recognised and preferred by societies, as well as the identification of indicators that are most meaningful to them. Stakeholders can also participate in community-based or citizen science-based monitoring of ecosystem services. Successfully integrating decision-makers in the assessment and valuation of services also allows for speedier adoption of the ecosystem services framework in practice, and the use of ecosystem service data into actual decision-making.

Emphasis has increasingly been put on the use of ecosystem service indicators towards agreed upon policy goals. That is the case of indicators that can inform on progress towards the Aichi Targets and more recently progress towards the Sustainable Development Goals. The challenge is to identify those indicators that are most relevant to measuring progress towards the goal, while at the same time being supported by actually available data, conceptual understanding and credibility.

Monitoring for ecosystem services at local to national and global scales needs to take into account how preferences and ecosystem services can change in space and time. Services that are most relevant at national to global scales could be monitored systematically, while locally relevant services could be assessed within particular locations.

Box 3.2. Monitoring Ecosystem Services for Coastal Planning in Belize

The coast of Belize includes hundreds of kilometres of mangrove forests, extensive seagrass beds, and the largest unbroken reef in the Western Hemisphere. 800,000 tourists visit the area for its renowned snorkelling and diving sites. Tourism, as well as commercial, recreational, and subsistence fisheries, contribute to income and livelihoods, but at the same time threaten the very ecosystems that make these activities possible. Efforts to put the Belize Barrier Reef on the United Nations Educational, Scientific and Cultural Organization's list of World Heritage Sites in Danger and the creation of a visionary legislation in 1998 calling for cross-sector, ecosystem-based management of coastal and marine ecosystems were insufficient to halt degradation. In 2010 The Natural Capital Project (www.naturalcapitalproject.org) partnered with the Coastal Management Authority and Institute to use ecosystem-service approaches and models to design a spatial plan (Arkema et al. 2015). Interactions with a range of stakeholders and government agencies led to the identification of different categories of human activities, a zoning scheme, and three alternative future scenarios. The supply and economic value of lobster fisheries, tourism, coastal protection and habitat (to support fisheries) were modelled for current and future scenarios using InVEST. Data sources included: (i) field assessments of lobster catch and revenue; (ii) high resolution land use cover maps developed from remote sensed data, (iii) model of lobster migration, (iv) current visitation data obtained from social media (e.g., flickr). Risk under alternative scenarios for individual services as well as trade-offs among services across zones were assessed using additional spatial data on human activities and habitats, as well as information from the peer- reviewed and grey literature on the expected impacts of human activities on the services and the habitats. The most desirable future scenario was identified and further refined to increase expected delivery of almost all services in all regions into 2025. The results from this future scenario were incorporated into the Coastal Zone Management plan for Belize in 2012. It was refined through further stakeholder involvement and expert review during 2013 and led to changes in national legislation such as the creation of marine reserves and the revocation of offshore drilling contracts issued earlier by the government of Belize.

3.17 Creating a Network for Observing and Managing Ecosystem Services

The ultimate goal of many efforts to monitor ecosystem services is to inform decision-makers and policy to ensure the long-term supply of services and the flow of benefits to societies. While progress has been made on the quantification and mapping of services, less attention has been given to the needs of decision-makers and resource users from local to global scales. Meaningful engagement with resource users and policy makers should occur early, explicitly and formally when monitoring services (Menzel and Teng 2010).

A network for monitoring ecosystem services is necessary to synergise work done by multiple partners, taking advantage of others' insights, increasing consistency, and reducing duplication of efforts. Creating such a network for monitoring ecosystem services at local to global scales will require significant effort from stakeholders from the research, policy and practice communities across the globe. National monitoring systems could create mechanisms by which local stakeholders can provide input and feed into the national system. City and regional governments may help facilitate the engagement with local stakeholders, and help assess the status of services at local scales. Stakeholder participation in monitoring activities will vary widely depending on many factors including local relevance of the services they are monitoring, and whether incentives are provided.

Local scale monitoring could dovetail into existing ecosystem services research which may have very different objectives but could contribute to an observation network. Examples of such on-going efforts include: the already existing networks associated with ARIES, and MIMES the Ecosystem Service Partnership (www.es-partnership.org/esp), the International Long-Term Ecological Research Network (www.ilternet.edu), the Natural Capital Project (www.naturalcapitalproject.org), the Program for Ecosystem Change and Society (PECS; www.pecs-science.org), the Sub-Global Assessment Network (www.unep-wcmc.org/sga-network_770.html), the Tropical Ecology Assessment and Monitoring Network (www.teamnetwork.org), the ESCom Scotland (http://escomscotland.wordpress.com/) and Vital Signs (http://vitalsigns.org/).

One major challenge to date is that multi-scale cross-site comparisons are only possible if comparable approaches and indicators are used. To date a wide diversity of approaches and indicators complicate such comparisons. Great emphasis has been given over the last decade to the development of new metrics, tools and approaches, which has fostered creative solutions. Yet, standard procedures will eventually need to be identified and practical examples be provided to opera-tionalise the ecosystem services concept (e.g., OPERAs; www.operas-project.eu/).

Efforts through the Group on Earth Observations Biodiversity Observation Network (GEO BON; www.geobon.org), to further develop and communicate standards and protocols for the collection of new ecosystem services observations to enhance comparability across scales and data sources, are on-going. Ecosystem Service tools are being incorporated into GEO BON developed toolkits, namely BON-in-a-Box.

Automated, remotely sensed Earth observations will increasingly be used in the future to assess ecosystem services as well as the drivers that modify their supply and delivery. Changes in environmental and socio-economic features are more available than ever with the new sensors, such as those in the Sentinel fleet. The critical issue is integration of the data in ways that make it readily usable for ecosystem service assessments (Cord et al. 2015).

3.18 Monitoring to Support Policy Design

Ecosystem services monitoring can be directly linked to on-going assessments that support policy design. Timely information from monitoring ecosystem services can be useful to the Intergovernmental Platform on Biodiversity and Ecosystem Services (IPBES; www.ipbes.net) that aims to strengthen the science policy inter-face for biodiversity and ecosystem services for the conservation and sustainable use of biodiversity, long-term human well-being and sustainable development. IPBES is aiming to establish strategic partnerships, such as with monitoring pro-grammes, to assist in the delivery of its work programme.

Similarly, National governments are also signatories to Multilateral Environmental Agreements. In most cases (for instance the CBD), these rely on technical and scientific bodies to assess progress towards implementation of agreed decisions. National progress reports and assessment of needs towards achieving targets rely on monitoring ecosystem services.

Agreements and commitments across different scales (national to global) on biodiversity and ecosystem services would benefit greatly from the extension and linking of various observing networks, which can promote the collection, access, packaging and communication of data. This often will require engagement with existing mechanisms such as the assessments to be performed by IPBES, CBD and individual nations.

3.19   Conclusions

Monitoring ecosystem services is vital for informing policy (or decision-making) to protect human well-being and the natural systems upon which it relies at different scales. While ecosystem services are linked to biodiversity, the social factors involved in their supply, delivery and value to human well-being implies that they cannot be predicted from biodiversity monitoring initiatives alone. Here we emphasise that monitoring systems for ecosystem services must take into account provisioning, regulating and cultural services as well as their components of supply, delivery, contribution to well-being and value. A wide variety of data sources is available and relevant to ecosystem services monitoring, including national statistics, field-based assessments, remote sensing and models. Their elaboration will help ensure monitoring at relevant (and where necessary multiple) scales of interest.

Outputs from monitoring a range of ecosystem services and their components at different spatial scales can actively support decision-making. Analyses of multiple services and biodiversity can inform decision-makers such as land managers as to trade-offs and synergies among them. Modelling and exploring future scenarios of ecosystem services can then clarify the impacts of alternative policies on such trade-offs and synergies.

Monitoring our life support systems and using this information in decision-making across all scales will be central to our endeavours to transform to more sustainable and equitable futures.

References

Altman, I., Boumans, R., Roman, J., Gopal, S. & Kaufman, L. (2014). An ecosystem accounting framework for marine ecosystem-based management. In: M. J. Fogarty & J. J. McCarthy (Eds.), *Marine ecosystem-based management. The sea: Ideas and observations on progress in the study of the seas* (Vol. 16, pp. 245–276). Cambridge, MA, USA: Harvard University Press.

Andrew, M. E., Wulder, M. A., Nelson, T. A., & Coops, N. C. (2015). Spatial data, analysis approaches, and information needs for spatial ecosystem service assessments: a review. *GIScience & Remote Sensing, 52*, 344–373.

Arkema, K. K., Verutes, G. M., Wood, S. A., Clarke-Samuels, C., Rosado, S., Canto, M., et al. (2015). Embedding ecosystem services in coastal planning leads to better outcomes for people and nature. *Proceedings of the National Academy of Sciences of the USA, 112*, 7390–7395.

Bagstad, K. J., Johnson, G. W., Voigt, B. & Villa, F. (2013a). Spatial dynamics of ecosystem service flows: A comprehensive approach to quantifying actual services. *Ecosystem Services, 4*, 117–125.

Bagstad, K. J., Semmens, D. J., Waage, S. & Winthrop, R. (2013b). A comparative assessment of decision-support tools for ecosystem services quantification and valuation. *Ecosystem Services, 5*, 27–39.

Bagstad, K. J., Semmens, D. J. & Winthrop, R. (2013c). Comparing approaches to spatially explicit ecosystem service modeling: A case study from the San Pedro River, Arizona. *Ecosystem Services, 5*, 40–50.

Bondeau, A., Smith, P. C., Zaehle, S., Schaphoff, S., Lucht, W., Cramer, W., et al. (2007). Modelling the role of agriculture for the 20th century global terrestrial carbon balance. *Global Change Biology, 13*, 679–706.

Cardinale, B. J., Duffy, J. E., Gonzalez, A., Hooper, D. U., Perrings, C., Venail, P., et al. (2012). Biodiversity loss and its impact on humanity. *Nature, 486*, 59–67.

Chan, K., Satterfield, T., & Goldstein, J. (2012). Rethinking ecosystem services to better address and navigate cultural values. *Ecological Economics, 74*, 8–18.

Cord, A. F., Seppelt, R., & Turner, W. (2015). Monitor ecosystem services from space. *Nature, 523*, 27–28.

Díaz, S., Demissew, S., Carabias, J., Joly, C., Lonsdale, M., Ash, N., et al. (2015). The IPBES conceptual framework—Connecting nature and people. *Current Opinion in Environmental Sustainability, 14*, 1–16.

Dinerstein, E., Varma, K., Wikramanayake, E., Powell, G., Lumpkin, S., Naidoo, R., et al. (2013). Enhancing conservation, ecosystem services, and local livelihoods through a wildlife premium mechanism. *Conservation Biology, 27*, 14–23.

Egoh, B., Dunbar, M. B., Maes, J., Willemen, L., & Drakou, E. G. (2012). *Indicators for mapping ecosystem services: A review*. Ispra, Italy: European Commission.

FAO. (2012). *The state of food and agriculture*. Rome, Italy: Food and Agriculture Organization of the United Nations.

Finkbeiner, M., Inaba, A., Tan, R. B. H., Christiansen, K., & Kluppel, H. J. (2006). The new international standards for life cycle assessment: ISO 14040 and ISO 14044. *International Journal of Life Cycle Assessment, 11*, 80–85.

Griggs, D., Stafford-Smith, M., Gaffney, O., Rockstrom, J., Ohman, M. C., Shyamsundar, P., et al. (2013). Sustainable development goals for people and planet. *Nature, 495*, 305–307.

Hein, L., Van Koppen, K., De Groot, R. S., & Van Ierland, E. C. (2006). Spatial scales, stakeholders and the valuation of ecosystem services. *Ecological Economics, 57*, 209–228.

Henrichs, T., Zurek, M., Eickhout, B., Kok, K., Raudsepp-Hearne, C., Ribeiro, T., et al. (2010). Scenario development and analysis for forward-looking ecosystem assessments. In: N. Ash, H. Blanco, C. Brown, K. Garcia, T. Henrichs, N. Lucas, C. Raudsepp-Hearne, R.D. Simpson, R. Scholes, T. P. Tomich, B. Vira, M. Zurek (Eds.), *Ecosystems and human well-being: A manual for assessment practitioners* (pp. 151–219). Washington, D.C., USA: Island Press.

Honey-Rosés, J., Acuna, V., Bardina, M., Brozovic, N., Marce, R., Munne, A., et al. (2013). Examining the demand for ecosystem services: The value of stream restoration for drinking water treatment managers in the Llobregat River, Spain. *Ecological Economics, 90*, 196–205.

Horning, N., Robinson, J. A., Sterling, E. J., Turner, W., & Spector, S. (2010). *Remote sensing for ecology and conservation: A handbook for techniques*. Oxford, UK: Oxford University Press.

IPBES (2016). Summary for policymakers of the methodological assessment of scenarios and models of biodiversity and ecosystem services of the Intergovernmental Science-Policy Platform on Biodiversity and Ecosystem Services. In S. Ferrier, K. N. Ninan, P. Leadley, R. Alkemade, L.A. Acosta, H. R. Akçakaya, L. Brotons, W. Cheung, V. Christensen, K. A. Harhash, J. Kabubo-Mariara, C. Lundquist, M. Obersteiner, H. Pereira, G. Peterson, R. Pichs-Madruga, N. H. Ravindranath, C. Rondinini, B. Wintle (Eds.). *Secretariat of the intergovernmental science-policy platform on biodiversity and ecosystem services* (pp. 1–32). Germany:Bonn. ISBN: 978-92-807-3570-3.

Kettunen, M., Bassi, S., Gantioler, S. & Ten Brink, P. (2009). *Assessing socio-economic benefits of Natura 2000: A toolkit for practitioners*. Brussels, Belgium: Institute for European Environmental Policy (IEEP). http://ec.europa.eu/environment/nature/natura2000/financing/docs/benefits_toolkit.pdf

MA (Millennium Ecosystem Assessment). (2005). *Ecosystems and human well-being: Synthesis*. Washington, DC: Island Press. http://www.millenniumassessment.org/documents/document.356.aspx.pdf

Mace, G. M., Norris, K., & Fitter, A. H. (2012). Biodiversity and ecosystem services: A multilayered relationship. *Trends in Ecology & Evolution, 27*, 19–26.

Mace, G. M., Reyers, B., Alkemade, R., Biggs, R., Chapin, F. S., III, Cornell, S. E., et al. (2014). Approaches to defining a planetary boundary for biodiversity. *Global Environmental Change, 28*, 289–297.

Maes, J., Liquete, C., Teller, A., Erhard, M., Paracchini, M. L., Barredo, J., et al. (2016). An indicator framework for assessing ecosystem services in support of the EU Biodiversity Strategy to 2020. *Ecosystem Services, 17*, 14–23.

Menzel, S., & Teng, J. (2010). Ecosystem services as a stakeholder-driven concept for conservation science. *Conservation Biology, 24*, 907.

Mulligan, M. (2015a). Trading off agriculture with nature's other benefits. In: C. A. Zolin, R. de A. R. Rodrigues (Eds.), *Impact of climate change on water resources in agriculture* (pp. 184–204). Boca Raton, FL, USA: CRC Press.

Mulligan, M. (2015b). WaterWorld: a self-parameterising, physically based model for application in data-poor but problem-rich environments globally. *Hydrology Research, 44*, 748–769.

Peh, K. S.-H., Balmford, A. P., Bradbury, R. B., Brown, C., Butchart, S. H. M., Hughes, F. M. R., et al. (2014). *Toolkit for ecosystem service site-based assessment* (*TESSA*). Version 1.2, Cambridge, UK. http://tessa.tools/

Reyers, B., Polasky, S., Tallis, H., Mooney, H. A., & Larigauderie, A. (2012). Finding common ground for biodiversity and ecosystem services. *BioScience, 62*, 503–507.

Scholes, R. J. (2009). Ecosystem services: Issues of scale and trade-offs. In: S. A. Levin (Ed.), *The Princeton guide to ecology* (pp. 579–583). Princeton, NJ, USA: Princeton University Press.

Tallis, H., Mooney, H., Andelman, S., Balvanera, P., Cramer, W., Karp, D., et al. (2012). A global system for monitoring ecosystem service change. *BioScience, 62*, 977–986.

Tallis, H. T., Ricketts, T., Guerry, A. D., Wood, S. A., Sharp, R., Nelson, E., et al. (2013). *InVEST 3.0.0 user's guide*. Stanford, USA: The Natural Capital Project. http://ncp-dev.stanford.edu/~dataportal/invest-releases/documentation/3_0_0/

Valero, F., & Arbós, R. (2010). Desalination of brackish river water using electrodialysis reversal (EDR) control of the THMs formation in the Barcelona (NE Spain) area. *Desalination, 253*, 170–174.

White, C., Halpern, B. S., & Kappel, C. V. (2012). Ecosystem service tradeoff analysis reveals the value of marine spatial planning for multiple ocean uses. *Proceedings of the National Academy of Sciences of the USA, 109*, 4696–4701.

Chapter 4
Monitoring Essential Biodiversity Variables at the Species Level

Henrique M. Pereira, Jayne Belnap, Monika Böhm, Neil Brummitt, Jaime Garcia-Moreno, Richard Gregory, Laura Martin, Cui Peng, Vânia Proença, Dirk Schmeller and Chris van Swaay

Abstract The Group on Earth Observations Biodiversity Observation Network (GEO BON) is developing a monitoring framework around a set of Essential Biodiversity Variables (EBVs) which aims at facilitating data integration, spatial scaling and contributing to the filling of gaps. Here we build on this framework to explore the monitoring of EBV classes at the species level: species populations, species traits and community composition. We start by discussing cross-cutting issues on species monitoring such as the identification of the question to be addressed, the choice of variables, taxa and spatial sampling scheme. Next, we discuss how to monitor EBVs for specific taxa, including mammals, amphibians,

H.M. Pereira (✉)
German Centre for Integrative Biodiversity Research (iDiv) Halle-Jena-Leipzig, Deutscher Platz 5e, 04103 Leipzig, Germany
e-mail: hpereira@idiv.de

H.M. Pereira
Institute of Biology, Martin Luther University Halle-Wittenberg, Am Kirchtor 1, 06108 Halle (Saale), Germany

H.M. Pereira
CIBIO/InBio—Research Network in Biodiversity and Evolutionary Biology, Universidade do Porto, Campus Agrário de Vairão, Rua Padre Armando Quintas, Vairão 4485–601, Portugal

J. Belnap
Southwest Biological Science Center, U.S. Geological Survey, Moab, UT 84532, USA
e-mail: jayne_belnap@usgs.gov

M. Böhm
Institute of Zoology, Zoological Society of London, Regent's Park, London NW1 4RY, UK
e-mail: monika.bohm@ioz.ac.uk

N. Brummitt
Department of Life Sciences, Natural History Museum, Cromwell Road, South Kensington, London SW7 5BD, UK
e-mail: n.brummitt@nhm.ac.uk

J. Garcia-Moreno
ESiLi Consulting, Het Haam 16, 6846 KW Arnhem, The Netherlands
e-mail: jaime_gm@yahoo.com

© The Author(s) 2017
M. Walters and R.J. Scholes (eds.), *The GEO Handbook on Biodiversity Observation Networks*, DOI 10.1007/978-3-319-27288-7_4

butterflies and plants. We show how the monitoring of species EBVs allows monitoring changes in the supply of ecosystem services. We conclude with a discussion of challenges in upscaling local observations to global EBVs and how indicator and model development can help address this challenge.

Keywords Species · EBV · Monitoring · Population abundance · Distribution

4.1 Introduction

People have monitored and managed species for thousands of years, but national and international biodiversity monitoring is a relatively recent phenomenon. By the end of the 1800s, some governments had established monitoring agencies, mostly taxon-specific. In the United States, for example, Congress established the U.S. Fish Commission in 1871 to recommend ways to manage the nation's food fishes, and the Division of Biological Survey in 1885 in order to promote 'economic ornithology, or the study of the interrelation of birds and agriculture.' In 1940, these divisions were combined into the U.S. Fish and Wildlife Service. Later, the U.S. Endangered Species Act of 1966 mandated species monitoring. At the international level, the multilateral CITES Treaty, established in 1973, required that the

R. Gregory
RSPB Centre for Conservation Science, The Lodge, Sandy, Bedfordshire SG19 2DL, UK
e-mail: richard.gregory@rspb.org.uk

L. Martin
Center for the Environment, Harvard University, Cambridge 02138, MA, USA
e-mail: laura.jane.martin@gmail.com

C. Peng
Center of Nature and Biodiversity Conservation, Nanjing Institute of Environmental Sciences, Ministry of Environmental Protection, 8 Jiangwangmiao Street, Nanjing 210042, Jiangsu Province, People's Republic of China
e-mail: cuipeng1126@163.com

V. Proença
MARETEC, Instituto Superior Técnico, University of Lisbon, Lisboa 1049-001, Portugal
e-mail: vania.proenca@tecnico.ulisboa.pt

D. Schmeller
Department of Conservation Biology, Helmholtz Centre for Environmental Research—UFZ, Permoserstrasse 15, 04318 Leipzig, Germany
e-mail: dirk.schmeller@ufz.de

C. van Swaay
Dutch Butterfly Conservation and Butterfly Conservation Europe, PO Box 506, 6700 AM Wageningen, The Netherlands
e-mail: chris.vanswaay@vlinderstichting.nl

international trade of potentially vulnerable species be monitored by countries. Starting in the 1960s and during the following decades, conservation-focussed Non-Governmental Organisations (NGOs) also became involved in monitoring schemes, such as the Common Bird Census of the British Trust for Ornithology. Since the 1990s, the Habitats and Birds directives further stimulated species monitoring in European countries, although even today major gaps remain (Schmeller 2008; Henle et al. 2013). The global change discourse has increased the demand for biological monitoring. The Aichi Targets for 2020 by Parties to the United Nations Convention on Biological Diversity affirm an international desire to curb the rate of biodiversity loss (Leadley et al. 2014) and their assessment requires an expansion of current species monitoring efforts (Pereira et al. 2012; Tittensor et al. 2014).

Ecological monitoring in the early 20th century was largely organised around estimating population sizes of specific species. Capture-recapture methods were developed for fish by the Danish biologist Carl Petersen in the 1890s. In the mid-20th century, technologies developed in the world wars, including radioisotopes and radio-tracking collars, revolutionised ecological monitoring, and broadened the scope of monitoring from individual populations to ecosystem level processes. Part of this trend was reflected in the development of the Long Term Ecological Research (LTER) network (Aronova et al. 2010). In the last few decades, the development of extensive monitoring schemes based on trained volunteers or citizen scientists has allowed for the tracking of entire taxonomic groups over national and continental scales, for example, the Breeding Bird Survey in the USA or the Pan-European Common Bird Monitoring Scheme (Pereira and Cooper 2006). At the same time, remote sensing technology has started to make incursions into species level monitoring (see Chap. 8), including population counts of birds and mammals or the detection of invasive species (Pettorelli et al. 2014). In the last decade, the development of websites, such as ebird.org, ispot.org, inaturalist.org and observado.org, which allow for the global recording and sharing of species observations, has led to a new wave of citizen science engagement (see Chap. 9).

Studies of biodiversity remain unevenly distributed across the globe. One review of papers published in ten leading journals from 2004 to 2009 found that approximately 75 % of studies are conducted in protected areas (Martin et al. 2012). Studies were also disproportionately conducted in temperate, wealthy countries. Similarly, Amano and Sutherland (2013) found that a country's wealth, language, geographical location, and security explain variation in data availability in four different types of biodiversity databases. At a global scale, biodiversity monitoring is also biased towards consideration of certain taxa. For example, systematic IUCN Red List assessments have been carried out for only a few taxonomic groups, and the proportion of species assessed in each group is unrelated to its representation in global diversity (Pereira et al. 2012). Such geographical biases and historical contingencies have led to mismatches between prioritisation and protection (Jenkins et al. 2013).

In the past, gathering data for biodiversity management involved querying colleagues and conducting extensive literature reviews. But in the past two decades, vast quantities of ecological data have been made digitally accessible. Nevertheless, aggregating relevant knowledge often remains difficult and inefficient. A key challenge for the future is the development of tools for aggregating local studies to generate broader-scale patterns. International conservation projects are seriously limited by spatial gaps in biodiversity monitoring data, and geographical biases must be taken into account when extrapolating from single-site studies.

The Group on Earth Observations Biodiversity Observation Network (GEO BON) is developing a monitoring framework around a set of Essential Biodiversity Variables (EBVs) which aims at facilitating data integration, spatial scaling and contributing to the filling of gaps. EBVs have been inspired by the Essential Climate Variables (ECVs) framework of the Global Climate Observing System developed by Parties to the UN Framework Convention on Climate Change (Pereira et al. 2013). Here we build on this framework to explore the monitoring of EBV classes at the species level: species populations, species traits and community composition. We start by discussing cross-cutting issues on species monitoring such as the identification of the question to be addressed, the choice of variables, taxa and spatial sampling scheme. Next, we discuss how to monitor EBVs for specific taxa, including mammals, amphibians, butterflies and plants. We show how the monitoring of species EBVs allows monitoring changes in the supply of ecosystem services. We conclude with a discussion of challenges in upscaling local observations to global EBVs and how indicator and model development can help address this challenge.

4.2 Defining the Scope of the Monitoring Program

When designing a monitoring scheme, one needs to keep in mind three main questions: why monitor, what to monitor, and how to monitor (Yoccoz et al. 2001)? Addressing the first question is important to define the monitoring goals. The second question leads to the identification of which biodiversity variables should be monitored. Finally, the third question leads to the assessment of different sampling schemes and methods (often taxon specific). This is a process that needs to be done with great care, as once a monitoring system is established, changing it can, in some instances, invalidate all the previous monitoring efforts.

4.2.1 Surveillance and Targeted Monitoring

We can classify monitoring in two broad categories: surveillance monitoring and targeted monitoring (Nichols and Williams 2006). In surveillance monitoring, the goal is to have baseline data for one or multiple biodiversity variables. For instance, one may want to know how species population abundances are changing across as many taxa as possible. There are no a priori specific questions to be addressed.

Instead the goal is to obtain as much data as possible about that biodiversity variable over time. Data obtained by surveillance monitoring can be used for a multitude of research and management questions, with many of them defined years after the monitoring program started.

In contrast, targeted monitoring addresses specific research or management questions. For example, if the main management goal of a reserve is the protection of a specific species, monitoring the population of that species, as well as vital forage and habitat for that species, will be a necessary part of any monitoring design. Another type of targeted monitoring addresses the impact of specific drivers on biodiversity change. For instance, one may want to compare areas that receive relatively low impacts from a driver of concern to those that receive high levels of impact from that same driver and to measure all the EBVs that are likely to change with exposure to that stressor. Thus, for example, if timber harvest is the driver of concern, comparing unlogged and logged areas is likely to show a difference in the abundance of tree and other plant or animal species.

4.2.2   Choosing Which Variables, Taxa and Metrics to Monitor

Based on the available list of candidate EBVs (see www.geobon.org), we chose seven variables to discuss in this chapter that are relevant at the species level (Table 4.1). Monitoring any of these variables requires that one or more particular taxonomic group is chosen (e.g., mammals). Next, for the variables in the species population class, a key sampling design question is how many species of a given taxonomic group shall be monitored for abundance or occurrence. For instance, one may be interested in monitoring as many species as possible and therefore choose methods that assess simultaneously a wide range of species in as many locations as possible. Monitoring species population variables across entire assemblages also provides a community level overview of biodiversity change (Dornelas et al. 2014). Such broad surveys may capture population trends of abundant species, but may fall short of providing precise abundances for rare species. Instead, rare species may require targeted sampling schemes both from the point of view of spatial sampling and field methodology (Thompson 2013).

For the community composition variables, the choice of metrics to measure taxonomic diversity or species interactions become paramount (Table 4.1). For instance taxonomic diversity can be measured by many metrics, including (Magurran 2004): species richness, Simpson's diversity index, phylogenetic diversity, functional diversity, beta diversity, among others. In some cases (e.g. richness), only the presence or absence of the species is needed to calculate the metric. In others, relative abundance is required (Simpson's index), or turnover over gradients (β diversity), or cladistic information (phylogenetic), or trait information (functional).

Table 4.1 Essential biodiversity classes, essential biodiversity variables, and associated sampling design questions

Essential biodiversity class	Essential biodiversity variable	Main design choice	Metrics or taxa groups (examples)
Species populations	Species abundance	How many taxa to monitor?	Common versus rare species
	Species distribution		
	Species age structure		
Species traits	Phenology	Which metrics and how many taxa to monitor?	Metrics are taxon-dependent: flowering time, migration time
	Body mass		Harvested versus non-harvested species
Community composition	Species interactions	Which metrics to monitor?	Connectedness, length of trophic chain, interaction strength
	Taxonomic diversity		Species richness, species α and β diversity, phylogenetic diversity, etc.

For variables in the species traits class, both the general identification of which variable should be measured, what particular metric of that variable, and which species should be monitored, have to be considered (Table 4.1).

In any case, metrics and taxa to be monitored should follow a range of required and desirable criteria. Required criteria include: (1) monitoring should have a low impact on the targeted organisms over time; (2) the monitoring protocol should be reliable and repeatable with different personnel; (3) for targeted monitoring, the variable should have a strong correlation with the driver of concern; and (4) the variable should be ecologically important, that is, impacts on the variable have meaning at an ecosystem level or localised impacts are significant enough to warrant concern. The variables or metrics that meet the four required criteria are then evaluated for the desired criteria. Desired criteria include: (1) a quick response to the stressor so that effects are detectable in a short time frame; (2) a quick response to management actions so the efficacy of actions can be determined in a short time frame; (3) minimal stochastic variability so sample number can be small and effects can be clearly connected to the stressor of concern; (4) ease of measurement; (5) extended sampling window so scheduling and staff time can be more effectively allocated; (6) cost effectiveness; (7) ease of training personnel; (8) baseline data is available so effects seen are known to be stressor-caused and not a natural fluctuation; and (9) a response to the stressor can be seen when the impacts are still relatively slight; if the change cannot be detected until a large decline in resource condition occurs, alteration to the systems may be impossible or difficult to repair. The metrics that meet all the required criteria and most of the desired criteria can be chosen and then ranked, based on the number of desirable criteria they meet.

If some metrics obtain similar rankings, budgetary considerations can be used to prioritise measures to be included in the final program. A two-tier system may be adopted: Tier 1 measures can be carried out more frequently (e.g., yearly) and are either very important or less expensive. Tier 2 metrics are done less frequently (e.g., every 5 years), generally because they are expensive, destructive (e.g., material has to be collected), or require expertise that is not readily available. In addition, Tier 2 indicators can act as a check on more simplistic Tier 1 indicators. One of the major challenges with this approach is finding a way to incorporate variables of both high ecological significance and low cost. It is also important to note that the frequency of the measurements depends on the taxa being studied. Taxa with shorter life spans often require more frequent monitoring.

4.2.3   Choosing a Spatial Sampling Scheme

Despite recent advances in remote sensing for particular species (Pettorelli et al. 2014), for most taxa it is impractical to monitor an entire region at the one to five year intervals sought by many programs. Therefore, a spatial sampling scheme needs to be adopted for each monitored variable. We can broadly divide spatial sampling schemes in two major groups, extensive and site-based monitoring schemes (Fig. 4.1; Couvet et al. 2011). In extensive monitoring schemes a variable is observed at numerous sites over a large territory at regular time intervals, often using volunteers or citizen scientists (e.g., Breeding Bird Survey in North America, or the Pan European Common Bird Monitoring Scheme). In contrast, site-based or intensive monitoring schemes observe a range of variables at a limited number of sites, often associated to field stations of universities or organisations (e.g., the International Long Term Ecological Research Network—ILTER, the National Ecological Observation Network in the USA—NEON). Therefore a trade-off exists between the number of sites in a monitoring scheme (that is, its extensiveness) and the number of variables to be monitored or even the time intervals for the sampling (that is, the intensity of the monitoring effort). While extensive monitoring schemes have been very successful in providing long-term data on biodiversity change across large areas in developed regions, much of the data coming from developing regions is associated with site-based monitoring schemes (Proença et al. in press). Where volunteer capacity exists, the development of extensive national monitoring programs can be done very rapidly and it has been proposed that this model could also be applied in some developing countries (Pereira et al. 2010).

For both extensive and site-based monitoring schemes, the question of where to place the monitoring sites arises. This can be done using a systematic sampling design such as a grid, a random sampling design or a stratified random design

(Elzinga et al. 2001). One of the most common stratification schemes used is environmental stratification based on important habitat variables (Metzger et al. 2013). Sometimes a mixed design is used, for instance by systematically defining a grid and then randomly sampling inside that grid or within each habitat stratum of the grid. de Kruijter et al. (2006) provide a comprehensive guide to designing sampling frames.

One type of spatial data that is becoming increasingly relevant is opportunistic data (Fig. 4.1c). Over the last century, much biodiversity data was collected for museums and natural history collections. For instance, the Global Biodiversity Information Facility (GBIF) indexed more than 500 million species occurence records as of 2015, many of them from such collections. More recently, the development of websites for recording and sharing species observations (Boakes et al. 2010) is mobilizing an impressive range of data almost in real-time. Despite opportunistic observations being vulnerable to multiple biases (e.g., they are often presence-only data, so it is difficult to distinguish true from false absences), Bayesian methods have been recently developed to use this data to track biodiversity change (van Strien et al. 2013). Furthermore, the interactive community features of the social web allows for mobilizing observers for biodiversity observations in novel ways.

4.3   Taxon-Specific and Driver-Specific Examples

In this section we discuss methods available to monitor species EBVs (Table 4.1), particularly species distributions (also referred to as species occupancy or species occurrences) and species abundances. We emphasise species distributions and species abundances since some other EBVs (e.g., taxonomic diversity) can be inferred from those when data is collected for entire species assemblages. We use taxon-specific examples for mammals, amphibians, butterflies, and plants. We also include an example for monitoring a specific driver: wildlife diseases.

4.3.1   Mammals

Harmonizing monitoring schemes is likely to be more challenging for mammals than for other taxa (e.g., birds), because observation techniques used for mammals are often very species-specific (Battersby and Greenwood 2004) and reliability of techniques is likely to be affected by habitat type. It is advantageous to monitor mammal species that are common and easily observed as part of a global harmonised observation system. However, at a national level, it is also important to monitor less common species, particularly those of conservation concern, because of reporting requirements from international policy agreements and to assess nationally set targets.

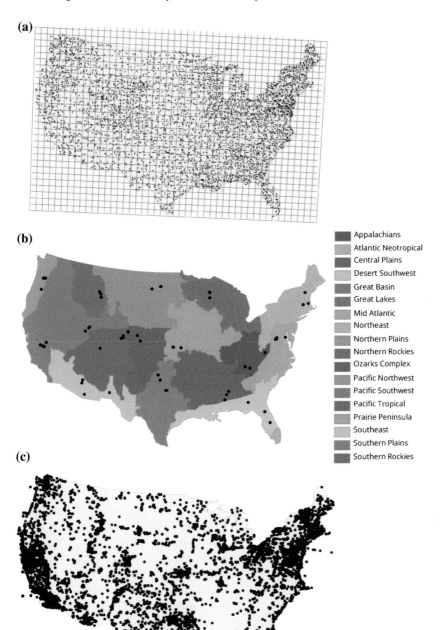

◀ **Fig. 4.1** Spatial sampling schemes for species data. **a** Extensive monitoring in the Breeding Bird Survey of the USA. Approximately 3000 routes are monitored yearly across the USA. The original routes were placed randomly for each $1° \times 1°$ cell, but the system has since expanded to take advantage of the proximity of cities with large numbers of observers. **b** Site-based monitoring in the National Ecological Observation Network. Each site was placed in order to be representative of an environmental domain. **c** Point species occurrence data from the iNaturalist portal, mostly opportunistic observations contributed by citizen scientists

The Tracking Mammals Partnership (TMP), established in 2005 by the Joint Nature Conservation Committee (JNCC), provides an interesting case study of a mammal monitoring programme developed at the national level. Despite a long history of natural history recording in the United Kingdom (Flowerdew 2004), reviews in the 1990s suggested a paucity of data on population, abundance, and distribution data for British mammals, prompting a call for an integrated monitoring programme to track the status of British mammals (Harris et al. 1995). The TMP is a collaborative effort between 25 organisations and uses a diverse programme of monitoring schemes, collecting data on a range of species in both urban and countryside environments, and covering a number of species relying on specialist survey methods. The TMP aims to detect changes in species distributions and abundance over time, by using stratified sampling to also provide regional trends, thus ensuring geographical representativeness (Battersby and Greenwood 2004).

Learning from monitoring efforts on bird populations was central to the development of the TMP, including through direct input from the ornithological community (Battersby and Greenwood 2004). For instance the British Trust for Ornithology (BTO) was involved in devising mammal tracking programmes such as the Winter Mammal Monitoring scheme. Specific lessons learnt included the importance of establishing long-term datasets of population indices through annual monitoring and the use of non-governmental conservation organisations and volunteers to collect data (Battersby and Greenwood 2004; Harris and Yalden 2004). While there is no single approach that suits all mammal species equally, it was suggested that a small number of monitoring techniques that can be applied to a large number of terrestrial mammal species could be integrated to form a multi-species monitoring programme (Harris and Yalden 2004). Most importantly, the chosen techniques should be applicable across a wide range of habitats to overcome biases established by past monitoring schemes focussing on specific habitat types (e.g., hedgerows, woodlands; Flowerdew 2004). By 2007, the TMP was reporting on annual trends for 35 species of terrestrial mammals.

While the TMP is less active at present, the constituent partner organisations are carrying out continued monitoring projects, some of which are run annually and cover multiple species (e.g., the National Bat Monitoring Programme run by the Bat Conservation Trust, the Breeding Bird Survey run by the BTO, and the Mini Mammal Monitoring run by the Mammal Society). Many of these monitoring schemes are based on line transects (for sightings of medium to large mammals and field signs) or live trapping transects (for small mammals) within specified grid squares (most often randomly selected 1 km^2 squares and involving two transects; e.g., Risely et al. 2012).

Transect counts are time-consuming. However, for large- to medium-sized mammals which occur at high densities in relatively open habitat, are relatively easily spotted (e.g., active at time of survey) or have field signs which are easily identifiable, transect counts can provide relatively robust estimates of species richness, relative abundances and habitat use. With help of specialist software such as DISTANCE (Thomas et al. 2010), estimates of absolute densities of species are also possible. Live trapping for small mammals has often been suggested as a key methodology for small mammal monitoring (Toms et al. 1999). Small annual changes in small mammal populations (e.g., 3–11 %) can be detected with 85 % power when monitoring is carried out for 10 years at a minimum of 50 sites (Flowerdew 2004). Other methodologies tested for use in the UK include road traffic casualties to monitor changes in relative abundance of several mammal species. With some refinement of the methodology, such as taking road type into account, the method may be sensitive enough to be used in national mammal monitoring schemes (Baker et al. 2004).

With the development of new technology, remotely monitoring mammals becomes more practical, often cutting down on man-hours spent in the field. In particular, camera trapping has been increasingly applied worldwide in monitoring and conservation (Fig. 4.2). It has been applied in a range of contexts from tracking specific species (e.g., the pygmy hippo in Sapo National Park; Collen et al. 2011), to multi-species monitoring, including tracking rare or elusive species in dense habitats such as tropical forests (Munari et al. 2011), monitoring small invasive

Fig. 4.2 Camera trapping is becoming one of the main methods to monitor medium to large mammals

mammals (Glen et al. 2013), and monitoring arboreal mammals (Cerbo and Biancardi 2012). Animal density estimation was previously only possible for species with individually recognisable markings; however, recent analytical developments have focussed on deriving methods and models to derive animal density estimates for species eliminating the requirement for individual recognition of animals (Rowcliffe et al. 2008; Chandler and Royle 2013). Methods have also been proposed to integrate data from camera trapping into biodiversity indicators (e.g., the Wildlife Picture Index; O'Brien et al. 2010; Beaudrot et al. 2016). Remote monitoring of mammals can result in large amounts of data, and the volunteer focus of traditional monitoring programmes is set to be turned into large-scale citizen scientist involvement to facilitate data processing (e.g., via species identification through mobile phone apps, such as Instant Wild; see www.edgeofexistence.org/instantwild/).

4.3.2 Amphibians

Assessing trends in amphibian populations can be challenging because they can fluctuate dramatically (Pechmann et al. 1989; Collins et al. 2009). In addition, many species often occur as meta-populations with some populations acting as 'sources' of individuals colonizing other places due to birth rates exceeding mortality rates, and some populations acting as 'sinks', receiving more animals than those that leave and where mortality rate exceeds birth rate. Therefore, it may be important to monitor the entire meta-population in order to produce meaningful results. Long-term studies have also shown that amphibian populations can vanish locally as a result of natural habitat changes that take place over decades (Collins et al. 2009).

As for other taxa, it is impossible to survey every habitat or catch every individual of a population, but ideally one should look to sample units that are separate and (statistically) independent. Sample units are usually individual animals for single population studies; they are quadrats, transects or habitat features like ponds and streams for community studies. Some monitoring programs focus on a handful of target species and report, in addition, all observations of rare species encountered during the surveys (e.g., Netherlands national monitoring scheme; Groenveld 1997).

A number of methods exist to survey species abundances and ranges for amphibians. Below we present very brief accounts of some of the most popular and promising ones:

- **Clutch counts** (also known as egg masses, spawn clumps, or batches) and **nest counts** are techniques that have been used to monitor population trends of some species and can also help to assess which factors are affecting populations. Egg mass counts have been used to assess population sizes of pool-breeding amphibians, particularly some explosive-breeding species, and they are relatively simple in that they only require surveying ponds repeatedly for clutches. Species whose eggs do not hatch very quickly (e.g., more than 10 days between laying and hatching) have higher detection probabilities (Crouch and Paton 2000). Nest counts have been used to estimate population size of some salamander species over long periods of time (e.g., Harris 2005).

- **Trapping** animals over time is a common method, either by using passive traps or by attracting animals to a trap (active traps). Nearly all passive traps for amphibians are either funnel traps or pitfall traps. Funnel traps have a funnel-shaped entrance that guides animals to a larger holding chamber, while pitfall traps consist of some type of container sunk into the ground with the rim level with the surface, and deep enough that the animals that fall into it cannot climb out (Gibbons and Semlitsch 1981). Traps are often used in combination with drift fences, which are vertical barriers that curtail the options of animals on the move and guide them towards a trap. The combination of drift fences and traps has proved very successful in some places (e.g., southern U.S.) but not in others (e.g., forests in NE Australia).

- **Area-based surveys** are used to estimate the abundance and density of a species or survey the amphibian fauna of a site. One needs to define small units within a larger area (plots or transects) that are sampled for amphibians, and, from the data collected, inferences are made about the larger area. The data can be used to compare species among habitats or to study how communities change over ecological gradients or over time. The literature indicates plots are generally square or rectangular, with median dimensions of 25×20 m (range 4–400×2–240 m); transects are narrow plots intended to be explored by a single person at a time, and their median dimensions are 100×2 m (range 7–2000×1–8 m) (Marsh and Haywood 2010). Though plots and transects are often surveyed visually, sometimes they can be sampled by registering calls. The final choice of the size, shape, and number of units to sample depends on the questions that the survey is intended to address.

- **Auditory monitoring** is a relatively efficient method for assessing frogs and toads. The method has proven a useful tool for anurans because many are more easily heard than seen and it is widely used in the U.S. and Canada (Weir and Mossman 2005). This is a good method for monitoring changes in anuran occupancy or for rough species inventories. Nevertheless, it has some limitations, as it relies on detecting singing males (and thus misses females and sub-adults), and cannot be applied to the non-singing salamanders and caecilians. More recently, automated systems, or frogloggers, are being used to collect data at single sites. Such automated systems may be the most efficient way to monitor threatened species or those with unpredictable breeding seasons in the future.

- **Environmental or e-DNA** is a promising technique that will likely be useful for detection of rare freshwater species (Ficetola et al. 2008; Thomsen et al. 2012). This technique relies on DNA obtained directly from small water samples of lakes, ponds and streams. It has been tested successfully in temperate systems for detection of amphibians, but to our knowledge is not yet being used for amphibian monitoring.

4.3.3   Butterflies

Contrary to most other groups of insects, butterflies are relatively well-documented, easy to recognise and popular with the general public. Butterflies use the landscape at a fine scale and react quickly to changes in management, intensification or abandonment. Furthermore, a sustainable butterfly population relies on a network of breeding habitats scattered over the landscape, where species exist in a meta-population structure. This makes butterflies especially vulnerable to habitat fragmentation. Moreover, as ecto-therm animals, many butterflies are highly sensitive to climate change.

At the national scale the following monitoring techniques can be used to monitor species ranges and species abundances of butterflies:

- Unvalidated, **opportunistic data** can only be used for coarse distribution maps. Species distribution modelling including habitat and climate variables can be used to refine the species ranges from opportunistic data (Jetz et al. 2012). If the quantity of observations is high enough and the quality of visits can be established, the Frescalo method (Hill 2012) and occupancy modelling can be used to establish distribution trends (Isaac et al. 2014).
- **Standardised day-lists** can be used for occupancy modelling (van Strien et al. 2011). An advantage of this method is that it can work with co-variates (e.g., the Julian date, as butterflies typically have a limited flight period). Occupancy modelling with day-lists also addresses the problem of detection probability. Occupancy modelling can also produce colonisation and persistence trends, population parameters that can be very helpful to identify the causes of observed occupancy changes. It is important to note that the statistical methods for occupancy modelling are data and computation intensive.
- **Standardised counts** following a protocol is ideal for population abundance monitoring. For instance, in Europe although field methods differ to some degree across countries, most counts are conducted along fixed transects of about 1 kilometre, consisting of smaller sections, each with a homogeneous habitat type (van Swaay et al. 2008). Visits are only conducted when weather conditions meet specified criteria. Site selection varies from random stratified designs (only in a few countries), to grid design (only in Switzerland), to free observer choice (most countries). Countries use a software package called TRIM to analyse and supply trend information at the national level. Trend data are then integrated to create European population indices for species and multi-species indicators.

4.3.4   Plants

Plants, as primary producers, are effectively the basis of life on earth, and funda-mental not only to many millions of species, known and unknown, but also our own. However, our knowledge of the world's flora remains limited, despite over 250 years of scientific research. In 1753 when Linnaeus published his Species Plantarum, some 5573 plant species were included; at that time, he was convinced the number would never exceed 10,000. Today, the total of known species stands at ~380,000 (Paton et al. 2008) out of a total of more than 890,000 published names for plant species, with almost 2000 newly described species published annually (www.ipni.org/stats.html). Centres of plant diversity (Davis et al. 1997) and hotspots of threatened plants (www.conservation.org/hotspots) have been identified. There are many permanent forest plots that have received one or more complete censuses (e.g., the CTFS network; www.ctfs.si.edu/plots). However, this is collectively only a very small proportion of the total land area of the Earth and for many individual species there is little available data beyond the natural history collections, herbarium specimens and their original description.

Recent attempts to consolidate existing knowledge, from which EBVs and hence global biodiversity indicators must be derived, have been largely driven by international policy objectives. The botanical community has galvanised around the Global Strategy for Plant Conservation, adopted by the Convention on Biological Diversity. This Strategy has a set of targets to be achieved by 2020, including Target 1 which is to produce '*an online flora of all known plants*' and Target 2 which is to undertake '*an assessment of the conservation status of all known plant species, as far as possible, to guide conservation action*'.

Formal assessments of the conservation status of most plant species are still lacking. Only 19 728 plant species have been assessed by the Red List (www.iucnredlist.org), totalling less than 5 % of the world's flora (as of November 2014). Of those assessed, about 54 % (10,584 plant species) have been classified as threatened. The assessment of extinction risk is based on objective and quantitative criteria that capture one or more EBVs (e.g., species distribution and species abundance). This can be based, in the first instance, on opportunistically-collected herbarium specimen data and published botanical literature (Brummitt et al. 2008; Rivers et al. 2011), followed by verification and validation in the field (Brummitt et al. 2015). It is important that assessments are based on a verifiable trail of data, from maintained long-term databases, preserved herbarium specimens, or published literature sources.

Field-based monitoring techniques for plant EBVs are many and varied, including:

- **Quadrats** can be used to survey plants, as it is a particularly effective method for sessile organisms. Quadrats can be of different sizes, depending on the size of the plants and the structure of the vegetation, but need to be consistent within the study. Typically they are a few times larger than the mean size of the organisms being monitored. Quadrats should be placed at random and should be permanently marked to allow repeated measures through time. In addition, there should be a sufficient number of replicates to ensure statistical power. Within each quadrat, species can be recorded as actual counts, as some measure of cover (see below) or density or frequency, or occasionally biomass (dry weight).

Species can be grouped into higher taxonomic units such as genera or families or as functional ecological units such as graminoids (grasses and grass-like plants), forbs (herbaceous plants), shrubs, trees, and climbers. The standardised plot surveys of the Centre for Tropical Forest Science, in which each individual tree is identified, tagged, and mapped on a repeated cycle, are perhaps some of the largest quadrats ($\sim$50 ha in size) being measured with standard protocols around the world.

- **Transects** of varying width, are often employed over longer distances, especially against an environmental gradient or gradient of disturbance that intentionally includes the range of floristic variation within the area. Along each transect, each species may be recorded including information on numbers of individuals, distance from transect, cover, biomass, density or frequency.

- Placement of **quadrats along transects** has several advantages. First, quadrats along a line can be easier to relocate than if scattered across an area. Second, quadrats allow for more vegetated space to be measured along the line than compared to points along a transect line. Finally, the advantage of a transect is maintained (i.e., covering more space, thus incorporating more variability, and enabling spatial analysis).

- **Cover** can be assessed using different methods, such as the DAFOR (Dominant, Abundant, Frequent, Occasional or Rare), Braun-Blanquet (5 classes up to 100 % cover, not of equal size) and Domin (10 classes up to 100 % cover, not of equal size) scales. Each can be used with existing sampling techniques such as quadrat or transect of defined length and width. The classes for the DAFOR scale can be interpreted by the user relative to the particular situation, as long as this is consistent and stated within each study. Assessments of extinction risk under IUCN Criterion A require estimates of population size and its change over time from 'an index of abundance appropriate to the taxon', using any of these cover assessment methods across the species range, as long as this is stated and applied consistently between time points.

- **Counts** of all individuals of conspicuous plants at low densities are possible, although this is time-consuming and it can be difficult to avoid double counting. Counts are particularly challenging for densely-growing plants and clonal plants. In those situations measures of cover, of numbers of ramets (modular, repeating, connected units of the plant) or numbers of stems or reproducing stems may be used instead. For Red List assessments under IUCN Criterion C, actual counts of numbers of individuals are required, but the thresholds for threatened categories are low in value. Therefore this is a feasible technique for species of known conservation concern, although it is not generally viable for widespread and less threatened species. Frequency of presence/absence in quadrats of known size can be related to population density.

- **Mapping** vegetation over larger areas is possible using GPS points or tracks and a pre-defined habitat classification such as the National Vegetation Survey of the UK, the Braun-Blanquet vegetation types, one specified by the user, or from remotely-sensed data. Available satellite imagery can detect fine spatial resolution and variation within vegetation, even detecting characteristic individual tree species with LIDAR data, to which image-recognition algorithms can be trained. Care needs to be given to seasonality for vegetation mapping, including the tropics where seasons tend to be defined by rainfall rather than temperature, even within apparently uniform rain forest. The combination of different methods is extremely useful in vegetation mapping, as remotely-sensed data needs validation and ground-truthing through on-the-ground observations from quadrats, transects or point surveys.

- **Environmental DNA (eDNA)** approaches, in which estimates of species richness and species abundances may be obtained from next-generation sequencing of leaf litter or soil samples, offer considerable promise for rapid ground-truthing of satellite imagery, if a suitable DNA library exists against which to compare the species.

Few plant species have sufficient data at the global or regional levels for the majority of the Essential Biodiversity Variables (Table 4.1). However, much is already known: there is a draft global species checklist (www.theplantlist.org), with synonymy and distributions for each species; species ranges are available for many vascular plants in some regions (e.g., Europe, USA); weight is one of the main traits compiled in the TRY database (Kattge et al. 2011); phenology, at least for flowering and often fruiting, can be inferred from herbarium specimens (collections are usually only made if a species is in flower or fruit, and collecting date is given on the label) and taxonomic literature; dispersal mode if not distance can be similarly inferred from fruit and seed morphology. What generally is not known for the overwhelming majority of plant species is how these variables are changing over time. Furthermore, data on local abundances and population structure is only being compiled at some research sites, such as the aforementioned forest plots (e.g., CTFS), and data on individual trophic interactions is even less available. Still, available plot data was recently used to provide a global assessment of changes in local species richness over the last few decades (Vellend et al. 2013), with the surprising result that no net change on species richness was found on the set of plots analysed.

The capacity for developing countries to undertake repeated measurements of the EBVs for which base data already exists, such as species ranges, populations, and phenology, is limited. Therefore measuring and monitoring EBVs for plants is inherently also a capacity-building exercise. Knowledge of the plants themselves and the ability to accurately identify them is of utmost importance. There is an ever-increasing availability of digital specimen data through GBIF (www.gbif.org) or other platforms, or crowd-sourced specimen databasing and georeferencing. Rapid, standardised satellite imagery can be used to monitor habitat loss and vegetation change. But it is essential to develop training workshops in assessment and monitoring techniques for local experts, provide easy-to-use identification tools and field guides, and develop long-term partnerships. Many of these approaches come together in work conducted for the IUCN Sampled Red List Index for Plants

(Brummitt and Bachman 2010) (www.threatenedplants.myspecies.info), where observable change in range size or population size is measured to re-assess the Red List status of a broadly representative sample of plant species from around the world.

4.3.5 Monitoring Diseases

Infectious wildlife diseases are emerging globally, and their adverse effects are becoming more and more visible (Fisher et al. 2012). It is therefore important to include disease surveillance or pathogen monitoring into global, regional, and national biodiversity monitoring strategies. The three main questions faced when designing a disease monitoring scheme, i.e. why, what, and how to monitor, are also relevant here. The answer to *why* to establish disease surveillance is straightforward: the adverse effects of non-native emerging infectious diseases can throw entire ecosystems out of balance and have major impacts on humans, livestock and crops (Keesing et al. 2010). The question of *what* to monitor is a bit more challenging, as one could monitor the symptoms of a disease, the disease itself, or the pathogen. Considering that disease monitoring should also be an early warning system, it might be suboptimal to monitor the symptoms of a disease or the disease itself. It is preferable to monitor the presence of a pathogen, but then, what are the EBVs needed to describe the status of a pathogen? Finally, the question of *how* to monitor pathogens needs to consider different sources of error such as the representativeness and detection probability. Random selection or stratified random selection of monitoring sites ensures that the sample will be representative for the larger area from which the sites are selected (Yoccoz et al. 2001). However, other questions might demand a different site selection strategy. Imperfect detection, or detection probability (Kéry and Schmidt 2008; Archaux et al. 2012), is of particular interest in pathogen monitoring, as pathogens are often difficult to detect (McClintock et al. 2010).

As pathogens depend on their host, pathogen monitoring often starts with monitoring of the host. In many cases, a pathogen is only detected after disease outbreaks and when negative effects on the host population become evident (Berger and Speare 1998; Blehert et al. 2009). Monitoring species distribution can detect a change in a host population linked to disease outbreaks and the presence of pathogens. Species abundance is more sensitive, but it is also more difficult to conduct over large regions. Pathogen monitoring should be conducted at the same sites (or a random subset of them) to establish the occurrence pattern of the pathogen in both space and in time and to track disease outbreaks. Once the occurrence of a pathogen has been detected, infection prevalence (the proportion of infected individuals in a population) needs to be recorded, followed by infection intensity. These two state variables will inform about the extent of the infection and will give information on the future dynamics of the disease, especially if prevalence is above a 5–10 % threshold (Knell et al. 1998). Above such a threshold, epidemics often occur. In case pathogen occurrence is clustered or when unusual mortality rates are observed, it is advisable to conduct more detailed surveys with more specific questions. This may include delineation of clusters, identification of areas of host population declines, determination of the involved variants of the pathogen, and investigating the taxonomic, seasonal and temporal variation of prevalence and infection intensity. Such information can then feed into a risk analysis for the host population(s).

Care needs to be taken that the same host species is monitored across different sites and different years to yield robust information on the pathogen. It is also important to have sufficient sample sizes when conducting detailed surveys, as otherwise false negatives may not allow delineating the distribution of the pathogen. The necessary sample size is dependent on the minimum prevalence expected if the population/specimen were infected. For example, the common prevalence of a resident disease in a population is approximately 5 %. With that level of prevalence, at least 90 specimens need testing for the likely detection of one or more positive individuals to reach 99 %. An approximation to the number of individuals that need to be tested to be 95 % certain of detecting at least 1 positive individual is $n = 3/p$ (for 99 % certainty it is $4.5/p$), where p is the prevalence expressed as a proportion (Walker et al. 2007). In case no visible symptoms of a disease can be detected, such as in the amphibian disease chytridiomycosis, detection and quantification of a pathogen might need quantitative molecular tools such as PCR (e.g., for *Batrachochytrium dendrobatidis*; see Boyle et al. 2004; Hyatt et al. 2007) or Next-Generation Sequencing.

4.4 From Species Monitoring to Ecosystem Services

Biodiversity plays several roles along the process chain that links ecosystems to human well-being and which includes ecosystem processes, final ecosystem services (i.e., services that directly underpin or give rise to goods), and the (material and non-material) goods generated by those services (Mace et al. 2012). As species may contribute to all these stages, the application of species monitoring data to ecosystem services should take into account their position in this process chain. Establishing these connections between species monitoring and ecosystem services is important to support the work of the Intergovernmental Platform on Biodiversity and Ecosystem Services (IPBES; Díaz et al. 2015).

If species constitute final ecosystem services or goods, that is, if species are directly linked to services, then species population data can be directly used to monitor ecosystem services. This is usually the case of provisioning services (i.e., material ecosystem outputs that can be directly used) and cultural services (i.e., non-material ecosystem outputs with cultural or spiritual significance). Examples of provisioning services provided directly by species include, among others, food (e.g., game birds, wheat, mushrooms), fibres (e.g., cork oak, timber trees, sheep) and medicines (e.g., *Aloe* spp., medicinal herbs, poison dart frogs). Examples of cultural services include, among others, charismatic species (e.g., monarch butterflies, primates, orchids) and species inspiring technology (e.g., *Morpho* butterflies, lotus plants). Therefore, a decrease in the species abundance or species range of a game bird or a primate species corresponds to a decrease in the supply of the associated provisioning or cultural service.

In other situations, species do not constitute final services or goods, but are known to play a facilitator or intermediary role in the ecosystem processes underpinning the services. This is particularly true for regulating services (i.e., non-material ecosystem outputs not directly used by people but that affect human well-being) such as water run-off regulation or pollination, but also for some provisioning or cultural services such as clean water provision and landscape character. While individual species may play a dominant role in ecosystem processes generating services, for example, fruit tree pollination by honey bees, in most cases, ecosystem processes are affected by multiple species in a community (Díaz et al. 2007; Hillebrand and Matthiessen 2009; Lavorel et al. 2011). In these situations, data on species abundance and distribution obtained through monitoring schemes can be complemented with data on species traits (i.e., morphological, physiological and life history attributes), in order to compute community-aggregated metrics that characterise the community regarding traits of interest for a particular function. For example, data on root size and architecture can be used to assess the contribution of plant communities to water regulation and soil stability, and data on body size and feeding habits can be used to assess the pollination potential of insect communities (de Bello et al. 2010).

Species traits can also be applied in the identification of species functional groups relevant to monitoring provisioning, cultural or regulating ecosystem services. For instance, protein content could be an indicator of plants' forage value in pastures (Lavorel et al. 2011), production of medicinally important compounds, such as antioxidants and alkaloids, could be an indicator of medicinal value (Canter et al. 2005), and structural complexity could be an indicator of existence value (Proença et al. 2008).

In addition to the traits determining species contribution to ecosystem processes, final services or goods (effect traits), species can also be characterised by traits shaping their responses to pressures (response traits). These two categories of traits provide complementary information regarding species interaction with their environment, that is, species responses to external drivers and species input to ecosystem processes and services. Response traits, such as fire response traits (e.g., resprouting ability, serotiny) and habitat specialisation, can be used to assess or predict the impacts of drivers of change or conservation measures on species populations and communities. The borderline between the two categories is not strict, as some effect traits may also be response traits. For example, leaf area has an effect on evapotranspiration, and hence on water regulation, but it can also respond to drought or nutrient availability. Response traits are not only reactive to pressures, providing a way of tracking their impacts on a certain area, but also to the variation of abiotic conditions across a landscape or region (Lavorel et al. 2011). Therefore, data on abiotic variables, such as climate and physiography, are also needed when monitoring ecosystem services using species data, since abiotic factors indirectly affect ecosystem processes through effects on species functional attributes. Moreover, the contribution of species or functional groups to the processes underpinning ecosystem services should be weighed against the direct influence of abiotic factors on these processes.

4.5 Scaling from Local Observations to the Global Monitoring of Biodiversity Change

Perhaps the main challenge facing the development of EBVs at the species level is the scaling from the temporally and spatially scattered local observations to the global level. Data collection, mobilisation, sharing and harmonisation are key steps in addressing this challenge, but two additional stages are important: the development of indicators and the development of models of EBV responses to drivers of biodiversity change (Akçakaya et al. 2016).

Over the last decade significant advances have been made in developing indicators of biodiversity change as assessment and communication tools (Sparks et al. 2011; Collen et al. 2013). Indicators are able to synthesise the wealth of data in a given EBV, for example, the abundance of each species i at time t in location $[x, y]$, into a single scalar number, such as geometric mean abundance at time t. This can confer statistical robustness to indicators: when individual observations are brought together, statistics such as means and variances can be calculated. Naturally the statistical power of indicators is completely dependent on the representativeness of the underlying data, and it has been argued that indicators used in recent assessments are spatially, temporally and taxonomically biased (Pereira et al. 2012; Akçakaya et al. 2016). Indicators also allow to communicate the evolution of a particular aspect of biodiversity (e.g., mean species abundance) to the public, which can be compared to targets set by managers and policy makers (Jones et al. 2011; Geijzendorffer et al. 2016). Several species based indicators where recently used to assess international progress towards the 2020 Aichi Targets of the Convention on Biological Diversity, including the Red List Index, the Living Planet Index, the number of mammal and bird extinctions, the Wild Bird Index, and the cumulative number of alien species introduction events (Tittensor et al. 2014).

Indicators are powerful communication tools that can help to transmit succinct information about the status of biodiversity, but they may be insufficient to uncover the drivers of biodiversity change. In order to understand what is driving biodiversity change, the indicators, or even better, the EBV data itself, needs to be analysed and modelled in relation to datasets on drivers of change such as land-use change, climate change, harvest or hunting pressure, and pollution. As an example, Rittenhouse et al. (2012) found a strong response of bird species richness and abundance to land-cover changes between 1992 and 2001, using correlative models. The PREDICTS project has reviewed studies of the impact of different types of land-use change on different metrics of biodiversity using over 1 million records of species abundance and over 300,000 records of species occurrence or richness (Newbold et al. 2015). They estimated a global reduction of 10 % in local species richness based on global models of land use in relation to a historical baseline (Newbold et al. 2015). An alternative approach is to develop indicators of the effect of a driver on biodiversity, such as the indicator of the impact of climate change on European Bird populations (Gregory et al. 2009) or the community temperature index (Devictor et al. 2012).

The development of models connecting responses of EBVs such as species distribution and species abundance to drivers such as land-use or other biophysical variables that can be measured using remote sensing is particularly important to address this upscaling challenge. Such models could allow the extrapolation of point observations resulting from *in situ* monitoring into continuous variables in space and time. Species distribution models are already capable of producing spatially explicit projections, at global scale, of how a species range might respond to climate change based on a limited number of point-based observations (Peterson et al. 2011) and wall-to-wall climate data. Similar correlative models have also been used to project species distributions for different scenarios of land-use change (Jetz et al. 2007; Rondinini et al. 2011).

With the support of CSIRO, Map of Life, PREDICTS and others, GEO BON is now developing several global biodiversity change indicators (GEO BON 2015) that build on the EBV framework concept (Pereira et al. 2013). The idea is that EBVs such as species distributions can be modelled continuously in space by integrating point-based species observations, remote-sensing of habitat cover, and other biophysical data such as elevation (Jetz et al. 2012). The availability of annual updates on the distribution of global forest cover, allows one to also estimate species ranges of forest dependent species over time. Finally, for any spatial region (e.g., a country or part of a country) an indicator of the total area of suitable habitat for each species can be calculated and averaged across a taxonomic group of interest (e.g., threatened birds).

As these examples illustrate, the collaboration between volunteers and professionals collecting biodiversity data, the scientists analysing the data, and the managers acting on the data, will be critical to address the on-going biodiversity crisis. We hope the EBV framework will help harmonise and integrate the work across these different communities.

References

Akçakaya, H. R., Pereira, H. M., Canziani, G., Mbow, C., Mori, A., Palomo, M. G., Soberon, J., et al. (2016). Improving the rigour and usefulness of scenarios and models through ongoing evaluation and refinement. In S. Ferrier, K. N. Ninan, P. Leadley, R. Alkemade, L. Acosta-Michlik, H. R. Akcakaya, L. Brotons, W. Cheung, V. Christensen, K. H. Harhash, J. Kabubo-Mariara, C. Lundquist, M. Obersteiner, H.M. Pereira, G. Peterson, R. Pichs, C. Rondinini, N. Ravindranath, & B. Wintle (Eds.). IPBES. (2016). *Methodological assessment of scenario analysis and modelling of biodiversity and ecosystem services.*

Amano, T., & Sutherland, W. J. (2013). Four barriers to the global understanding of biodiversity conservation: wealth, language, geographical location and security. *Proceedings of the Royal Society B: Biological Sciences, 280,* 1756.

Archaux, F., Henry, P.-Y., & Gimenez, O. (2012). When can we ignore the problem of imperfect detection in comparative studies? *Methods in Ecology and Evolution, 3,* 188–194.

Aronova, E., Baker, K. S., & Oreskes, N. (2010). Big science and big data in biology: From the International Geophysical Year through the International Biological Program to the Long Term Ecological Research (LTER) Network, 1957–Present. *Historical Studies in the Natural Sciences, 40,* 183–224.

Baker, P. J., Harris, S., Robertson, C. P. J., Saunders, G., & White, P. C. L. (2004). Is it possible to monitor mammal population changes from counts of road traffic casualties? An analysis using Bristol's red foxes *Vulpes vulpes* as an example. *Mammal Review, 34,* 115–130.

Battersby, J. E., & Greenwood, J. J. D. (2004). Monitoring terrestrial mammals in the UK: Past, present and future, using lessons from the bird world. *Mammal Review, 34,* 3–29.

Beaudrot, L., Ahumada, J. A., O'Brien, T., Alvarez-Loayza, P., Boekee, K., Campos-Arceiz, A., et al. (2016). Standardized assessment of biodiversity trends in tropical forest protected areas: The end is not in sight. *PLoS Biology, 14,* e1002357.

Berger, L., & Speare, R. (1998). Chytridiomycosis—a new disease of amphibians. *ANZCCART News, 11,* 1–3.

Blehert, D. S., Hicks, A. C., Behr, M., Meteyer, C. U., Berlowski-Zier, B. M., Buckles, E. L., et al. (2009). Bat white-nose syndrome: An emerging fungal pathogen? *Science, 323,* 227.

Boakes, E. H., McGowan, P. J. K., Fuller, R. A., Chang-qing, D., Clark, N. E., O'Connor, K., et al. (2010). Distorted views of biodiversity: Spatial and temporal bias in species occurrence data. *PLoS Biology, 8,* e1000385.

Boyle, D. G., Boyle, D. B., Olsen, V., Morgan, J. A. T., & Hyatt, A. D. (2004). Rapid quantitative detection of chytridiomycosis (*Batrachochytrium dendrobatidis*) in amphibian samples using real-time Taqman PCR assay. *Diseases of Aquatic Organisms, 60,* 141–148.

Brummitt, N. A., & Bachman, S. P. (2010). *Plants under pressure—A global assessment: the first report of the IUCN Sampled Red List Index for Plants.* Kew, Richmond, UK: Royal Botanical Gardens.

Brummitt, N. A., Bachman, S. P., & Moat, J. (2008). Applications of the IUCN Red List: towards a global barometer for plant diversity. *Endangered Species Research, 6,* 127–135.

Brummitt, N. A., Bachman, S. P., Aletrari, E., Chadburn, H., Griffiths-Lee, J., Lutz, M., et al. (2015). The sampled red list index for plants, phase II: ground-truthing specimen-based conservation assessments. *Philosophical Transactions of the Royal Society of London B: Biological Sciences, 370,* 20140015.

Canter, P. H., Thomas, H., & Ernst, E. (2005). Bringing medicinal plants into cultivation: Opportunities and challenges for biotechnology. *Trends in Biotechnology, 23,* 180–185.

Cerbo, A. R. D., & Biancardi, C. M. (2012). Monitoring small and arboreal mammals by camera traps: Effectiveness and applications. *Acta Theriologica, 58,* 279–283.

Chandler, R. B., & Royle, J. A. (2013). Spatially explicit models for inference about density in unmarked or partially marked populations. *The Annals of Applied Statistics, 7,* 936–954.

Collen, B., Howard, R., Konie, J., Daniel, O., & Rist, J. (2011). Field surveys for the endangered pygmy hippopotamus *Choeropsis liberiensis* in Sapo National Park, Liberia. *Oryx, 45,* 35–37.

Collen, B., Pettorelli, N., Baillie, J. E., & Durant, S. M. (2013). *Biodiversity monitoring and conservation: bridging the gap between global commitment and local action.* Cambridge, UK: Willey-Blackwell.

Collins, J. P., Crump, M. L., & Lovejoy, T. E., III. (2009). *Extinction in our times: Global amphibian decline.* New York, USA: Oxford University Press.

Couvet, D., Devictor, V., Jiguet, F., & Julliard, R. (2011). Scientific contributions of extensive biodiversity monitoring. *Comptes Rendus Biologies, 334*, 370–377.

Crouch, W. B., & Paton, P. W. C. (2000). Using egg-mass counts to monitor wood frog populations. *Wildlife Society Bulletin, 28*, 895–901.

Davis, S. D., Heywood, V. H., & Hamilton, A. C. (Eds.). (1997). Centres of plant diversity: A guide and strategy for their conservation (Vol. 3, Americas). Gland, Switzerland: IUCN and WWF.

de Bello, F., Lavorel, S., Díaz, S., Harrington, R., Cornelissen, J. H. C., Bardgett, R. D., et al. (2010). Towards an assessment of multiple ecosystem processes and services via functional traits. *Biodiversity and Conservation, 19*, 2873–2893.

de Kruijter, J., Brus, D., Bierkens, H., & Knotters, M. (2006). *Sampling for natural resource monitoring.* Berlin, Germany: Springer.

Devictor, V., van Swaay, C., Brereton, T., Brotons, L., Chamberlain, D., Heliola, J., et al. (2012). Differences in the climatic debts of birds and butterflies at a continental scale. *Nature Climate Change, 2*, 121–124.

Díaz, S., Demissew, S., Carabias, J., Joly, C., Lonsdale, M., Ash, N., et al. (2015). The IPBES conceptual framework—Connecting nature and people. *Current Opinion in Environmental Sustainability, 14*, 1–16.

Díaz, S., Lavorel, S., de Bello, F., Quétier, F., Grigulis, K., & Robson, T. M. (2007). Incorporating plant functional diversity effects in ecosystem service assessments. *Proceedings of the National Academy of Sciences of the USA, 104*, 20684.

Dornelas, M., Gotelli, N. J., McGill, B., Shimadzu, H., Moyes, F., Sievers, C., et al. (2014). Assemblage time series reveal biodiversity change but not systematic loss. *Science, 344*, 296–299.

Elzinga, C., Salzer, D., Willoughby, J., & Gibbs, J. (2001). *Monitoring plant and animal populations.* Oxford, UK: Blackwell Science.

Ficetola, G. F., Miaud, C., Pompanon, F., & Taberlet, P. (2008). Species detection using environmental DNA from water samples. *Biology Letters, 4*, 423–425.

Fisher, M. C., Henk, D. A., Briggs, C. J., Brownstein, J. S., Madoff, L. C., McCraw, S. L., et al. (2012). Emerging fungal threats to animal, plant and ecosystem health. *Nature, 484*, 186–194.

Flowerdew, J. R. (2004). Advances in the conservation of British mammals, 1954–2004: 50 years of progress with The Mammal Society. *Mammal Review, 34*, 169–210.

Geijzendorffer, I. R., Regan, E. C., Pereira, H. M., Brotons, L., Brummitt, N. A., Gavish, Y., Haase, P., et al. (2016). Bridging the gap between biodiversity data and policy reporting needs: An Essential Biodiversity Variables perspective. *Journal of Applied Ecology, 53*, 1341–1350.

GEO BON. (2015). *Global biodiversity change indicators: Model-based integration of remote-sensing & in situ observations that enables dynamic updates and transparency at low cost.* Leipzig, Germany: GEO BON Secretariat.

Gibbons, J. W., & Semlitsch, R. D. (1981). Terrestrial drift fences with pitfall traps: An effective technique for quantitative sampling of animal populations. *Brimleyana*, 1–6.

Glen, A. S., Cockburn, S., Nichols, M., Ekanayake, J., & Warburton, B. (2013). Optimising camera traps for monitoring small mammals. *PLoS ONE, 8*, e67940.

Gregory, R., Willis, S., Jiguet, F., Vorisek, P., Klvanova, A., van Strien, A., et al. (2009). An indicator of the impact of climatic change on European bird populations. *PLoS ONE, 4*, e4678.

Groenveld, A. (1997). *Handleiding voor het monitoren van amfibieën in Nederland.* Nijmegen, Netherlands: Stichting RAVON.

Harris, R. N. (2005). *Hemidactylium scutatum.* In M. Lannoo (Ed.), *Amphibian declines: The conservation status of United States species* (pp. 780–781). Berkeley, CA, USA: University of California Press.

Harris, S., Morris, P., Wray, S., & Yalden, D. (1995). *A review of British mammals: Population estimates and conservation status of British mammals other than cetaceans.* Peterborough, UK: JNCC.

Harris, S., & Yalden, D. W. (2004). An integrated monitoring programme for terrestrial mammals in Britain. *Mammal Review, 34,* 157–167.

Henle, K., Bauch, B., Auliya, M., Kuelvik, M., Pe'er, G., Schmeller, D. S., et al. (2013). Priorities for biodiversity monitoring in Europe: A review of supranational policies and a novel scheme for integrative prioritization. *Ecological Indicators, 33,* 5–18.

Hillebrand, H., & Matthiessen, B. (2009). Biodiversity in a complex world: Consolidation and progress in functional biodiversity research. *Ecology Letters, 12,* 1405–1419.

Hill, M. O. (2012). Local frequency as a key to interpreting species occurrence data when recording effort is not known. *Methods in Ecology and Evolution, 3,* 195–205.

Hyatt, A. D., Boyle, D. G., Olsen, V., Boyle, D. B., Berger, L., Obendorf, D., et al. (2007). Diagnostic assays and sampling protocols for the detection of *Batrachochytrium dendrobatidis*. *Diseases of Aquatic Organisms, 73,* 175–192.

Isaac, N. J. B., van Strien, A. J., August, T. A., de Zeeuw, M. P., & Roy, D. B. (2014). Statistics for citizen science: extracting signals of change from noisy ecological data. *Methods in Ecology and Evolution, 5,* 1052–1060.

Jenkins, C. N., Pimm, S. L., & Joppa, L. N. (2013). Global patterns of terrestrial vertebrate diversity and conservation. *Proceedings of the National Academy of Sciences of the USA, 110,* E2602–E2610.

Jetz, W., McPherson, J. M., & Guralnick, R. P. (2012). Integrating biodiversity distribution knowledge: Toward a global map of life. *Trends in Ecology & Evolution, 27,* 151–159.

Jetz, W., Wilcove, D. S., & Dobson, A. P. (2007). Projected impacts of climate and land-use change on the global diversity of birds. *PLoS Biology, 5,* e157.

Jones, J. P. G., Collen, G., Atkinson, G., Baxter, P. W. J., Bubb, P., Illian, J. B., et al. (2011). The why, what, and how of global biodiversity indicators beyond the 2010 target. *Conservation Biology, 25,* 450–457.

Kattge, J., Díaz, S., Lavorel, S., Prentice, I. C., Leadley, P., Bönisch, G., et al. (2011). TRY - a global database of plant traits. *Global Change Biology, 17,* 2905–2935.

Keesing, F., Belden, K., Daszak, P., Dobson, A., Harvell, C. D., Holt, R. D., Hudson, P., et al. (2010). Impacts of biodiversity on the emergence and transmission of infectious diseases. *Nature* 468, 647–652.

Kéry, M., & Schmidt, B. R. (2008). Imperfect detection and its consequences for monitoring for conservation. *Community Ecology, 9,* 207–216.

Knell, R. J., Begon, M., & Thompson, D. J. (1998). Host-pathogen population dynamics, basic reproductive rates and threshold densities. *Oikos, 81,* 299–308.

Lavorel, S., Grigulis, K., Lamarque, P., Colace, M.-P., Garden, D., Girel, J., et al. (2011). Using plant functional traits to understand the landscape distribution of multiple ecosystem services: Plant functional traits and provision of multiple ecosystem services. *Journal of Ecology, 99,* 135–147.

Leadley, P. W., Krug, C. B., Alkemade, R., Pereira, H. M., Sumaila, U. R., Walpole, M., et al. (2014). *Progress towards the Aichi biodiversity targets.* Montréal, Canada: Secretariat of the Convention on Biological Diversity.

Mace, G. M., Norris, K., & Fitter, A. H. (2012). Biodiversity and ecosystem services: a multilayered relationship. *Trends in Ecology & Evolution, 27,* 19–26.

Magurran, A. E. (2004). *Measuring biological diversity.* Malden, MA, USA: Blackwell.

Marsh, D. M. & Haywood, L. M. B. (2010). Area-based surveys. *Amphibian Ecology and Conservation: A Handbook of Techniques* (pp. 247–262). Oxford University Press, New York, USA.

Martin, L. J., Blossey, B., & Ellis, E. (2012). Mapping where ecologists work: Biases in the global distribution of terrestrial ecological observations. *Frontiers in Ecology and the Environment, 10,* 195–201.

McClintock, B. T., Nichols, J. D., Bailey, L. L., MacKenzie, D. I., Kendall, W., Franklin, A. B., et al. (2010). Seeking a second opinion: uncertainty in disease ecology. *Ecology Letters, 13,* 659–674.

Metzger, M. J., Brus, D. J., Bunce, R. G. H., Carey, P. D., Gonçalves, J., Honrado, J. P., et al. (2013). Environmental stratifications as the basis for national, European and global ecological monitoring. *Ecological Indicators, 33,* 26–35.

Munari, D. P., Keller, C., & Venticinque, E. M. (2011). An evaluation of field techniques for monitoring terrestrial mammal populations in Amazonia. *Mammalian Biology - Zeitschrift für Säugetierkunde, 76,* 401–408.

Newbold, T., Hudson, L. N., Hill, S. L. L., Contu, S., Lysenko, I., Senior, R. A., et al. (2015). Global effects of land use on local terrestrial biodiversity. *Nature, 520,* 45–50.

Nichols, J. D., & Williams, B. K. (2006). Monitoring for conservation. *Trends in Ecology & Evolution, 21,* 668–673.

O'Brien, T. G., Baillie, J. E. M., Krueger, L., & Cuke, M. (2010). The wildlife picture index: Monitoring top trophic levels. *Animal Conservation, 13,* 335–343.

Paton, A. J., Brummitt, N. A., Govaerts, R., Harman, K., Hinchcliffe, S., Allkin, B., et al. (2008). Towards target 1 of the global strategy for plant conservation: A working list of all known plant species, progress and prospects. *Taxon, 57,* 602–611.

Pechmann, J. H., Scott, D. E., Gibbons, J. W., & Semlitsch, R. D. (1989). Influence of wetland hydroperiod on diversity and abundance of metamorphosing juvenile amphibians. *Wetlands Ecology and Management, 1,* 3–11.

Pereira, H. M., Belnap, J., Brummitt, N. A., Collen, B., Ding, H., Gonzalez-Espinosa, M., et al. (2010). Global biodiversity monitoring. *Frontiers in Ecology and the Environment, 8,* 459–460.

Pereira, H. M., & Cooper, H. D. (2006). Towards the global monitoring of biodiversity change. *Trends in Ecology & Evolution, 21,* 123–129.

Pereira, H. M., Ferrier, S., Walters, M., Geller, G. N., Jongman, R. H. G., Scholes, R. J., et al. (2013). Essential biodiversity variables. *Science, 339,* 277–278.

Pereira, H. M., Navarro, L. M., & Martins, I. S. (2012). Global biodiversity change: The bad, the good, and the unknown. *Annual Review of Environment and Resources, 37,* 25–50.

Peterson, A. T., Soberón, J., Pearson, R. G., Anderson, R. P., Martínez-Meyer, E., Nakamura, M., et al. (2011). *Ecological niches and geographic distributions.* Princeton, NJ, USA: Princeton University Press.

Pettorelli, N., Laurance, W. F., O'Brien, T. G., Wegmann, M., Nagendra, H., & Turner, W. (2014). Satellite remote sensing for applied ecologists: Opportunities and challenges. *Journal of Applied Ecology, 51,* 839–848.

Proença, V., Martin, L. J., Pereira, H. M, Fernandez M., McRae, L., Belnap, J., Böhm, M., et al. (in press). Global biodiversity monitoring: From data sources to essential biodiversity variables. *Biological Conservation.* doi:10.1016/j.biocon.2016.07.014.

Proença, V. M., Pereira, H. M., & Vicente, L. (2008). Organismal complexity is an indicator of species existence value. *Frontiers in Ecology and the Environment, 6,* 298–299.

Risely, K., Massimino, D., Johnston, A., Newson, S. E., Eaton, M. A., Musgrove, A. J., et al. (2012). *The breeding bird survey 2011.* Thetford, UK: BTO.

Rittenhouse, C. D., Pidgeon, A. M., Albright, T. P., Culbert, P. D., Clayton, M. K., Flather, C. H., et al. (2012). Land-cover change and avian diversity in the conterminous United States. *Conservation Biology, 26,* 821–829.

Rivers, M. C., Taylor, L., Brummitt, N. A., Meagher, T. R., Roberts, D. L., & Lughadha, E. N. (2011). How many herbarium specimens are needed to detect threatened species? *Biological Conservation, 144,* 2541–2547.

Rondinini, C., Marco, M. D., Chiozza, F., Santulli, G., Baisero, D., Visconti, P., et al. (2011). Global habitat suitability models of terrestrial mammals. *Philosophical Transactions of the Royal Society B: Biological Sciences, 366,* 2633–2641.

Rowcliffe, J. M., Field, J., Turvey, S. T., & Carbone, C. (2008). Estimating animal density using camera traps without the need for individual recognition. *Journal of Applied Ecology, 45,* 1228–1236.

Schmeller, D. S. (2008). European species and habitat monitoring: Where are we now? *Biodiversity and Conservation, 17*, 3321–3326.

Sparks, T. H., Butchart, S. H. M., Balmford, A., Bennun, L., Stanwell-Smith, D., Walpole, M., et al. (2011). Linked indicator sets for addressing biodiversity loss. *Oryx, 45*, 411–419.

Thomas, L., Buckland, S. T., Rexstad, E. A., Laake, J. L., Strindberg, S., Hedley, S. L., et al. (2010). Distance software: Design and analysis of distance sampling surveys for estimating population size. *Journal of Applied Ecology, 47*, 5–14.

Thompson, W. (2013). *Sampling rare or elusive species: Concepts, designs, and techniques for estimating population parameters*. New York, USA: Island Press.

Thomsen, P. F., Kielgast, J., Iversen, L. L., Wiuf, C., Rasmussen, M., Gilbert, M. T. P., et al. (2012). Monitoring endangered freshwater biodiversity using environmental DNA. *Molecular Ecology, 21*, 2565–2573.

Tittensor, D. P., Walpole, M., Hill, S. L. L., Boyce, D. G., Britten, G. L., Burgess, N. D., et al. (2014). A mid-term analysis of progress towards international biodiversity targets. *Science, 346*, 241–244.

Toms, M. P., Siriwardena, G. M., Greenwood, J. J. D., & Freeman, S. N. (1999). *Developing a mammal monitoring programme for the UK*. Thetford, UK: BTO.

van Strien, A. J., van Swaay, C. A. M., & Kery, M. (2011). Metapopulation dynamics in the butterfly *Hipparchia semele* changed decades before occupancy declined in the Netherlands. *Ecological Applications, 21*, 2510–2520.

van Strien, A. J., van Swaay, C. A. M. & Termaat, T. (2013). Opportunistic citizen science data of animal species produce reliable estimates of distribution trends if analysed with occupancy models. *Journal of Applied Ecology*, 1450–1458.

van Swaay, C. A. M., Nowicki, P., Settele, P., & van Strien, A. (2008). Butterfly monitoring in Europe: methods, applications and perspectives. *Biodiversity and Conservation, 17*, 3455–3469.

Vellend, M., Baeten, L., Myers-Smith, I. H., Elmendorf, S. C., Beauséjour, R., Brown, C. D., et al. (2013). Global meta-analysis reveals no net change in local-scale plant biodiversity over time. *Proceedings of the National Academy of Sciences of the USA, 110*, 19456–19459.

Walker, S. F., Baldi Salas, M., Jenkins, D., Garner, T. W., Cunningham, A. A., Hyatt, A. D., et al. (2007). Environmental detection of *Batrachochytrium dendrobatidis* in a temperate climate. *Diseases of Aquatic Organisms, 77*, 105.

Weir, L. A., & Mossman, M. J. (2005). North American Amphibian Monitoring Program (NAAMP). In M. Lannoo (Ed.), *Amphibian declines: The conservation status of United States species* (pp. 307–313). Berkeley, CA, USA: University of California Press.

Yoccoz, N. G., Nichols, J. D., & Boulinier, T. (2001). Monitoring of biological diversity in space and time. *Trends in Ecology & Evolution, 16*, 446–453.

Chapter 5
Monitoring Changes in Genetic Diversity

Michael W. Bruford, Neil Davies, Mohammad Ehsan Dulloo, Daniel P. Faith and Michele Walters

Abstract DNA is the most elemental level of biodiversity, drives the process of speciation, and underpins other levels of biodiversity, including functional traits, species and ecosystems. Until recently biodiversity indicators have largely overlooked data from the molecular tools that are available for measuring variation at the DNA level. More direct analysis of trends in genetic diversity are now feasible and are ready to be incorporated into biodiversity monitoring. This chapter explores the current state-of-the-art in genetic monitoring, with an emphasis on new molecular tools and the richness of data they provide to supplement existing approaches. We also briefly consider proxy approaches that may be useful for many-species, global scale monitoring cases.

Keywords Biodiversity · Molecular tools · Genetic diversity · Variation · Monitoring

---

M.W. Bruford (✉)
School of Biosciences and Sustainable Places Institute,
Cardiff University, Cardiff, Wales CF10 3AX, UK
e-mail: brufordmw@cardiff.ac.uk

N. Davies
Richard P Gump South Pacific Research Station, University of California Berkeley,
Moorea, French Polynesia
e-mail: ndavies@moorea.berkeley.edu

M.E. Dulloo
Biodiversity International, Maccarese, 00054 Fiumicino, Italy
e-mail: e.dulloo@cgiar.org

D.P. Faith
The Australian Museum Research Institute, The Australian Museum, Sydney, Australia
e-mail: dan.faith@austmus.gov.au

M. Walters
Natural Resources and Environment, Council for Scientific and Industrial Research,
PO Box 395, Pretoria 0001, South Africa
e-mail: mwalters@csir.co.za

M. Walters
Centre for Wildlife Management, University of Pretoria, Pretoria 0002, South Africa

5.1   Introduction

As the most elemental level of biodiversity, DNA is part of the software on which all life operates. Life has thrived in many different environments over the billions of years, encoding its solutions into DNA—the heredity material. Thanks to this genetic patrimony, many species are equipped with sufficient evolutionary resilience to overcome rapid environmental change (Hughes et al. 2008). Genetic divergence drives the process of speciation. Genetic variation, within and among species, plays an important role in ecosystem structure and function (Whitham et al. 2008). Genetic diversity therefore underpins other levels of biodiversity, including functional traits, species and ecosystems (see Fig. 1.1 in Chap. 1). Life's capacity to adapt relies on genetic variation, and we should thus value it as a major way of mitigating the ecological degradation threatened by growing human impacts on the Earth system. Genetic variation within species is not only the currency of natural selection, it also underpins animal and plant breeding. As raw material for biotechnology, global genomic biodiversity provides a rich source of 'parts' for synthetic biology fuelling the new bio-economy. Molecular solutions discovered over the eons will help humanity address grand societal challenges of the 21st century regarding food, energy, water, and health. For example, crop genetic diversity has a critical role in addressing food and nutrition security, continually increasing yield from crops and livestock (on smaller land space), and instilling resilience to climate change (Dulloo et al. 2014; Hajjar et al. 2008; FAO 2015).

The value of genetic resources includes their capacity to generate ecosystem services, including supporting landscape-level ecosystem resilience (Hajjar et al. 2008; Narloch et al. 2011), maintaining socio-cultural traditions, local identities and traditional knowledge, and allowing plants and animals to undergo natural evolutionary processes, which in turn generate broad genetic variation essential for adaptation to change (Bellon 2009). Genetic variation contributes directly to agriculture by providing a range of valuable traits and genes that are used by modern day breeders for improvement, in particular those species which are closely related to domesticated forms (Hajjar and Hodgkin 2007). Genetic variation also enhances resilience to climate change by providing the traits that are key to the efficiency and adaptability of production systems. It underpins the efforts of local communities and researchers to improve the quality and output of food production (FAO 2015).

This chapter focuses on monitoring of changes in genetic diversity. In that context it is important to ask what is the definition and scope of genetic diversity? The Convention on Biological Diversity (CBD, article 2; www.cbd.int/sp/) defines biodiversity as: '*the variability among living organisms from all sources. This includes diversity within species, between species and of ecosystems*'. For example, the Intergovernmental Platform on Biodiversity and Ecosystem Services (IPBES; Díaz et al. 2015) has defined biodiversity as variation, but also included in the definition '*changes in abundance and distribution over time and space within and among species, biological communities and ecosystems.*'

These and other global efforts highlight the need to clarify the scope and meaning of terms such as 'variability' and 'variation' particularly when we are concerned with monitoring change over time. Many studies adopt the full range of indices from ecology that have been, or might be, referred to as 'diversity' indices, and equate these with 'biodiversity' (Faith 2016). McGill et al. (2015) recognised 15 kinds of trends in biodiversity (including genetic diversity), and it is important to consider whether these define the scope of concerns for monitoring within-species genetic diversity. The authors also considered four spatial scales (local, meta-community, biogeographical, and global) and four 'classes of biodiversity metrics' (alpha diversity, spatial beta diversity, temporal beta diversity, and abundance). In principle all of these categories could be relevant to genetic variation. However, this expanded notion of biodiversity—which includes change over time, spatial variation and abundance—is a relatively recent development for studies of within-species genetic diversity, which has tended to focus on estimating the number of different genetic units of some kind at a range of possible geographic scales. Homogenisation is also an important kind of genetic change. Other estimates (including many referred to in ecology as diversity indices) can be made, but are not by themselves complete descriptions of biodiversity and do a poor job at representing genetic diversity. Generally, we do not know which genetic units are most crucial to species and ecosystems, and so variability in itself is valued.

This focus on variation helps understand the value of genetic diversity referred to above. Genetic diversity provides 'option value'—the value that variation has in potentially providing unanticipated benefits for humans in the future and the evolutionary potential of species (Faith 1992). While the relevance and role of genetic diversity was recognised in the Convention on Biological Diversity (CBD), its importance was largely overlooked during the following two decades (Laikre 2010). However, genetic diversity has been given more visibility since the release of the Aichi Biodiversity Targets in 2010, particularly in Target 13: '*By 2020 the genetic diversity of cultivated plants and farmed and domesticated animals and of wild relatives, including other socio—economically as well as culturally valuable species is maintained and strategies have been developed and implemented for minimizing genetic erosion and safeguarding their genetic diversity.*' This bold and wide-ranging goal poses a major challenge for the scientific community because a globally coordinated approach to monitoring genetic diversity, whether for agricultural species or wildlife, is currently lacking (Hoban et al. 2013; Dulloo et al. 2010). A recent analysis of progress towards the Aichi Biodiversity Targets was unable to adequately assess progress towards Aichi Target 13 due to lack of time series data sources (Tittensor et al. 2014). To discern and compare trends, we need fit-for-purpose genetic monitoring tools that can be easily applied and replicated (Brown 2008; Pinsky and Palumbi 2014). The recently formed Genomic Observatories Network (GOs Network; see Box 5.1) is an example of one concerted, international attempt to respond to these needs. The GOs Network encourages major long-term research sites (e.g., International Long Term

Ecological Research network; ILTER), whether in natural or agricultural ecosystems, to integrate genomics into their longitudinal (time-series) studies and to make these data available according to global data standards.

Until now biodiversity indicators have largely overlooked data from the molecular tools that are available for measuring variation at the DNA level, partly due to their limited availability, high expense, and inaccessibility, focusing instead on proxies, such as trends in the number of domestic livestock breeds and their wild relatives (see Tittensor et al. 2014). While such indicators may be useful in capturing higher order biodiversity trends (it is debatable whether number of breeds is an appropriate measure), they do not account for the genetic distinctiveness of the populations they assess; for instance, some breeds are more distinct than others. More direct analysis of trends in genetic diversity using molecular data are now feasible and are ready to be incorporated into biodiversity monitoring. To mobilise molecular genetic information in monitoring programs, standardised estimates of molecular genetic diversity within and among taxa at specific georeferenced points over time need to be implemented to enable spatial (among site) and temporal (within site) genetic variation to be compared. This chapter explores the current state-of-the-art in genetic monitoring, with an emphasis on new molecular tools and the richness of data they provide to supplement existing approaches. We will also briefly consider complementarity proxy approaches that still may be useful for the many-species, global scale monitoring cases.

5.2 Brief Overview of Developments in the Monitoring of Genetic Diversity

During the last 40 years, studies of genetic diversity have been transformed from simple statistical comparisons of allele frequencies of a handful of soluble enzymes (allozymes) for a few individuals within and among populations. It is now possible, and increasingly affordable, to analyse genome-wide sequence variation (thousands to millions of locations across a genome) across many (hundreds to thousands) individuals of any species, even from non-living remains like faeces or feathers. In parallel with the advances in DNA sequencing and related technologies, many sophisticated bioinformatics tools, software architectures and frameworks have been developed, driven by the need to analyse the huge amounts of data that these studies can generate. Some generally accepted standards are also now emerging from the many kinds of data suitable for monitoring of genetic diversity (Table 5.1). For example, DNA 'barcodes' (Hebert et al. 2003) allow building a library of sequences of the same gene across many different taxa linked to museum specimens and an authoritative taxonomic identification. Sequencing the barcode

Table 5.1 A list of molecular tools that may be used to analyse genetic diversity, including applications, advantages and disadvantages of each

Molecular marker	Application	Advantages	Disadvantages
Microsatellites			
Microsatellites are short repeated sequences, commonly 1–10 bp long, used for DNA profiling. Microsatellites are amplified by PCR using unique flanking DNA sequences. They reflect the ploidy level of the organism studied (e.g. up to two alleles for diploid species) and alleles vary according to the length of repeat sequence. They are a form of Copy Number Variant (CNV) sequence	Microsatellites are used where high levels of polymorphism is desirable—for individual identification, forensics, population genetics studies and genome mapping	• High numbers of markers per genome (typically in the thousands), easily identified using next generation shotgun sequencing • Can use low quantity and quality DNA (for non-invasive samples) • Multiple microsatellites can be amplified and scored during PCR and gel electrophoresis, decreasing costs • High information content of the markers provides statistical reliability	• No amplification of the intended PCR may occur (null alleles), leading to errors in screening • Allelic dropout in poor quality DNA • 'Stutter bands' may occur due to DNA slippage during PCR amplification, complicating the interpretation of band profiles • High mutation rates of microsatellite loci make them unsuitable for higher taxonomic studies
Mitochondrial and chloroplast DNA			
Mitochondrial and chloroplast sequences are organelle-based plastid sequences, uniparentally inherited in a single copy, with well-known gene orders and standard PCR primers for genes and non-genic regions	Used to analyse taxonomic diversity (DNA barcoding), interspecific phylogeny and intraspecific phylogeography	• High copy number per cell and per tissue make these markers robust and easy to amplify • Known evolutionary rates for many genes in many taxonomic groups • Rapid evolution compared to nuclear DNA gives high resolution • Different sequences evolve at different rates, allowing multiple levels of taxonomic diversity to be investigated	• Uniparental mode of inheritance renders population-level inference complex—usually maternally inherited • Nuclear copies can complicate PCR and inference of sequence divergence • Heteroplasmy, where more than one copy is inherited, can occasionally occur, limiting haplotype inference
Single nucleotide polymorphisms			
Single Nucleotide polymorphisms (SNPs) occur throughout the nuclear and organelle genomes and normally refer to single base mutations. Can be detected by direct sequencing, or indirectly by restriction enzymes (e.g. RAD sequencing or Genotyping by Sequencing)	SNPs are rapidly replacing microsatellites as the marker of choice for population genetics but can be applied to genome mapping, genome wide association studies and when used in genotyping arrays, for identifying individuals, populations and even species	• Ubiquitously present, often at millions of locations in the genome • Easily automated using bead chip arrays (and other methods) for efficient genotyping of thousands to hundreds of thousands of markers simultaneously for <100 dollars	• Are rarely trans-specific, so specific assays need to be modified for almost all species • Suffer from ascertainment bias—markers identified in one population as being polymorphic may not be in another population (or may be less so)

(continued)

Table 5.1 (continued)

Molecular marker	Application	Advantages	Disadvantages
		• Can be identified in all classes of DNA and on all chromosomes, including usually less variable sex chromosomes	• Polymorphism is usually limited to two alleles per locus, so use of a limited number of SNPs can suffer lack of resolution at the desired taxonomic level
Direct sequencing			
It is now possible, using high throughput, low cost sequencing platforms to directly sequence whole genomes, transcriptomes or epigenomes of individual organisms at relatively low cost. This approach can be used to produce de novo reference genomes, or to re-sequence the genomes of multiple individuals	This method can be used to provide complete information on the elemental, modified or expressed nucleotide sequences for studies of diversity, evolutionary adaptation, phenotypic plasticity and physiological response of almost any organism in any environment	• The ultimate way to gain information on genetic diversity, huge statistical power • Very easy to carry out and automate • Can be applied to many biological materials • Cost has decreased dramatically, and so many sequence reads are generated that individual genomes can be pooled for each sequencing run, further decreasing costs • Many new statistical tools have been and are being produced to analyse whole 'omic data	• While much cheaper than previously, cost per individual genome can still be prohibitive ($\sim$\$1000) • Some new generation sequencing technologies have high sequencing error rates • Enormous computational requirements to process and analyse whole genome data • Sequencing capacity has rapidly outpaced bioinformatics tools to analyse the data, although this imbalance is being addressed • Viewed by some as an analytical 'black box' where almost any result is possible with little understanding of the underlying data

gene of any biological sample (including eggs, larvae, or parts of an organism such as legs or leaves) leads to rapid identification if the species has already been catalogued in a reference library (e.g., Barcode of Life Database; Ratnasingham and Hebert 2007).

In addition to species identification and studies of phylogenetic relationships, the variation in DNA sequences also enables refined estimates of genetic diversity at the species level and above (Faith 1992). These can be applied to specific taxa (e.g., an endangered species) and/or places (e.g., a national park or farm) and monitored over time, and these are the units we focus on for genetic monitoring. Within-species genetic diversity estimation has been transformed by the use of various genetic profiling methods since the late 1980s involving the use of DNA sequencing and DNA fragment analysis (Sunnucks 2000) and an ever-expanding range of statistical frameworks in which to analyse the data (Beaumont et al. 2002). Notably we can not only analyse levels of genetic variation but use this information to infer population parameters and demographic trajectories, often from a single point sample. Furthermore, the advent of metagenomics through environmental shotgun sequencing (Tyson et al. 2004; Venter et al. 2004) opened up the microbial world, heralding a new age of biotic exploration documenting what constitutes the overwhelming majority of life in both biomass and variation terms. Some of these advances have not reached the conservation monitoring literature and seemingly went almost unnoticed by the CBD and its associated bodies until recently. This oversight can perhaps be attributed to the astonishing pace of DNA sequencing capacity, increasing at a much faster rate than Moore's Law since the mid 2000s. Perhaps the explosion in technologies and analytical methods made it difficult to settle on standardised genomics-based approaches for biodiversity monitoring. Of course, attention has also been largely focused on the more established (and visible) levels of biodiversity (e.g., CBD-related efforts on the global taxonomy initiative and the so-called 'ecosystem approach').

Ignoring the power and promise of genomics seems increasingly anachronistic. The public is increasingly aware of the benefits the 'new age of genomics' offers for personal and public health, and food and energy production (Field and Davies 2015). Genetics is likely to become increasingly important in biodiversity monitoring with rapid molecular assessment of species and ecosystems now feasible using high throughput DNA sequencing in a fraction of the time and cost of previous approaches (Whitham et al. 2008). Simultaneously, with the establishment of the IPBES, the recognition that understanding and maintaining genetic diversity within and among species may be key to ecosystem (and therefore ecosystem service) resilience in the face of climate change and other anthropogenic stressors, has raised the profile of genetic diversity substantially (Sgro et al. 2011; Mace et al. 2012; Pereira et al. 2013) leading to its incorporation into the Aichi Targets.

5.3 Spatio-Temporal Considerations in Genetic Monitoring

While genetic monitoring is a tool that has global relevance for the maintenance of biodiversity, like other monitoring techniques it can be costly and time-consuming. In particular, DNA cannot be read at a distance in contrast to ecosystems (e.g., remote sensing) or species (e.g., visual observation); rather, all genetic analyses require access to biodiversity and its physical sampling (Davies et al. 2012b). Genetic approaches are thus unlikely to be applicable in all cases where monitoring is required and may not be the most cost-effective option in some.

> **Box 5.1. The Genomic Observatories Network**
>
> Genomic Observatories (GOs) are sites where genomic information is collected alongside social-ecological, environmental and/or other biological data, ensuring co-location of observations and much-needed context for such genomic information (Davies et al. 2012a). GOs show commitment to the long-term collection of data, now and into the future, as well as to the depositing of such data in suitable repositories (Field 2011). GOs should be based on a subset of sites of 'utmost scientific importance' (Davies et al. 2012b) and be supported by field stations, universities, museums or similar organisations or institutions (Davies et al. 2012a), allowing for long-term observations and thus change detection.
>
> The first published calls for the establishment of a GOs Network (Davies et al. 2012a, b) highlighted the fact that DNA sequences should be part of the data collected to monitor life on earth and that, whilst the costs of collecting and processing such samples remains high, the establishment of GOs could consolidate these monitoring efforts.
>
> By hosting workshops and meetings on the side-lines of various conferences (see www.genomicobservatories.org/ for more information), the efforts in building a community around the GOs Network concept, culminated in the publication of the founding charter of the GOs Network and agreement on the network's mission as working towards 'Biocoding the Earth; integrating DNA data into Earth observing systems and eventually building a global Genomic Observatory within the Global Earth Observation System of Systems (GEOSS)' (Davies et al. 2014; GEO Secretariat n.d.).
>
> The GOs Network, which is a collaboration between the Group on Earth Observations Biodiversity Observation Network (GEO BON) and the Genomic Standards Consortium (GSC), held its first coordinated action in the form of Ocean Sampling Day (OSD) on 21 June 2014 and repeated it on the same day in 2015 (Field and Davies 2015; Kopf et al. 2015). The effort was joined by a number of GOs Network (marine) sites with the purpose of coordinated, standardised collection and sequencing of seawater throughout the world's oceans (Field and Davies 2015; see Fig. 5.1).

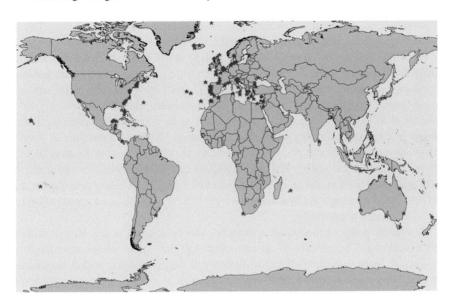

Fig. 5.1 Map showing registered sites for Ocean Sampling Day, 21 June 2014. *Source* Kopf et al. (2015)

This emerging network has not only 'site members' but recently New Zealand launched its Genomic Observatory project (see http://data. genomicobservatory.cs.auckland.ac.nz/) aimed at the characterisation (environmental and phylogenetically) of terrestrial species in a selected ecosystem, as a first national contribution to the GOs Network (Drummond et al. 2015).

Furthermore, other scientific communities have started work on supporting the efforts of the GOs Network, for example through the creation of the Biological Collections Ontology (BCO), which is to provide the informatics stack for the network (Walls et al. 2014).

A list of the scientific community members involved in the development of the network can be found at http://wiki.gensc.org/index.php?title=GOs_ Network_Membership. Parties interested in joining the GOs Network may contact the coordinators of the initiative through their website at http://www. genomicobservatories.org/ and those interested in participating in upcoming OSD events can visit https://www.microb3.eu/osd.

One possibility is that specific sites can act as genetic biodiversity observatories where special efforts are made to document and monitor genetic biodiversity. The GOs Network is promoting just such an approach at the best-studied sites around the world (Davies et al. 2012a, 2014). Apart from the scientific and technical aspects of genetic monitoring, the need to access genetic resources in situ and carry out downstream analyses in laboratories and museums around the globe raises important legal and social concerns that must be addressed at each site. These include compliance with international legal instruments, such as the CBD's Nagoya protocol on Access and Benefit Sharing (ABS) and CITES, as well as national and/or local regulations and sensitivities, which were often not written with genetic monitoring in mind, but for other reasons like protecting species or preserving sovereignty. Sampling and associated costs appear relatively incompressible (Davies et al. 2012b) and careful thought needs to be given to the spatio-temporal design of genetic monitoring or genetic assessments (defined as multiple or single sampling events, respectively, by Schwartz et al. (2007). A combination of the Global Genome Biodiversity Network (GGBN) and GOs Network, in conjunction with initiatives such as the ILTER (International Long Term Ecosystem Research) network, offers perhaps the best hope for establishing a coordinated global effort to monitor genetic biodiversity.

A wide diversity of molecular approaches can be adopted at any site, or within any ecosystem. The concept of an ecosystem can now be extended to individual organisms, including establishing the diversity of the microbiome within organisms or to the environment using metagenomics (Tringe et al. 2005). Community level diversity (e.g., species richness) might correlate with within-species genetic diversity of ecosystem-defining taxa (Zytynska et al. 2012), however, recent studies attempting to establish whether one diversity estimate might act as a reasonable proxy for the other, have been inconclusive (e.g., Evanno et al. 2009; Struebig et al. 2011; Taberlet et al. 2012). Within-species genetic diversity studies have proliferated during the last 20 years to the point that they are now routinely carried out globally and represent a huge, largely untapped resource for ecosystem evaluation. They have recently begun to be augmented by studies at the genome, epigenome and transcriptome level (Shafer et al. 2015). The domestic animal and plant community has led the way in within-species molecular biodiversity assessment (e.g., Boettcher et al. 2010; Cheng et al. 2011), including making recommendations on common tools for measuring variation in the same species at a global scale. Considering this vast amount of genetic data being generated anyway, another option is to evaluate genetic variation of species expected to have undergone erosion (e.g., due to harvest) and compare this to 'reference' or 'control' species, those having experienced no impact. This analysis of existing data could give an overall picture of genetic erosion (Hoban, *pers comm*). Indeed, Pinsky and Palumbi (2014) used this approach for more than 100 species of fish and found identifiable genetic erosion in harvested fish.

Although tools for genetic monitoring are now almost universally available, the statistical approaches needed to compare data, evaluate trends and provide indicators of genetic health are less well developed. This is partly because temporally spaced sampling of the same species at the same site for population genetic evaluation is rare, with the possible exception of commercial species (Hutchinson et al. 2003) where genetic material (e.g., fish fin clips) has been collected since population monitoring began. In the absence of sampling a population over time, analysis of genetic data from a single 'point sample' can still provide insights into recent demographic change (Goossens et al. 2006), although different estimators can have wide confidence intervals and provide inconsistent values depending on the methods chosen or model assumptions (Barker 2011). Recently, Hoban et al. (2014) carried out an assessment of temporal indicators of genetic erosion (sensu Aichi Target 13) to assess which metric and sampling would be the most sensitive to detecting short-term declines in genetic diversity. The number of alleles per genetic locus outperformed all other potential indicators (such as heterozygosity) across all scenarios tested. Sampling 50 individuals at as few as two time points with 20 microsatellite (DNA profiling) markers could reliably detect genetic erosion even in cases where 80–90 % of diversity remained. Power increased substantially with more samples or markers, with, for example, 2500 single nucleotide polymorphisms (SNPs) being extremely effective at detecting relatively subtle demographic declines. The latter observation is of particular relevance because since about 2010, biodiversity assessment has routinely used tens of thousands of SNPs (genome scale, or 'next generation' sequencing) in domestic animals and plants (e.g., Kijas et al. 2012) and the increased application of genome resequencing (sequencing whole genomes of multiple individuals of the same species) in non-model organisms (e.g., Lamichhaney et al. 2015). These methods can also be used in soil or water (marine, freshwater) samples to analyse 'environmental DNA' (eDNA), which includes ancient and modern genetic material from animals and plants as well as microbes (Pedersen et al. 2015; Kelly et al. 2014). Thus global capacity to perform molecular genetic monitoring with very high detail and precision is available and is being deployed in environmental assessment worldwide. Hoban et al. (2014) also concluded that there is 'high potential for using historic collections in monitoring programs'. They found that statistical power to detect change was significantly improved if samples were available before the onset of decline—so that archived and museum collections clearly could play an important role as part of the monitoring program. Hoban et al. (2014) made another interesting conclusion relevant to the design of monitoring programs 'many schemes were sufficient, and strict adherence to a particular sampling protocol seems unnecessary'.

5.4 What to Monitor?

While Schwartz et al. (2007) broadly defined genetic monitoring as the tracking of neutral genetic markers through time to estimate demographic and/or population genetic parameters, this Chapter is focused on changes in within-species genetic diversity. This focus is much closer to the studies addressing conservation strategies, at the level of within species genetic diversity. Eizaguirre and Baltazar-Soares (2014) summarised the key challenge: '*which genetic metrics, temporal sampling protocols, and genetic markers are sufficiently sensitive and robust to be informative on conservation-relevant timescales?*' Dornelas et al. (2013) argued that, for genetic diversity, '*sources of error are associated with the processes of selection of the genes of interest, amplifying and sequencing genes, and (especially for microbes) determining the boundaries of operational taxonomic units.*' These make sense, however microbial work has largely side-stepped the operational taxonomic units problem by using indices defined at the phylogenetic level (for discussion, see Faith et al. 2009). Unfortunately, these estimators are also the most sensitive to sampling bias (Leberg 2002). Consistent sampling is required to eliminate error or correction for the lowest sample size will often be necessary, resulting in a loss of information and sensitivity (but see Dornelas et al. 2013). Recent advances have provided robust methods that correct for sampling biases in estimates of phylogenetic diversity (e.g., Chao et al. 2015). van Zonneveld et al. (2014) concluded that the number of 'locally common' alleles (defined as alleles restricted to a limited area of a species' distribution, but having high frequencies in these areas) may also be good indicators of overall genetic diversity. The question 'What to monitor?' also involves choosing which species or groups of species (including environmental sampling) are of highest priority given the substantial costs of physically sampling biodiversity over time. Target 13 of the CBD cites '*domesticated plants and animals, their wild relatives, socio-economically important species and culturally valuable species*'. We will consider these three categories separately although there is much overlap between them.

5.4.1 Domesticated Species

Domesticated species are relatively easy to define, they are largely found in agri-ecosystems. In such ecosystems, the key commercial crops and livestock must form an integral part of genetic monitoring protocols. We will focus here primarily on crop genetic diversity. To date there is poor knowledge of the distribution of genetic diversity that exists at the global level for all crops. Crop genetic diversity is distributed in space and time and occurs in farmer's fields, wild habitats, market places and as conserved in ex situ collections such as genebanks and botanic gardens. It is widely believed that crop genetic diversity is being lost in farmers' fields at an accelerated rate mainly due to the replacement of the heterogeneous (highly diverse) traditional varieties by uniform high-yielding improved varieties, as well as contributing factors that also affect natural ecosystems, such as climate change, habitat change, invasive alien species, overexploitation and pollution (FAO 2010; MA 2005). An additional problem, given the current development of genebank methodologies and management, is that of potential genetic erosion within genebanks, which should be monitored as well (Schoen and Brown 2001).

There is conflicting evidence for the erosion of crop genetic diversity (Dulloo et al. 2010; Bonneuil et al. 2012). There are many examples that have shown the loss of genetic diversity in cultivated plants. One of the classic examples is the study by the Rural Advancement Fund International (RAFI) that compared the number of varieties of different commercial crops known to the U. S. Department of Agriculture in 1903 to the number of varieties of these crops for which seeds existed in the National Seed Storage Laboratory (NSSL) in 1983 (Fowler and

Mooney 1990). Considering about 75 different vegetables together, the RAFI study found that approximately 97 % of the varieties on the 1903 lists are now extinct. In China, of almost 10,000 varieties of wheat in use in 1949, only about 1000 remained by the 1970s (Arunachalam 1999). Further evidence for genetic erosion is provided by Hammer et al. (1996), who analysed differences between collecting missions in Albania (1941 and 1993) and in southern Italy (in 1950 and the late 1980s) and claimed high losses in genetic variability—genetic erosion of 72.4 and 72.8 %, respectively.

A major challenge is that there is no consensus on what type of data (genetic or proxy) can effectively be used to monitor genetic diversity at the global level (Dulloo et al. 2010; Pereira et al. 2013; Graudal et al. 2014). Although the molecular tools for measuring genetic diversity are well advanced and the cost becoming more affordable, as mentioned earlier, a major challenge remains to develop simple, inexpensive, and standardised means to monitor genetic diversity at a global scale (Frankham 2010). Up to now, monitoring of crop genetic diversity has focused mainly on existing data and the use of proxy indicators for genetic diversity (Brown 2008; Jarvis et al. 2008; Last et al. 2014). It has also been driven by the FAO's country-led processes in developing State of the World Reports on plant genetic resources for food and agriculture (FAO 1997, 2010) and in monitoring the implementation of Global Plan of Action for the Conservation and Sustainable Utilization of Plant Genetic Resources for Food and Agriculture (PGRFA). Based on the occurrences and capacity to conserve and use PGRFA, a set of 63 indicators as well as three targets (Conservation of PGRFA, Sustainable use, and institutional and human capacities) for monitoring the status and trends of conservation and use of PGRFA has been developed and adopted (CGRFA 2013). Unfortunately, none of the indicators directly use genetic diversity metrics for assessing status of crop genetic diversity, but proxy indicators for in situ conservation, ex situ conservation, sustainable use and existing capacities are used instead. Further the FAO is developing a composite index based on the 63 indicators for each of the three targets. Brown (2008) discussed many diversity variables and argued that practical ones are based on number of individuals, area occupied in situ, number of accessions and number in gene banks ex situ. He provided a set of 22 genetic indicators for cultivated and wild plants.

At the European level, a pan-European collaborative initiative, Streamlining European Biodiversity Indicators (SEBI) was started in 2005 to provide a workable set of biodiversity indicators for Europe to measure progress towards the target of halting the loss of biodiversity in Europe by 2010 (Biała et al. 2012). The SEBI aim was to build on current monitoring and available data to avoid duplication of efforts and to complement other activities to describe, model and understand biodiversity and the pressures upon it. Within this context, Last et al. (2014) proposed five indicators for the estimation of genetic diversity, by which they meant crop accession or breed diversity at the farm level. These were 'Crop-Species Richness', 'Crop-Cultivar Diversity', 'Type of Crop Accessions', 'Livestock-Species Richness' and 'Breed Diversity'. Additionally, they evaluated the potential role of such indicators for developing strategies to conserve or increase crop cultivars and livestock breeds in agro-ecosystems.

One of the best examples of monitoring crop diversity at the genetic level was by Bonneuil et al. (2012). They carried out a literature review to assess bread wheat diversity, as well as a range of general studies on the assessment of crop genetic diversity (see list in Goffaux et al. 2011), according to the level of genetic diversity they target (varietal or allelic) and in which pool it is measured (ex situ collections, registered varieties, or in situ (on-farm). They showed that most studies assess diversity within registered varieties or genebanks, while only a few considered the diversity actually cultivated. They also argued that the different indices (the number of varieties; the Shannon index, the Simpson index and the Piélou's hierarchical diversity index) of diversity for in situ (on farm) diversity, account for the richness and evenness of varieties spatial distribution, while Nei's index accounts for allelic diversity across a pool of varieties. However, there was no study that has combined intra varietal allelic diversity. Bonneuil et al. (2012) therefore developed a composite indicator that integrates richness, evenness and inter-variety allelic diversity as well as within-variety allelic diversity and proposed the H_T^* index as an integrated indicator for crop diversity on farms. They tested the index on a comprehensive historical dataset of bread wheat varieties dating back to 1878 from a French territory, Eure-et-Loire department. The study revealed that more varieties (the varietal richness factor) can mean less diversity when their genetic structure is more similar (the effect of between-variety genetic diversity), or when more diverse landraces are replaced by many homogeneous lines (the effect of within-variety genetic diversity), or else when one or a few varieties become hegemonic in the landscape (the spatial evenness effect). Furthermore, increased evenness in variety distribution (varietal evenness) can also mean less diversity when varieties are genetically related.

The domestic animal community fully embraced molecular characterisation within and among livestock breeds and species since the early 1990s, thanks to the proactive role of the International Association for Animal Genetics and the FAO, who established and revised guidelines for the use of genetic markers (initially microsatellites; see Hoffmann et al. 2004; Groeneveld et al. 2010). As a result, many studies have used the same marker systems for the more common domestic livestock species. More recently, microsatellites have been largely superceded by medium to high-density Single Nucleotide Polymorphism (SNP) arrays, which provide comprehensive, genome-wide surveys of genetic diversity at an affordable cost (Matukumalli et al. 2009; for a review see Bruford et al. 2015).

5.4.2   Socioeconomically (and Ecologically) Important Species

Socio-economically important species may include those that humans exploit for food, shelter, medicines, fuel and ecotourism income but may also include those that are ecologically important providing other key ecosystem services such as pollination, nutrient cycling and pest regulation (Bailey 2011). The genetics of 'foundation species', those that structure their environment, such as trees and corals, could be particularly important as heritable changes in these species could affect entire ecosystems. The field of community genetics (Whitham et al. 2008) has tended to focus on genetic surveys of key indicator species of ecosystems, often including vegetation (Bailey 2011; Zytynska et al. 2012). Indicators or foundation species may be a first priority to monitor. An alternative approach may be to choose representatives from key functional groups within ecosystems, standard practice in microbial community genetics (Nannipieri et al. 2003; He et al. 2007), and routinely applied in animal and plant ecology.

Wild species that are commercially relevant such as marine and freshwater fish, timber trees, crop pests and large predators have seen an explosion in genetic studies during the last three decades, focusing on DNA barcoding for species identification and wildlife forensics (e.g., Minhós et al. 2013), delineation of commercial stocks using population genetics approaches (e.g., Nielsen et al. 2012), genetic assignment of individuals or their products to source population (e.g., Lowe and Cross 2011) and assessing the impacts of anthropogenic barriers such as roads or dams to the movement of individuals (e.g., Keller and Largiader 2003). Such approaches are directly influencing policy and management, enabling law enforcement helping authorities to prioritise their direct interventions and redefine populations for exploitation management.

5.4.3   Monitoring Genetic Diversity in Culturally Valued Species

The definition of culturally valuable species is even more open to interpretation, but could reasonably include locally important domesticated breeds and landraces, wild species of emblematic significance, wild species of medicinal value (e.g., Shivaprakash et al. 2014) and those in immediate danger of extinction that attain cultural significance. The field of conservation genetics has traditionally focused on emblematic and endangered species and local breeds that might be regarded as culturally valuable, however, a wider definition of cultural value may be needed to ensure all elements of this category can be encapsulated in a comprehensive manner (Hoban et al. 2013).

5.5   Proxies for Reporting Changes in Genetic Diversity

Earlier we expressed some caution about the use of simple proxies for within species genetic diversity. The increased capacity to capture genomics information for many species at many places will gradually reduce reliance on proxy approaches. However, interest in human impacts on biodiversity at the level of within-species genetic variation includes not only poorly studied species but also those still unknown to science. Thus, well-designed proxies may still fill a gap in providing broad-brush 'report cards' on change in within-species genetic diversity, for many species at broad scales (e.g., globally; see the section on PGRFA above). Hoban et al. (2014) cautioned against some simplistic uses of proxies for genetic diversity change, and this seems particularly relevant when considering a single target species. In contrast, proxies may serve well as a complement to these direct approaches, when a broad brush report card on all species is needed. What sorts of proxies may be useful? Good candidates will build on information that is already widely available through existing monitoring efforts at the species level. For example, information on the range extent, distribution and abundance for many species is available, and this information is often complemented by associated information on key environmental variables.

There are two fundamental geographic scales for such proxies—proxies may be developed within one area, or for a collection of many areas (thus, regional or global). As an example of a localised proxy, Taberlet et al. (2012) assessed how well the estimated species richness of geographic areas corresponded to the area's average within-species genetic diversity. These proxies were judged as not useful, countering conventional assumptions that patterns in species richness among areas may be informative about genetic diversity. More effective proxies may operate among-areas, and take advantage of changes in species' range extent and/or occupancy of 'environmental space'. Such broad-brush approaches can take advantage of, and add value to, the well-developed regional-to-global monitoring systems at the species level (e.g., Map of Life; https://www.mol.org/). One broad-brush approach can assess the loss of genetic diversity, using models that link loss of geographic range for a given species to its loss of genetic diversity. Previous work has largely focussed on comparisons among species to make predictions about their relative levels of genetic diversity. For example, Frankham (1996) showed that genetic variation will be greater in those species with wider ranges. Vellend and Geber (2005) treated species diversity–genetic diversity relationships by looking at correlation coefficients between species diversity and the genetic diversity of a focal species among localities. The key information gap is about how loss of the geographic or environmental range within a given species relates to loss of its genetic diversity. Recently, Mimura et al. (in revision) provided some evidence that patterns of genetic diversity, covering the range of various well-sampled species, provide support for a 'power curve' relationship (analogous to the well-known species-area relationship) linking range loss to loss of genetic diversity. They also argued that the exact relationship for a given species may vary in a predictable way according to factors such as the general dispersal ability of the species. This may allow a small number of proxy-models based on power curves to infer genetic diversity losses for a wide range of species.

A closely-related approach can use changes in a species' coverage of its environmental range or 'environmental space' to infer its consequent loss in genetic diversity. When a population of a given species exists in predictable environmental space, changes in area of that space may suggest changes in genetic diversity thus loss of environmental range may correlate with loss of genetic diversity. There is some empirical support for these proxies. Zhang et al. (2013) showed that both geographic and environmental distances are significant correlates of genetic differentiation among locations. Congruence between geographic and environmental distances and genetic distances supports the potential for genetic diversity proxy models that assess loss of coverage of environmental space (Faith 2015).

The challenge to produce global report cards on the loss of within-species genetic diversity is timely. For example, the Planetary Boundaries framework (Rockström et al. 2009) has proposed that loss of global genetic diversity is one of the key variables for understanding whether society is within a 'safe operating space' for sustainability. Such assessments could be augmented by monitoring the loss of geographic and/or environmental range, for a representative sample of species. Mimura et al. (in revision) argue that the indicator value for a representative subset of species can provide a general indicator of within-species genetic diversity loss for all species and outline a procedure to derive this subset, based on the available distribution information. They define three steps: (1) For any two species, calculate their 'dissimilarity' based on the difference in their locations in geographic (or environmental) space; (2) Use the dissimilarities to derive an arbitrary but pre-chosen k number of clusters of species. For example, k-means clustering algorithms can directly use dissimilarities or genetic data to derive k clusters. Choose a member of each cluster to form the subset of k representative species; (3) For the k species, apply the proxy model to infer loss of genetic diversity based on loss of geographic (or environmental) range extent.

Proxy indicators are a potentially efficient approach to bridging the evidence gap on genetic diversity within species. However, the relationships mentioned above in terms of genetic diversity and ecological space are dependent on various assumptions including demographic history, natural and/or artificial selection signatures and the ability to disperse. Therefore, we would advocate these approaches as primarily extension mechanisms and would not suggest they can replace genetic data nor do they provide the rich information available from the genomes of the planet's species.

References

Arunachalam, V. (1999). *Genetic erosion in plant genetic resources and early warning system: a diagnosis dilating genetic conservation.* M. S. Swaminathan Research Foundation, Chennai, India. http://59.160.153.185/library/node/403

Bailey, J. K. (2011). From genes to ecosystems: A genetic basis to ecosystem services. *Population Ecology, 53*, 47–52.

Barker, J. S. F. (2011). Effective population size of natural populations of Drosophila buzzatti, with a comparative evaluation of nine methods of estimation. *Molecular Ecology, 20,* 4452–4471.

Beaumont, M. A., Zhang, W., & Balding, D. J. (2002). Approximate Bayesian computation in population genetics. *Genetics, 162*, 2025–2035.

Bellon, M. R. (2009). Do we need crop landraces for the future? Realizing the global option value of in situ conservation. In A. Kontoleon, U. Pascual, & M. Smale (Eds.), *Agrobiodiversity and economic development* (pp. 51–59). London and New York: Routledge.

Biała, K., Condé, S., Delbaere, B., Jones-Walters, L., & Torre-Marín, A., (2012). *Streamlining European biodiversity indicators 2020: Building a future on lessons learnt from the SEBI 2010 process.* EEA Technical Report No. 11/2012. EEA, Copenhagen, Denmark.

Boettcher, P. J., Tixier-Boichard, M., Toro, M. A., Simianer, H., Eding, H., Gandini, G., et al. (2010). Objectives, criteria and methods for using molecular genetic data in priority-setting for conservation of animal genetic resources. *Animal Genetics, 41*(s1), 64–77.

Bonneuil, C., Goffaux, R., Bonnin, I., Montalent, P., Hamon, C., Balfourier, F., et al. (2012). A new integrative indicator to assess crop genetic diversity. *Ecological Indicators, 23,* 280–289.

Brown, A. H. D. (2008). *Indicators of Genetic Diversity, Genetic Erosion and Genetic Vulnerability for Plant Genetic Resources for Food and Agriculture.* Food and Agriculture Organisation of the United Nations, Rome, Italy.

Bruford, M. W., Ginja, C., Hoffmann, I., Joost, S., Orozco-terWengel, P., Alberto, F. J., et al. (2015). Prospects and challenges for the conservation of farm animal genomic resources 2015-2025. *Frontiers in Genetics, 6*, 314.

CGRFA. (2013). *Report of the fourteenth regular session of the Commission on genetic resources for food and agriculture, Rome Italy 15-19 April 2013. CGRFA- 14/13/Report.* Food and Agriculture Organisation of the United Nations, Rome, Italy.

Chao, A., Chiu, C. H., Hsieh, T. C., Davis, T., Nipperess, D. A., & Faith, D. P. (2015). Rarefaction and extrapolation of phylogenetic diversity. *Methods in Ecology and Evolution, 6*, 380–388.

Cheng, F., Liu, S., Wu, J., Fang, L., Sun, S., Liu, B., et al. (2011). BRAD: The genetics and genomics database for brassica plants. *BMC Plant Biology, 11*, 1.

Davies, N. & Field, D., and the Genomic Observatories Network (2012a). Sequencing data: A genomic network to monitor Earth. *Nature 481*, 145.

Davies, N., Meyer, C., Gilbert, J. A., Amaral-Zettler, L., Deck, J., Bicak, M., Rocca-Serra, P., et al. (2012b). A call for an international network of genomic observatories (GOs). *Gigascience 1*, 5.

Davies, N., Field, D., Amaral-Zettler, L., Clark, M., Deck, J., Drummond, A., et al. (2014). The founding charter of the Genomic Observatories Network. *GigaScience, 3*, 2.

Díaz, S., Demissew, S., Carabias, J., Joly, C., Lonsdale, M., Ash, N., et al. (2015). The IPBES Conceptual Framework—Connecting nature and people. *Current Opinion in Environmental Sustainability, 14*, 1–16.

Dornelas, M., Magurran, A. E., Buckland, S. T., Chao, A., Chazdon, R. L., Colwell, R. K., et al. (2013). Quantifying temporal change in biodiversity: Challenges and opportunities. *Proceedings of the Royal Society of London B: Biological Sciences, 280*, 20121931.

Drummond, A. J., Newcomb, R. D., Buckley, T. R., Xie, D., Dopheide, A., Potter, B. C., et al. (2015). Evaluating a multigene environmental DNA approach for biodiversity assessment. *GigaScience, 4*, 46.

Dulloo, M. E., Hunter, D., & Borelli, T. (2010). Ex situ and in situ conservation of agricultural biodiversity: Major advances and research needs. *Notulae Botanicae Horti Agrobotanici Cluj-Napoca, 38*, 123–135.

Dulloo, M. E., Hunter, D., & Leaman, D. (2014). Plant diversity in addressing food, nutrition and medicinal needs. In A. Gurib-Fakim (Ed.), *Novel plant bioresources: applications in food, medicine and cosmetics* (pp. 1–21). Chichester, UK: Wiley.

Eizaguirre, C., & Baltazar-Soares, M. (2014). Evolutionary conservation—Evaluating the adaptive potential of species. *Evolutionary Applications, 7*, 963–967.

Evanno, G., Castella, E., Antoine, C., Paillat, G., & Goudet, J. (2009). Parallel changes in genetic diversity and species diversity following a natural disturbance. *Molecular Ecology, 18*, 1137–1144.

Faith, D. P. (1992). Conservation evaluation and phylogenetic diversity. *Biological Conservation, 61*, 1–10.

Faith, D. P. (2015). Phylogenetic diversity, function trait diversity and extinction: avoiding tipping points and worst-case losses. *Philosophical Transactions of the Royal Society of London B: Biological Sciences, 370*, 1–10.

Faith, D. P. (2016). A general model for biodiversity and its value. In J. Garson, A. Plutynski, & S. Sarkar (Eds.), *The routledge handbook on the philosophy of biodiversity.* Routledge.

Faith, D. P., Lozupone, C. A., Nipperess, D., & Knight, R. (2009). The cladistic basis for the phylogenetic diversity (PD) measure links evolutionary features to environmental gradients and supports broad applications of microbial ecology's "phylogenetic beta diversity" framework. *International Journal of Molecular Sciences, 10*, 4723–4741.

FAO. (1997). *Report on the state of the world's plant genetic resources.* Food and Agriculture Organisation of the United Nations, Rome, Italy.

FAO. (2015). *Coping with climate change—The roles of genetic resources for food and agriculture.* Food and Agriculture Organisation of the United Nations, Rome, Italy.

FAO. (2010). *The second report on the state of the world's plant genetic resources for food and agriculture.* Food and Agriculture Organisation of the United Nations, Rome, Italy.

Field, D. (2011). *What is a Genomic Observatory?* Retrieved November 7, 2015, from http://genomicobservatories.blogspot.co.za/2011/08/what-is-genomic-observatory.html

Field, D., & Davies, N. (2015). *Biocode: The new age of genomics.* Oxford, UK: Oxford University Press.

Fowler, C., & Mooney, P. (1990). *Shattering food, politics, and the loss of genetic diversity.* Tucson, USA: The University of Arizona Press.

Frankham, R. (1996). Relationship of genetic variation to population size in wildlife. *Conservation Biology, 10*, 1500–1508.

Frankham, R. (2010). Challenges and opportunities of genetic approaches to biological conservation. *Biological Conservation, 143*, 1919–1927.

GEO Secretariat. (n.d.). *GEOSS: access—Connecting users.* Retrieved November 7, 2015, from http://www.earthobservations.org/geoss.php

Goffaux, R., Goldringer, I., Bonneuil, C., Montalent, P., & Bonnin, I. (2011). *Quels indicateurs pour suivre la diversité génétique des plantes cultivées. Le cas du blé tendre cultivé en France depuis un siècle.* Rapport FRB, Série Expertise et synthèse, 44.

Goossens, B., Chikhi, L., Ancrenaz, M., Lackman-Ancrenaz, I., Andau, P., & Bruford, M. W. (2006). Genetic signature of anthropogenic population collapse in orang-utans. *PLoS Biology, 4*, 285–291.

Graudal, L., Aravanopoulos, F., Bennadji, Z., Changtragoon, S., Fady, B., Kjær, E. D., et al. (2014). Global to local genetic diversity indicators of evolutionary potential in tree species within and outside forests. *Forest Ecology and Management, 333*, 35–51.

Groeneveld, L. F., Lenstra, J. A., Eding, H., Toro, M. A., Scherf, B., Pilling, D., et al. (2010). Genetic diversity in farm animals—A review. *Animal Genetics, 41*(s1), 6–31.

Hajjar, R., & Hodgkin, T. (2007). The use of wild relatives in crop improvement: A survey of developments over the last 20 years. *Euphytica, 156*, 1–13.

Hajjar, R., Jarvis, D. I., & Gemmill-Herren, B. (2008). The utility of crop genetic diversity in maintaining ecosystem services. *Agriculture, Ecosystems & Environment, 123*, 261–270.

Hammer, K., Knüpffer, H., Xhuveli, L., & Perrino, P. (1996). Estimating genetic erosion in landraces—Two case studies. *Genetic Resources and Crop Evolution, 43*, 329–336.

He, Z., Gentry, T. J., Schadt, C. W., Wu, L., Liebich, J., Chong, S. C., et al. (2007). GeoChip: A comprehensive microarray for investigating biogeochemical, ecological and environmental processes. *The ISME journal, 1*, 67–77.

Hebert, P. D., Cywinska, A., & Ball, S. L. (2003). Biological identifications through DNA barcodes. *Proceedings of the Royal Society of London B: Biological Sciences, 270*, 313–321.

Hoban, S., Arntzen, J. A., Bruford, M. W., Godoy, J. A., Rus Hoelzel, A., Segelbacher, G., et al. (2014). Comparative evaluation of potential indicators and temporal sampling protocols for monitoring genetic erosion. *Evolutionary Applications, 7*, 984–998.

Hoban, S. M., Hauffe, H. C., Pérez-Espona, S., Arntzen, J. W., Bertorelle, G., Bryja, J., et al. (2013). Bringing genetic diversity to the forefront of conservation policy and management. *Conservation Genetics Resources, 5*, 593–598.

Hoffmann, I., Marsan, P. A., Barker, J. S. F., Cothran, E. G., Hanotte, O., Lenstra, J. A., Milan, D., et al. (2004). New MoDAD marker sets to be used in diversity studies for the major farm animal species: recommendations of a joint ISAG/FAO working group. In *Proceedings of the 29th International Conference on Animal Genetics* (Vol. 123), Meiji University, Tokyo, Japan. Food and Agriculture Organisation of the United Nations, Rome, Italy.

Hughes, A. R., Inouye, B. D., Johnson, M. T., Underwood, N., & Vellend, M. (2008). Ecological consequences of genetic diversity. *Ecology Letters, 11*, 609–623.

Hutchinson, W. F., van Oosterhout, C., Rogers, S. I., & Carvalho, G. R. (2003). Temporal analysis of archived samples indicates marked genetic changes in declining North Sea cod (*Gadus morhua*). *Proceedings of the Royal Society of London B: Biological Sciences, 270*, 2125–2132.

Jarvis, D. I., Brown, A. H., Cuong, P. H., Collado-Panduro, L., Latournerie-Moreno, L., Gyawali, S., et al. (2008). A global perspective of the richness and evenness of traditional crop-variety diversity maintained by farming communities. *Proceedings of the National Academy of Sciences, 105*, 5326–5331.

Keller, I., & Largiader, C. R. (2003). Recent habitat fragmentation caused by major roads leads to reduction of gene flow and loss of genetic variability in ground beetles. *Proceedings of the Royal Society of London B: Biological Sciences, 270*, 417–423.

Kelly, R. P., Port, J. A., Yamahara, K. M., Martone, R. G., Lowell, N., Thomsen, P. F., et al. (2014). Harnessing DNA to improve environmental management. *Science, 344*, 1455–1456.

Kijas, J. W., Lenstra, J. A., Hayes, B., Boitard, S., Neto, L. R. P., San Cristobal, M., et al. (2012). Genome-wide analysis of the world's sheep breeds reveals high levels of historic mixture and strong recent selection. *PLoS Biology, 10*, e1001258.

Kopf, A., Bicak, M., Kottmann, R., Schnetzer, J., Kostadinov, I., Lehmann, K., Fernandez-Guerra, A., et al. (2015). The ocean sampling day consortium. *GigaScience, 4*, 1–5. http://doi.org/10.1186/s13742-015-0066-5

Laikre, L. (2010). Genetic diversity is overlooked in international conservation policy implementation. *Conservation Genetics, 11*, 349–354.

Lamichhaney, S., Berglund, J., Almén, M. S., Maqbool, K., Grabherr, M., Martinez-Barrio, A., et al. (2015). Evolution of Darwin's finches and their beaks revealed by genome sequencing. *Nature, 518*, 371–377.

Last, L., Arndorfer, M., Balázs, K., Dennis, P., Dyman, T., Fjellstad, W., et al. (2014). Indicators for the on-farm assessment of crop cultivar and livestock breed diversity: A survey-based participatory approach. *Biodiversity Conservation, 23*, 3051–3071.

Leberg, P. L. (2002). Estimating allelic richness: Effects of sample size and bottlenecks. *Molecular Ecology, 11*, 2445–2449.

Lowe, A. J., & Cross, H. B. (2011). Application of DNA methods to timber tracking and origin verification. *IWA Journal, 32*, 251–262.

MA [Millennium Ecosystem Assessment]. (2005). *Ecosystems and human wellbeing: Synthesis.* Washington, DC, USA: Island Press.

Mace, G. M., Norris, K., & Fitter, A. H. (2012). Biodiversity and ecosystem services: A multilayered relationship. *Trends in Ecology & Evolution, 27*, 19–26.

Matukumalli, L. K., Lawley, C. T., Schnabel, R. D., Taylor, J. F., Allan, M. F., Heaton, M. P., et al. (2009). Development and characterization of a high density SNP genotyping assay for cattle. *PLoS ONE, 4*, e5350.

McGill, B. J., Dornelas, M., Gotelli, N. J., & Magurran, A. E. (2015). Fifteen forms of biodiversity trend in the Anthropocene. *Trends in Ecology & Evolution, 30*, 104–113.

Minhós, T., Wallace, E., da Silva, M. J. F., Sá, R. M., Carmo, M., Barata, A., et al. (2013). DNA identification of primate bushmeat from urban markets in Guinea-Bissau and its implications for conservation. *Biological Conservation, 167*, 43–49.

Nannipieri, P., Ascher, J., Ceccherini, M., Landi, L., Pietramellara, G., & Renella, G. (2003). Microbial diversity and soil functions. *European Journal of Soil Science, 54*, 655–670.

Narloch, U., Pascual, U., & Drucker, A. G. (2011). Cost-effectiveness targeting under multiple conservation goals and equity considerations in the Andes. *Environmental Conservation, 38*, 417–425.

Nielsen, E. E., Cariani, A., Mac Aoidh, E., Maes, G. E., Milano, I., Ogden, R., et al. (2012). Gene-associated markers provide tools for tackling illegal fishing and false eco-certification. *Nature Communications, 3*, 851.

Pedersen, M. W., Overballe-Petersen, S., Ermini, L., Der Sarkissian, C., Haile, J., Hellstrom, M., et al. (2015). Ancient and modern environmental DNA. *Philosophical Transactions of the Royal Society of London B: Biological Sciences, 370*, 20130383.

Pereira, H. M., Ferrier, S., Walters, M., Geller, G. N., Jongman, R. H. G., Scholes, R. J., et al. (2013). Essential biodiversity variables. *Science, 339*, 277–278.

Pinsky, M. L., & Palumbi, S. R. (2014). Meta-analysis reveals lower genetic diversity in overfished populations. *Molecular Ecology, 23*, 29–39.

Ratnasingham, S., & Hebert, P. D. (2007). BOLD: The Barcode of Life Data System (http://www.barcodinglife.org). *Molecular Ecology Notes 7*, 355–364.

Rockström, J., Steffen, W., Noone, K., Persson, Å., Chapin, F. S., Lambin, E. F., et al. (2009). A safe operating space for humanity. *Nature, 461*, 472–475.

Schoen, D. J., & Brown, A. H. D. (2001). The conservation of wild plant species in seed banks. *Biosciences, 51*, 960–966.

Schwartz, M. K., Luikart, G., & Waples, R. (2007). Genetic monitoring as a promising tool for conservation and management. *Trends in Ecology and Evolution, 22*, 25–33.

Sgro, C., Lowe, A. J., & Hoffmann, A. A. (2011). Building evolutionary resilience for conserving biodiversity under climate change. *Evolutionary Aspplications, 4*, 326–337.

Shafer, A. B. A., Wolf, J. B. W., Alves, P. C., Bergström, L., Bruford, M. W., Brännström, I., et al. (2015). Genomics and the challenging transition into conservation practice. *Trends in Ecology and Evolution, 30*, 78–87.

Shivaprakash, K. N., Ramesha, B. T., Shaanker, R. U., Dayanandan, S., & Ravikanth, G. (2014). Genetic structure, diversity and long term viability of a medicinal plant, *Nothapodytes nimmoniana* Graham. (Icacinaceae), in protected and non-protected areas in the Western Ghats biodiversity hotspot. *PloS One, 9*, e112769.

Struebig, M., Kingston, T., & Petit, E. J. (2011). Parallel declines in species and genetic diversity in tropical forest fragments. *Ecology Letters, 14*, 582–590.

Sunnucks, P. (2000). Efficient markers for population biology. *Trends in Ecology and Evolution, 15*, 199–203.

Taberlet, P., Zimmermann, N. E., Englisch, T., Tribsch, A., Holderegger, R., Alvarez, N., et al. (2012). Genetic diversity in widespread species is not congruent with species richness in alpine plant communities. *Ecology Letters, 15*, 1439–1448.

Tittensor, D. P., Walpole, M., Hill, S. L. L., Boyce, D. G., Britten, G. L., Burgess, N. D., et al. (2014). A mid-term analysis of progress towards international biodiversity targets. *Science, 346*, 241–244.

Tringe, S. G., Von Mering, C., Kobayashi, A., Salamov, A. A., Chen, K., Chang, H. W., et al. (2005). Comparative metagenomics of microbial communities. *Science, 308*, 554–557.

Tyson, G. W., Chapman, J., Hugenholtz, P., Allen, E. E., Ram, R. J., Richardson, P. M., et al. (2004). Community structure and metabolism through reconstruction of microbial genomes from the environment. *Nature, 428*, 37–43.

van Zonneveld, M., Dawson, I., Thomas, E., Scheldeman, X., van Etten, J., Loo, J., et al. (2014). Application of molecular markers in spatial analysis to optimize in situ conservation of plant genetic resources. In R. Tuberosa, A. Graner, & E. Frison (Eds.), *Genomics of plant genetic resources* (pp. 67–91). Netherlands: Springer.

Vellend, M., & Geber, M. A. (2005). Connections between species diversity and genetic diversity. *Ecology Letters, 8*, 767–781.

Venter, J. C., Remington, K., Heidelberg, J. F., Halpern, A. L., Rusch, D., Eisen, J. A., et al. (2004). Environmental genome shotgun sequencing of the Sargasso Sea. *Science, 304*, 66–74.

Walls, R. L., Deck, J., Guralnick, R., Baskauf, S., Beaman, R., Blum, S., et al. (2014). Semantics in support of biodiversity knowledge discovery: an introduction to the biological collections ontology and related ontologies. *PLoS One, 9*, e89606.

Whitham, T. G., DiFazio, S. P., Schweitzer, J. A., Shuster, S. M., Allan, G. J., Bailey, J. K., et al. (2008). Extending genomics to natural communities and ecosystems. *Science, 320*, 492–495.

Zhang, Q. X., Shen, Y. K., Shao, R. X., Fang, J., He, Y. Q., Ren, J. X., et al. (2013). Genetic diversity of natural *Miscanthus sinensis* populations in China revealed by ISSR markers. *Biochemical Systematics and Ecology, 48*, 248–256.

Zytynska, S. E., Khudr, M. S., Harris, E., & Preziosi, R. F. (2012). Genetic effects of tank-forming bromeliads on the associated invertebrate community in a tropical forest ecosystem. *Oecologica, 170*, 467–475.

Chapter 6
Methods for the Study of Marine Biodiversity

Mark J. Costello, Zeenatul Basher, Laura McLeod, Irawan Asaad, Simon Claus, Leen Vandepitte, Moriaki Yasuhara, Henrik Gislason, Martin Edwards, Ward Appeltans, Henrik Enevoldsen, Graham J. Edgar, Patricia Miloslavich, Silvia De Monte, Isabel Sousa Pinto, David Obura and Amanda E. Bates

Abstract Recognition of the threats to biodiversity and its importance to society has led to calls for globally coordinated sampling of trends in marine ecosystems. As a step to defining such efforts, we review current methods of collecting and managing marine biodiversity data. A fundamental component of marine biodiversity is knowing what, where, and when species are present. However, monitoring methods are invariably biased in what taxa, ecological guilds, and body sizes they collect.

M.J. Costello (✉) · Z. Basher · L. McLeod · I. Asaad
Institute of Marine Science, University of Auckland, Auckland 1142, New Zealand
e-mail: m.costello@auckland.ac.nz

Z. Basher
e-mail: zbasher@gmail.com

L. McLeod
e-mail: lmcl513@aucklanduni.ac.nz

I. Asaad
e-mail: irawan.asaad@gmail.com

S. Claus · L. Vandepitte
Flanders Marine Institute, Ostend, Belgium
e-mail: simon.claus@vliz.be

L. Vandepitte
e-mail: leen.vandepitte@vliz.be

M. Yasuhara
School of Biological Sciences, Swire Institute of Marine Science,
and Department of Earth Sciences, The University of Hong Kong, Hong Kong, China
e-mail: moriakiyasuhara@gmail.com

H. Gislason
National Institute of Aquatic Resources, Technical University of Denmark,
Charlottenlund Slot, Jægersborg Allé 1, 2920, Charlottenlund, Denmark
e-mail: hg@aqua.dtu.dk

© The Author(s) 2017
M. Walters and R.J. Scholes (eds.), *The GEO Handbook on Biodiversity Observation Networks*, DOI 10.1007/978-3-319-27288-7_6

In addition, the data need to be placed, and/or mapped, into an environmental context. Thus a suite of methods will be needed to encompass representative components of biodiversity in an ecosystem. Some sampling methods can damage habitat and kill species, including unnecessary bycatch. Less destructive alternatives are preferable, especially in conservation areas, such as photography, hydrophones, tagging, acoustics, artificial substrata, light-traps, hook and line, and live-traps. Here we highlight examples of operational international sampling programmes and data management infrastructures, notably the Continuous Plankton Recorder, Reef Life Survey, and detection of Harmful Algal Blooms and MarineGEO. Data management infrastructures include the World Register of Marine Species for species nomenclature and attributes, the Ocean Biogeographic Information System for distribution

M. Edwards
Sir Alister Hardy Foundation for Ocean Science, Marine Institute, Plymouth University,
Plymouth, UK
e-mail: maed@sahfos.ac.uk

W. Appeltans
Ocean Biogeographic Information System, IODE, Intergovernmental Oceanographic
Commission of UNESCO, Ostend, Belgium
e-mail: w.appeltans@unesco.org

H. Enevoldsen
Intergovernmental Oceanographic Commission of UNESCO, IOC Science and
Communication Centre on Harmful Algae at University of Copenhagen, Copenhagen,
Denmark
e-mail: h.enevoldsen@bio.ku.dk

G.J. Edgar
Institute for Marine and Antarctic Studies, University of Tasmania, Hobart, Australia
e-mail: g.edgar@utas.edu.au

P. Miloslavich
Departamento de Estudios Ambientales, Universidad Simon Bolivar, Caracas, Venezuela
e-mail: pmilos@usb.ve

S. De Monte
Institut de Biologie de l'Ecole Normale Supérieure, Ecole Normale Supérieure, PSL Research
University, Paris, France
e-mail: demonte@biologie.ens.fr

I.S. Pinto
Centre for Marine and Environmental Research Ciimar, University of Porto, R. Dos Bragas,
289, 4050-123 Porto, Portugal
e-mail: isabel.sousa.pinto@gmail.com

D. Obura
CORDIO East Africa, P.O. BOX 10135, 80101 Mombasa, Kenya
e-mail: davidobura@gmail.com

A.E. Bates
Ocean and Earth Science, National Oceanography Centre Southampton,
University of Southampton, SO14 3ZH Southampton, UK
e-mail: A.E.Bates@soton.ac.uk

data, Marine Regions for maps, and Global Marine Environmental Datasets for global environmental data. Existing national sampling programmes, such as fishery trawl surveys and intertidal surveys, may provide a global perspective if their data can be integrated to provide useful information. Less utilised and emerging sampling methods, such as artificial substrata, light-traps, microfossils and eDNA also hold promise for sampling the less studied components of biodiversity. All of these initiatives need to develop international standards and protocols, and long-term plans for their governance and support.

Keywords Marine · Sampling · Methods · Biodiversity · Monitoring

6.1 Introduction

Current concerns about the Earth's ecosystems and the loss of biodiversity drives the need to measure spatial and temporal variation in biodiversity from local to global scales (Costello 2001; Andréfouët et al. 2008a; Ash et al. 2009). In the ocean, over-fishing and other threats to species' populations reduce resources for society, have altered ecosystems, and put many mammals, birds, reptiles, and fish in danger of extinction (e.g., Costello and Baker 2011; Hiscock 2014; Costello 2015; Webb and Mindel 2015). Global and regional scale assessments need data that are either collected by similar methods and procedures, or produce variables that can be integrated for analyses (Pereira et al. 2013). For example the EU Marine Strategy Framework Directive (MSFD) requires extensive measures of biodiversity and ecosystem functioning to monitor the health of European marine waters and to guide measures that ensure that they achieve a Good Environmental Status by 2021 (Boero et al. 2015). The World Ocean Assessment will emphasise the need for more standardised reporting of information (Inniss et al. 2016). To that end, variables that are 'essential' for the monitoring of biodiversity and understanding ecosystem change are being developed (Box 6.1). As yet, how to measure these variables, and manage and analyse the data, has not been elaborated. Here, we review methods used for field observations and sampling marine biodiversity, provide examples of methods and operational global monitoring programmes, and how data systems have emerged to assist in data publication and analysis. It cannot be assumed that established or popular methods are the most cost-effective and suitable for monitoring biodiversity. Thus we outline the potential of less prominent methods as well as those considered more conventional. This synthesis thus provides an introduction to how marine biodiversity may be monitored and assessed into the future.

Box 6.1. Essential Ocean Variables (EOVs)

Under the leadership of the Intergovernmental Oceanographic Commission (IOC) of UNESCO, the Global Ocean Observing System (GOOS) has proposed to develop an integrated framework for sustained ocean observing based on Essential Ocean Variables (EOVs). An EOV, should have by definition, a high impact in responding to scientific and societal issues and a high feasibility of sustained observation. These will include biogeochemical and biological variables (ecosystem EOVs), to help understand marine ecosystems, in addition to the existing physical ocean variables. At the same time GEO BON has been developing the Essential Biodiversity Variables (Pereira et al. 2013). GOOS in collaboration with GEO BON, has established the GOOS Panel on Biology and Ecosystems (GOOS BioEco), which is responsible for the development and assessment of ecosystem EOVs. This includes documentation, best practice, readiness, implementation strategies, coordination of activities, and fitness-of-purpose of data and information streams resulting from observations to improve their recommendations to policy-making. GOOS BioEco is also considering societal needs and human pressures affecting marine biodiversity and ecosystems to identify the EOVs. The first GOOS Biology technical expert workshop in Townsville, Australia in November 2013, resulted in a preliminary list of 42 candidate ecosystem EOVs. From these, 10 were selected for high impact and feasibility within four major areas identified as key for a healthy and productive ocean: (1) Productivity, (2) Biodiversity, (3) Ecosystem Services, and (4) Human activities and pressures. Some of the candidate EOVs that meet these requirements were chlorophyll, harmful algal blooms (HAB), zooplankton biomass and abundance, and the extent and live cover of marine communities such as coral reefs, mangroves, seagrasses, and salt marshes.

6.2 Sampling Methods

An impressive variety of methods have been used to sample marine species, including observations, nets, hooks, traps, grabs, sediment collection, sound, chemicals and electricity (Table 6.1) (e.g., Santhanam and Srinivasan 1994; Kingsford and Battershill 1998; Tait and Dipper 1998; Elliott and Hemingway 2002; Eleftheriou 2013; Hiscock 2014). All methods are selective, at least for body size by excluding smaller and/or larger organisms. Such bias should be explicitly recognised in the design and interpretation of field data. Because of methodological biases a comprehensive sampling of marine biodiversity across habitats, body sizes and trophic levels would need to use a variety of complementary methods. Such a

Table 6.1 General methods of sampling marine biodiversity and their biases

	Methods	Bias
Pelagic	Nets: sieve, gill, trammel (tangle), pelagic and demersal trawl, fyke (hoop), drop, push, dip, trap	Body size based on size of net opening (gape), mesh size, towing speed, and dimensions of trap
	Visual by observer on boat	Larger megafauna at or above water surface
	Visual by underwater video and scuba, aerial (aircraft) surveys	Larger species that swim close to observation point, or at sea surface
	Plankton pumps, powerstation screens	Capture depends on body size, agility, and flow rates
	Hooks, long-lines	Bait selective, and body size related to hook size
	Acoustic (echo-sound)	Species level recognition only for some larger fish species with distinct reflectance
Benthos	Dredges, benthic (beam, otter) trawls, sledges	Body size based on net opening and mesh size
	Baited traps and pots	Only animals attracted to bait and contained within mesh size of trap
	Artificial substrata (panels, mesh)	Taxa sampled depend on substratum used and time period of deployment
– Epifauna	Visual census and inspections by scuba, snorkel, video, submersible, Remotely Operated Vehicle (ROV), photographs	Larger taxa identified to species, cover of hard substrata, and tubes, tracks and burrows observed on sediments
– Sediment infauna	Grabs, cores, suction samples	Body size captured within sample and sieve. Some animals may escape capture as they are too large or mobile
Mobile macrofauna, especially in complex habitats, reefs (rock, coral), kelp forests, seagrass	Visual census and inspections and hand collection by snorkel, scuba, video, stereophotography	Larger species identifiable by eye in field or on video and still photograph images
	Poisons and anaesthetics (sometimes combined with suction samplers)	Collection of affected animals biased by collection method (e.g. if by hand then body size)
	Light-traps	Capture plankton and mobile benthos that are attracted to light. Body size of catch depends on trap size

(continued)

Table 6.1 (continued)

	Methods	Bias
	Emergence traps, sediment traps	Select benthic animals that move up or down in water column
Other	Hydrophone	Species that produce distinctive sounds and when they do so
	Gut contents, faeces	Prey that can be identified from samples of animal gut contents or faeces
Marking	Tags: plastic, dyes, chemical, branding, tattooing, fin clips, ultrasonic, satellite, loggers	Tag suitability depends on animal body size and anatomy
Intertidal	Visual, hand, shovel, rake, photography, electric current (sediments)	Limited to animals remaining on seashore when tide is out, and by body size if hand-collected, by visual counts, or if sediments are sieved

suite of methods can produce an inventory of species present that reflect the environment, habitats, and ecology of an area.

A species inventory provides the evidence of what species are present and an estimate of species richness. Knowing which species are present is essential to distinguish those that are of socio-economic or ecological importance, endemic, threatened with extinction, introduced, or considered pests (McGeoch et al. 2016). Indeed, species richness is by far the most common measure of 'diversity' used in science and conservation management (Gotelli and Colwell 2001; Costello et al. 2004).

For microbes, species identification can be impractical and so ad hoc 'metagenomic' and 'barcoding' style guidance on molecular 'Operational Taxonomic Units' (OTU) are used as indicators of diversity. However, OTU are not standardised between studies, and values vary due to different resolution of the genes analysed for different taxa. For some taxa they may indicate genus level and others population level differences. For bacteria, the species concept used for eukaryotes is doubtfully applicable, and while they have high genetic diversity, the number of formally named 'species' is relatively low (Costello et al. 2013a, b). Thus, while an indicator of genetic diversity, OTU should not be equated with 'species'.

Various methods and metrics have been used to characterise the relative abundance of species, including numbers of individuals, areal cover, and/or biomass within samples (Hiscock 2014). Assessments of measures of biodiversity thus need to consider that every sampling method is biased, and that different methods are required to sample different components of biodiversity. Thus quantitative sampling is best focused on measuring dynamics of particular species populations rather than measuring biodiversity across species. Instead, the relative abundance of species may be compared on semi-quantitative (e.g., log 10) abundance scales (e.g., Davies et al.

2001; Haegeman et al. 2013; Hiscock 2014). The more abundant and/or conspicuous species define communities and biotopes and indicate how an ecosystem functions in terms of habitat, productivity, and food-webs. Changes in the identity of the dominant species can indicate changes in the community present in space and time, and thus changes to the ecosystem. However, often the ecosystem effects of species are unrelated to their abundance or body size. For example, top-predators are typically low in abundance and density but large in body size. Thus a range of species of different guilds and body sizes should be sampled to monitor ecosystems.

In addition to the bias of how samples are taken, results will depend on when and where sampling takes place. The design of field surveys thus needs to be clear which habitats, body sizes and taxa it has focused on, and what has been excluded; i.e., how it has 'stratified' sampling. Perhaps the most effective way to place the data into an environmental context is to map the geographic distribution of environmental variables (e.g., depth, salinity, temperature, substratum, topography) and habitats (Costello 1992; Costello and Emblow 2005; Costello et al. 2005, 2010a; Hiscock 2014). These environmental variables can be mapped through 'remote sensing' from satellites, aircraft and ships (Andréfouët et al. 2008b, 2011) and can include: seabed depth, topography, and roughness; surface water colour (an estimate of phytoplankton biomass and dominance) and temperature; depth-profiles of density (salinity) and temperature; acoustic signatures of zooplankton and pelagic megafauna; and the distribution and extent of intertidal and shallow-water habitats such as coral reefs, kelp and seagrass beds, mangrove forests, and salt-marshes. As the technology improves and cost reduces, it is likely that 'remotely operated' and 'autonomous' vehicles (ROV, AUV) will become more commonly used for underwater and aerial surveillance. The potential of sound signatures in the marine environment as indicators of biodiversity is also being researched (Harris et al. 2015). Although sensors borne on satellites and aircraft may have limited ability to identify species they provide an invaluable environmental context for biodiversity, and may indicate global large-scale patterns in biodiversity (De Monte et al. 2013). They thus complement in situ observations and enable mapping of habitats and biotopes (e.g., Neilson and Costello 1999; Connor et al. 2006; Leleu et al. 2012; Remy-Zephir et al. 2012; Hiscock 2014). Other methods may identify species from images, such as video and still photography (Table 6.1). Techniques for unsupervised image processing continue to improve and may lead to an increased use of automated image systems for large and microscopic species. Crowd-sourcing is also increasingly assisting the digitisation of large ecological image libraries (Edgar et al. 2016).

6.2.1 Bottom Trawl Surveys

In many countries bottom trawl surveys are used for monitoring commercially important fish stocks. Although originally designed to provide fisheries independent information for fish stock assessment and management, they are now increasingly being used to analyse trends in the abundance, distribution and diversity of both commercial and non-commercial species of fish and epibenthos (e.g., Bianchi et al. 2000; Shackell and Frank 2003; Daan et al. 2005; Perry et al. 2005; Atkinson et al. 2011).

Bottom trawls come in different designs suited for catching fish on different types of seabed. Beam trawls use a horizontal metal beam to keep the mouth of the trawl open, and target flatfish and other near-bottom species. They sometimes have 'tickler chains' attached to the front bottom part of the gear to scare shrimps or flatfish up from the seabed and into the net. Otter trawls use otter boards (trawl doors) attached to the trawl net by wires to keep the mouth of the net horizontally open. Over fine grained sediments the otter boards generate clouds of suspended material on each side of the trawl net which helps to herd the fish into the mouth of the trawl. Often wings of netting are attached to both sides of the trawl mouth to increase the herding effect further. Vertically the mouth of an otter trawl is held open by floats and by a footrope to which weights, rollers or bobbins are attached. These vary from small rubber discs used on sandy or muddy bottoms to large metal balls that can roll over rocks or larger stones and prevent the footrope from becoming snagged on rougher and harder grounds. The body of the trawl is funnel-shaped and narrows from the mouth towards the cod end where the fish accumulate during the tow. It is the mesh size of the cod end that determines the size of the fish that are retained. In commercial trawl gears minimum mesh size regulations are often used to reduce the catch of juvenile undersized fish. However, in research surveys the mesh size in the cod end is usually small enough to ensure that the smaller species and individuals are retained. Pelagic trawls target fish such as anchovies, mackerels, and sardines in the water column.

The catch efficiency of a bottom trawl is defined as the proportion of the fish in the area swept by the gear that is retained in the cod end. The area swept equals the length of the tow multiplied by the width of the gear, where the latter often is assumed to correspond either to the spread of the wings or to the distance between the otter boards during fishing to account for the herding effect of the boards and bridles. However, the catch efficiency is influenced by a multitude of factors including the escape behaviour of the fish species, properties of the gear, and the fishing operation (Benoît and Swain 2003; Fraser et al. 2007, 2008; Queirolo et al. 2012; Weinberg and Kotwicki 2008; Winger et al. 2010; Sistiaga et al. 2015). Fish may escape by burrowing in the seabed, by swimming under the footrope, by escaping over the head-rope of the gear, or by passing through the meshes in the front part of the trawl. The size of the vertical and horizontal opening is often monitored during the tow by sensors attached to the gear and has been found to depend on the warp length and towing speed as well as the weight of the catch accumulating in the cod end. During fishing, fish accumulate in the mouth of the trawl where they try to keep pace with the gear. As individuals tire they fall back towards the cod end. How fast a fish will get tired, and whether it can outswim the gear is species and size dependent. The amount caught per area swept may also depend on the time of day because this can influence how close to the seabed the fish are found (Kotwicki et al. 2009). To ensure that catch rates can be compared across years, much is therefore done to standardise the trawling operation, the gear and the procedures for sampling and for analysing the catch (e.g., Miller 2013).

In some parts of the world standardised bottom trawl surveys have now been conducted for more than 50 years and some of the resulting data are publicly available or available upon request. ICES provides online access to a database with trawl survey data from the north eastern Atlantic (www.ices.dk/marine-data/data-portals/Pages/DATRAS.aspx) and similar databases are available for other areas such as the Eastern Bering Sea and Gulf of Alaska (www.afsc.noaa.gov/RACE/groundfish/survey_data). Additional data can be downloaded from international data portals such as OBIS (Table 6.2), but much data still reside in the custody of national fisheries research institutions. These data constitute a so far underutilised source of information on the distribution, abundance, and diversity of marine fishes on the world's continental shelves.

Table 6.2 Examples of marine biodiversity data management systems

Resource	Objectives	Content	Output
World Register of Marine Species (WoRMS) www.marinespecies.org Host: Flanders Marine Institute	To provide an expert validated and comprehensive list of names of all marine organisms	Species names including information on higher classification, synonymy, images and links to other species information	Database of over 240,000 accepted species names
Marine Regions www.marineregions.org Host: Flanders Marine Institute	To provide a standard list of marine georeferenced place names and areas	A data system of geographic marine areas from different national and global marine gazetteers and databases	Spatial information of 264 different physical and administrative boundaries; over 30,000 unique marine geographic places
Ocean Biogeographic Information System www.iobis.org Host: Intergovernmental Oceanographic Commission (IOC) of UNESCO	A global science alliance that facilitates free and open access to data and information on marine biodiversity	Database of the diversity, distribution and abundance of marine life	Over 1900 datasets that covers more than 45 million observations of 114,000 marine species
Global Marine Environment Datasets (GMED) http://gmed.auckland.ac.nz Host: University of Auckland	To provide standardised global marine environment datasets of climatic, biological and geophysical environmental layers	Environmental datasets featuring present, past and future environmental conditions to a common spatial resolution	60 datasets of climatic, biological and geophysical environmental layers ready to use for species distribution modelling and data visualisation software

6.2.2   Light Traps

Light traps are commonly used for collecting insects as a means for monitoring pest species. The American Center for Disease Control has had a standardised light trap for mosquito monitoring for over 50 years (Sudia and Chamberlain 1988). Moths, beetles and other crop pests are also commonly surveyed this way (Szentkiralyi 2002). However, light traps have a shorter history of use in the aquatic environment. The earliest uses were in freshwater for capturing insects and they were soon found to be excellent for collecting young fish (Hungerford et al. 1955) and zooplankton (Meekan et al. 2001; Øresland 2007), but also collect many benthic species that emerge from the benthos at night. They have been used extensively around coral reefs where the structural complexity of the reef system makes other methods susceptible to damage (Doherty 1987). There can be species, gender and ontogenetic specific responses to light traps making them more useful for some organisms than others. Species may vary in their abundance at different times of the night and lunar cycle. A benefit of light-trapping is that the animals are not harmed during collection, and have thus proved useful for sampling of museum specimens and laboratory animals (Doherty 1987; Holmes and O'Connor 1988). However, light trap catches may not work well in areas of high current or excessive turbidity. The potential of light traps for monitoring mobile benthic and demersal organisms, mostly crustaceans, has yet to be adequately explored. This 'fish food' component of biodiversity forms an important trophic link in many ecosystems, and has been overlooked in marine biodiversity monitoring.

6.2.3   Artificial Substrata

A problem in sampling the natural environment is that it is variable at every spatial scale, and thus the abundance of species sampled varies because of micro-habitat variation as well as changes in species abundance in space and over time. Advantages of artificial substrata are that they provide a standard replicable physical habitat and thus low variation between replicate samples. In addition their use avoids damage to natural habitat, and they can be low cost, amenable to experimental manipulation, easily deployed and retrieved, and rapidly processed (reviewed in Costello and Thrush 1991). Because the date and duration of deployment of artificial substrata is known their community can also be standardised for successional age. They can be hard panels, balls of plastic mesh, sediment trays, and made of a variety of materials. They can also capture species otherwise difficult to sample, such as mobile epifaunal macroinvertebrates that nestle into plastic mesh. Species composition and community structure has been found to be similar and comparable to natural substrata (Costello and Myers 1996). Artificial substrata have been long used in freshwater environments as a standard method of monitoring biodiversity, especially in large rivers and lakes where other methods may be difficult (APHA et al. 2007). They have had widespread use in experiments in the marine environment, such as looking at colonisation, succession, competition, and community stability on plastic mesh (e.g., Costello and Myers 1996) and flat panels (e.g., Atalah et al. 2007a, b; Wahl et al. 2011). Recently, hundreds of Autonomous Reef Monitoring Structures (ARMS) have become deployed on coral reefs and other habitats around the world (e.g., http://www.pifsc. noaa.gov/cred/arms.php; Leray and Knowlton 2015). ARMS are a stack of hard plastic plates that capture crevice living invertebrates otherwise difficult to collect without damaging reefs. Artificial substrata merit wider use in marine biodiversity monitoring considering their benefits of standardisation and lack of damage to natural habitat.

6.2.4   Microfossils

Microfossils are microscopic sized organisms that have hard parts with high fossilisation potential (e.g., calcareous or siliceous shells), including foraminiferans, ostracods, diatoms, radiolarians and coccolithophores, or are microscopic sized hard parts of larger organisms, including ichthyoliths. Microfossils can be a proxy for biodiversity patterns across a broader range of organisms, because they have excellent fossil records, occupy a wide range of ecological niches, and are abundant even in a small amount of sediment. Marine sediment cores available from almost the entire ocean through national and international drilling projects (e.g., International Ocean Discovery Program; IODP) include abundant microfossils and provide long-term continuous time-series sedimentary records at decadal, centennial, millennial, and multi-millennial time scales covering the entire Cenozoic Era. Thus microfossils in sediment cores are an archive that enables reconstruction of long-term time-series beyond the temporal coverage of recent biological monitoring (Yasuhara et al. 2015).

Sample procedures involve physical and chemical treatments of sediment subsamples to disaggregate consolidated sediment, clean up microfossils, concentrate specimens and remove extraneous material, for example, by freeze drying, hydrogen peroxide treatment, wet sieving, centrifugation, and acid treatment. The resulting sample can be mounted on a glass slide (e.g., for diatoms, radiolarians and coccolithophores) or manually picked from treated material onto a paper slide (e.g., for foraminiferans, ostracods and ichthyoliths) for counting under stereo and compound microscopes respectively.

For example, North Atlantic deep-sea ostracod diversity has been found to track global climate change for the last 500,000 years, being less during glacial and high during interglacial periods (Yasuhara et al. 2009). Climatic control of deep-sea ostracod diversity has also been shown for shorter, decadal-centennial time scales (Yasuhara et al. 2008). Latitudinal species diversity gradients of deep-sea ostracods in the North Atlantic Ocean were distinct during interglacials (including present day) but indistinct or collapsed during glacials (Yasuhara et al. 2009). These deep-sea diversity patterns in space and time in North Atlantic microfossil records are explained by temperature control of deep-sea biodiversity (Hunt et al. 2005; Yasuhara and Cronin 2008; Yasuhara et al. 2009, 2014). Further applications of microfossils as a model system for biodiversity research are found in Yasuhara et al. (2015).

6.2.5 Molecular Observations of Microbial Communities

Genomic analysis of marine microbes has become common both at the local (marine stations, localised cruises) and at the global scale. After the Global Ocean Sampling expedition (Venter et al. 2004) proved that high-throughput molecular approaches were able to reveal an unprecedented diversity of bacterial sequences, several other programs have quantified the molecular diversity and biogeography of planktonic communities. The Tara Oceans missions (http://oceans.taraexpeditions.org/en) have sampled coastal and open oceans worldwide (Bork et al. 2015), including eddies, upwellings, oxygen-minimum zones, coral reefs, regions of natural iron fertilisation, and lately the Arctic and Mediterranean regions. These missions are uncovering marine planktonic communities from viruses to protists, up to metazoan larvae. The Malaspina project (http://scientific.expedicionmalaspina.es) complements these observations with samples of the deep seas at the global scale, and 'Ocean Sampling Day' with about 150 stations globally sampled on the same day (Kopf et al. 2015). An increasing number of cruises include molecular high-throughput analyses of genes, transcripts, and metabolites of planktonic organisms, together with environmental variables such as physical and biochemical parameters (e.g., Atlantic Meridional Transect http://www.amt-uk.org).

6.3 Case Studies

6.3.1 The Continuous Plankton Recorder (CPR)

The Continuous Plankton Reorder (CPR) survey is the longest sustained and geographically most extensive marine biological survey in the world, covering ~1000 taxa over multi-decadal periods since 1931 (Edwards et al. 2010). It samples phytoplankton and zooplankton in oceans and shelf seas using ships of opportunity from ~30 different shipping companies, at monthly intervals on ~50 trans-ocean routes. In this way the survey autonomously collects biological and physical data from ships covering ~20,000 km of the ocean per month, ranging from the Arctic to the Southern Ocean. The survey is operated by the Sir Alister Hardy Foundation for Ocean Science (SAHFOS), an internationally funded charity with a wide consortium of stakeholders. Since the first tow of a CPR more than 6 million nautical miles of sea have been sampled and over 100 million data entries have been recorded. Plankton are collected on a band of silk and subsequently visually identified by experts. Additionally, over the last decade the CPRs have been equipped with modern chemical and physical sensors as well as molecular probes. The database and sample archive together provide a resource that can be utilised in a wide range of environmental, ecological and fisheries related research, e.g., molecular analyses of marine pathogens, modelling for forecasting and data for incorporation in new approaches to ecosystem and fishery management.

In 2011 SAHFOS, along with 12 other research organisations using the CPR from around the world formed a Global Alliance of CPR surveys (GACs) with the aim of developing new surveys and a global database, and producing a global ocean status report (Edwards et al. 2012). This global network of CPR surveys now routinely monitors the North Sea, North Atlantic, Arctic, North Pacific and Southern Ocean. New surveys are underway in Australian, New Zealand, Japanese and South African waters with a Brazilian and an Indian Ocean survey under development. These surveys provide coverage of large parts of the world's oceans but many gaps still exist particularly in the South Atlantic, Indian and Pacific Oceans. This global network also brings together the expertise of approximately 60 plankton specialists, scientists and technicians from 14 laboratories around the world. Working together, centralising the database and working in close partnership with the maritime shipping industry, this global network of CPR surveys with its low costs and new technologies makes the CPR an ideal tool for an expanded and comprehensive marine biological sampling programme.

6.3.2 Tropical Coral Reefs

Monitoring of tropical shallow reefs is conducted with near-global coverage using methods described by English et al. (1994). Considerable effort has been invested in comparing the accuracy and agreement among different methods (e.g., Leujak and Ormond 2007; Facon et al. 2016). The emerging consensus is to focus on the output variables from monitoring, rather than the methods: e.g., proportional cover for sessile taxa, abundance or density per unit area for mobile taxa and biomass, particularly for fishes. This is consistent with emerging guidance on observation and indicator systems (UNESCO 2012).

The principal framework for aggregating coral reef data to global levels has been the Global Coral Reef Monitoring Network (GCRMN) of the International Coral Reef Initiative (ICRI), which was initiated in 1995. The establishment of the GCRMN coincided with the largest global impact to reefs ever recorded, the 1997–98 El Niño event, giving strong impetus for global reporting for a decade. However funding for this level of reporting has been difficult to sustain, forcing the GCRMN to focus on regional level reporting, such as in the Caribbean (Jackson et al. 2014) and currently underway in the Western Indian Ocean. The GCRMN regions closely match those of the UNEP Regional Seas programmes, and inform countries regarding fisheries and food security. The GCRMN provides guidance for three levels of monitoring effort: citizen volunteer-focused, 'intermediate' and 'expert' (Wilkinson and Hill 2004). The challenges across these levels include data reliability and quality, replication and representation, and taxonomy, the latter exacerbated by the high diversity of coral reef taxa. The intermediate level of monitoring is most frequently applied and is implemented through technical staff (e.g., marine rangers), students and experienced volunteers, and focused on functional group or genus-level identifications for principal benthic taxa (e.g., hard corals, algae) and family or genus level identification for fish. The basic sampling unit recommended by the GCRMN has been line transects or photoquadrats for benthic cover, 50 m belt transects (2 or 5 m width) for fish and narrower belt transects or quadrats for mobile invertebrates. The configuration of these samples varies greatly among programmes. Expert-level monitoring has been the domain of professional researchers, often with genus-level identification for corals and species-level identification for fish. Due to the popularity of coral reefs for SCUBA diving, sampling by volunteers has been feasible, with the most widespread methods being those of Reef Check (Hodgson 1999), REEF (Francisco-Ramos and Arias-González 2013), and the Reef Life Survey (see below). In volunteer programs, assessments are generally restricted to indicator species and more rapid estimates of variables such as benthic cover, and lower levels of replication are accepted than in intermediate and expert monitoring. Though variable in quality and coverage, the resulting data can be invaluable in broad scale scientific assessments of reef status (Bruno and Selig 2007).

The urgency for accurate and reliable monitoring of coral reefs, that can serve both national (local) and international (global) needs is high, due to the poor performance of coral reef targets in the mid-term assessment of Aichi Target performance (GBO 2014). The GCRMN is developing with involvement from GEO BON and GOOS to become a mature observation network (UNESCO 2012), to better report on global targets (Aichi Target 10 on climate-sensitive ecosystems, and 14 on Oceans), and to feed into management, such as through the IUCN Red Lists of species and ecosystems. At the same time, extending citizen science contributions, and establishing a more open-data philosophy for monitoring data to maximise its accessibility, for example, through OBIS (Table 6.2), are emerging priorities.

6.3.3   The Reef Life Survey (RLS)

The Reef Life Survey (RLS) was established in 2007 to test the concept that a rigorous scientific approach to marine biodiversity monitoring could be developed within a citizen science framework (Edgar and Stuart-Smith 2014). The primary aim was to engage recreational divers to obtain scientific data from biodiversity observations that spanned geographic, temporal and taxonomic scales too costly for scientists to collect. It also aimed to extend other citizen science programs such as

Reef Check and the REEF (see Sect. 6.3.2) that collected less detailed data (Edgar et al. 2016). Following establishment of the charitable Reef Life Survey Foundation (www.reeflifesurvey.com) to oversee field activities, appropriate data collection methodology, training, data entry and management procedures were developed, and different mechanisms for data collection were tested. Field survey methods were based on those applied over two decades by University of Tasmania researchers in Marine Protected Area (MPA) monitoring studies (Edgar and Barrett 1999; Barrett et al. 2009).

Three coincident elements of biodiversity are documented along 50 m long underwater transect lines. Divers record abundances and sizes of all fish, and abundances of all large (>2.5 cm length) mobile invertebrates (echinoderms, crustaceans and gastropods) and cryptic fishes. The area covered by sessile invertebrates, macrophytes and abiotic habitat is quantified through digitisation of photoquadrats (e.g., using Coral Point Count; Kohler and Gill 2006). Divers are trained on a one-on-one basis, each novice diver following behind a trained diver and duplicating transect blocks until the required level of expertise is reached. A comparison of data collected by trained volunteers and experienced scientists at the same sites showed that the variation attributable to diver experience was not significant, and negligible (<1 %) relative to differences between sites and regions (Edgar and Stuart-Smith 2009). The RLS program possesses a degree of self-regulation, where the keenest volunteers tend to also collect the best data, participate most frequently and persist longest (Edgar and Stuart-Smith 2009). A network of over 100 active RLS divers has now been established worldwide.

Application of RLS methods has allowed the first global analyses using standardised site-based procedures that are quantitative, species-level and cover multiple higher taxa. Data have been obtained for over 4500 species, 2800 sites, 600,000 species abundance records, 43 countries, and 83 marine ecoregions including Antarctica (e.g., Stuart-Smith et al. 2013, 2015). Many sites have been surveyed on multiple occasions, in some cases annually since 2007. These data add enormous contextual value to local surveys, and provide sufficient replication to disentangle many interactive and non-linear threats to marine biodiversity, including impacts of climate change, fishing and invasive species. For example, Edgar et al. (2014) included an order of magnitude more MPAs than any previously attempted using standardised field data. They found no detectable differences between fish communities present in most of the 87 MPAs investigated when compared with comparable fished communities (i.e., most MPAs were 'paper parks'). However, some MPAs were extremely effective, with many large fishes and high conservation success. The RLS data are expected to be increasingly useful for (i) assessing ecosystem impacts of global threats to species at all levels of the food web from primary producers to higher predators, (ii) quantifying population trends for threatened species, and (iii) tracking international commitments associated with marine biodiversity in shallow reef ecosystems.

6.3.4   Harmful Algal Blooms (HAB)

Proliferation of microalgae in marine or brackish waters can cause massive fish kills, contaminate seafood with toxins, and alter ecosystems in ways that humans perceive as harmful. These phenomena are referred to as harmful algal blooms (HAB). Data on the distribution of toxic and harmful microalgae are collected through national surveillance programmes aimed at protecting public health, wild and cultured fish and shellfish, and bathing water quality. Sampling methods include plankton net hauls, water samples and molecular tools to detect species or genus-specific algal toxins in fish and shellfish. Benthic HAB species are collected from sediment, corals, seaweed or standardised screens. The detection of HAB species is challenging as many are difficult or impossible to identify even by using a light microscope. The challenge of maintaining a consistent microalgal taxonomy is addressed in the IOC Taxonomic Reference List of Toxic Plankton Algae within the World Register of Marine Species (Moestrup et al. 2009). The Intergovernmental Oceanographic Commission (IOC) of UNESCO has for two decades facilitated research to improve observations of harmful algae, provided training opportunities for their improved monitoring, as well as supported regional and global networks for knowledge and data sharing. The provision of method manuals and guides is central to observations of HAB species. The manual on HAB (Hallegraeff et al. 2003) is a base reference for methods and has been complemented by Babin et al.'s (2008) monograph on real-time observation systems, Karlson et al.'s (2010) intercomparison of quantative methods, and Reguera et al.'s (2011) sampling and analysis manual.

Global data on HAB species occurrences and their impacts are stored in the Harmful Algae Event Data Base (HAEDAT) in OBIS (Table 6.2). This international compiling and sharing of HAB data was initiated in the 1980s and is now accelerating and will provide the basis for a 'Global HAB Status Report' with the aims of compiling an overview of HAB events and their societal impacts; providing a worldwide appraisal of the occurrence of toxin-producing microalgae; and assessing the status and probability of change in HAB frequencies, intensities, and distribution resulting from environmental changes at the local and global scale. Linkages will be established with the International Panel on Climate Change (IPCC) reporting on the biological impacts of climate change. The Status report will provide the scientific community as well as decision makers with a reference on HAB occurrence and impacts on ecosystem services. IOC UNESCO project partners include the International Atomic Energy Agency (IAEA), the International Council for Exploration of the Sea (ICES), the North Pacific Marine Science Organization (PICES) and the International Society for the Study of Harmful Algae (ISSHA).

6.4 Data Management

Field data may be mapped to geographic areas, seascapes, habitats and against environmental parameters. Similar, globally applicable systems for the classification of marine habitats have been developed in Europe (Connor et al. 2004; Costello and Emblow 2005; Anon. 2014) and USA (Anon. 2012). The former leads to species-level biotopes (i.e., habitat + community), while the latter does not go to biotope level but does include seascape features (reviewed by Costello 2009). These can be presented as hierarchical lists and two-dimensional matrices (Fig. 6.1). The term habitat is highly context dependent and loosely used. Strictly speaking habitats are the immediate physical environment repeatedly associated with a species or distinct assemblage (or community) of species. The lowest level of habitat classifications are thus characterised by particular species. In contrast, related concepts of seascapes (landscapes, topographic features) and ecosystems will contain a variety of habitats (Costello 2009). These can be mapped over larger areas using remote sensing methods, whereas habitats usually need in situ sampling to identify their characteristic species, although exceptions exist in locations with biogenic habitat structure (e.g., seagrass beds, mangrove forests) (e.g., Andréfouët et al. 2001).

Knowing which species are present at a place and time is fundamental to biodiversity studies. Usually species are classified taxonomically because this is convenient and closely related species tend to have similar functional roles in ecosystems. However, ecologists may also classify species by their ecological traits (e.g., Wahl et al. 2013). Thus WoRMS (see Sect. 6.4.1) is developing a standardised approach to apply biological and ecological traits to marine species (Costello et al. 2015a).

A necessary step in organizing marine biodiversity data in integrated information systems is the development of appropriate thesauri and classification systems, as well as implementing quality control and feedback mechanisms. When integrating quantitative and qualitative natural history and distributional data, the use of both authoritative taxonomic and geographical hierarchical schema is essential. Here we introduce the leading taxonomic and geographic standards databases for the marine environment (Table 6.2).

6.4.1 World Register of Marine Species (WoRMS)

WoRMS is an open-access online database that provides an authoritative and comprehensive list of names of all marine organisms, including information on higher classification, synonymy, images and links to other information (Costello et al. 2013c). It currently contains over 240,000 accepted species names (Boxshall et al. 2015). While highest priority goes to valid names, other names in use are included so that this register is a guide to interpret taxonomic literature. Automated

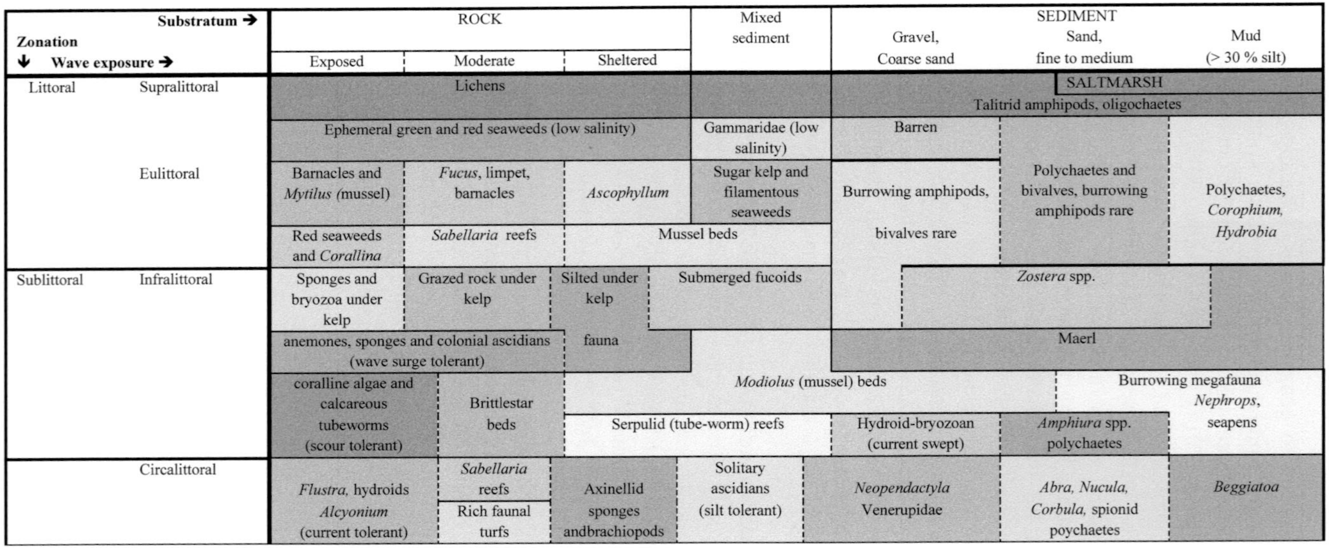

Zonation / Wave exposure →	ROCK			SEDIMENT			
	Exposed	Moderate	Sheltered	Mixed sediment	Gravel, Coarse sand	Sand, fine to medium	Mud (>30 % silt)
Littoral — Supralittoral	Lichens			SALTMARSH			
				Talitrid amphipods, oligochaetes			
Eulittoral	Ephemeral green and red seaweeds (low salinity)			Gammaridae (low salinity)	Barren		Polychaetes, Corophium, Hydrobia
	Barnacles and Mytilus (mussel)	Fucus, limpet, barnacles	Ascophyllum	Sugar kelp and filamentous seaweeds	Burrowing amphipods, bivalves rare	Polychaetes and bivalves, burrowing amphipods rare	
	Red seaweeds and Corallina	Sabellaria reefs	Mussel beds				
Sublittoral — Infralittoral	Sponges and bryozoa under kelp	Grazed rock under kelp	Silted under kelp fauna	Submerged fucoids		Zostera spp.	
	anemones, sponges and colonial ascidians (wave surge tolerant)			Modiolus (mussel) beds		Maerl	
				Serpulid (tube-worm) reefs			
Circalittoral	coralline algae and calcareous tubeworms (scour tolerant)	Brittlestar beds	Axinellid sponges and brachiopods	Solitary ascidians (silt tolerant)	Hydroid-bryozoan (current swept)	Amphiura spp. polychaetes	Abra, Nucula, Corbula, spionid poychaetes
	Flustra, hydroids Alcyonium (current tolerant)	Sabellaria reefs / Rich faunal turfs			Neopendactyla Venerupidae	Burrowing megafauna Nephrops, seapens	Beggiatoa

Fig. 6.1 A biotope matrix with the most important physical habitat features on the axes. This illustrates the relationship of shore height (littoral) and sea depth (sublittoral, offshore), with substratum (rock and grades of sediment), and the exposure of rocky habitats to wave action. These factors distinguish biotopes at the *upper levels* of the habitat classification. Within the matrix the characteristic species of the communities occurring in the habitats are indicated. Adapted from Costello and Emblow (2005)

tools allow users to upload their species lists and match and classify their names against WoRMS. WoRMS makes use of the Aphia infrastructure which is designed to capture taxonomic and related data and information (Vandepitte et al. 2015a). WoRMS was a development from the European Register of Marine Species (ERMS) (Costello 2000; Costello et al. 2001), and thus its content is controlled by an Editorial Board of taxonomic and thematic experts who elect a governing and steering committee, and invite colleagues to assist them. A permanent host institution provides professional computational support for the database, including monthly archiving. As of January 2016, there were 393 editors from 273 institutions in 50 countries actively involved in the management and quality control of the WoRMS content. Through this editorial community, communication and collaboration within and beyond this community is facilitated (e.g., Appeltans et al. 2012), which can lead to increased rates of species discoveries and synonym names, which in turn can lead to a reduced rate of creating new synonyms and homonyms. WoRMS uses Life Science Identifiers (LSIDs) as persistent, location-independent, resource identifiers for each species name (Costello et al. 2013a). WoRMS forms the taxonomic backbone for OBIS, meaning that each taxon name in OBIS is matched against WoRMS to verify its validity and spelling (Vandepitte et al. 2011, 2015a, b). WoRMS is also a major contributor to the Catalogue of Life, Encyclopedia of Life, and LifeWatch Marine Virtual Research Environment (http:// marine.lifewatch.eu). Species can be grouped with WoRMS to form Global, Regional and Thematic Databases. For example, the World Register of Introduced Marine Species (WRIMS) provides an entry point and experts to manage information on alien species (Pagad et al. 2015).

6.4.2   Marine Regions

Marine Regions (www.marineregions.org) hierarchically organises over 30,000 geographic areas from national and global marine gazetteers and databases (Claus et al. 2014). It contains spatial information of 264 different physical (e.g., sandbank, seamount, island, bay) and administrative (e.g., Exclusive Economic Zones, Marine Protected Area, Fisheries Zones or Biogeographic Regions) kinds of places. Both marine (e.g., seamounts, canyons, guyots, fracture zones, banks, ridges, basins) and coastal features (e.g., bays, fjords, cliffs, lagoons, beaches) are included. In order to preserve the identity of the marine geographic objects from the database, and to name and locate the geographic resources on the web, each geographic object is allocated a Marine Region Identifier, or MRGID. This unique persistent resource identifier is comparable to a LSID, being a unique identifier to locate the item on the World Wide Web.

6.4.3   Ocean Biogeographic Information System (OBIS)

OBIS is the world's largest database on the distribution and abundance of marine life. In 2009, IOC Member States recognised the importance of knowledge of the ocean's biodiversity to national and global environmental policies when they adopted it from the Census of Marine Life (Costello and Vanden Berghe 2006; Costello et al. 2007; O'Dor et al. 2012). OBIS operates through a network of national, regional and thematic nodes, and a secretariat based at the IOC's International Oceanographic Data and Information Exchange (IODE) programme office in Oostende, Belgium. This office provides training and technical assistance, guides new data standards and technical developments, and encourages international cooperation to foster the group benefits of the network.

OBIS is a global science alliance that facilitates free and open access to data and information on marine biodiversity. It provides a single access point to over 45 million observations of 114,000 marine species, collected on 4.6 million sampling events from 3.2 million sampling stations, integrated from over 1900 datasets provided by nearly 500 institutions in 56 countries, It grows by about 3 million records per year. Data are subject to a series of quality control steps, including for taxonomic nomenclature and geography (Vandepitte et al. 2011, 2015a, b; IODE Steering Group for OBIS 2013).

Communities associated with OBIS include OBIS-SEAMAP (Spatial Ecological Analysis of Megavertebrate Populations) focusing on megafauna, and MICROBIS (http://icomm.mbl.edu/microbis) on microbes. The latter collects molecular observations of marine microbial organisms at taxonomic ranks from phyla to genus, together with their contextual physical and biochemical data measured in situ or from remote sensing. It has developed tools for extracting diversity measures, as well as other ecologically relevant statistics, from molecular datasets (Giongo et al. 2010; Buttigieg and Ramette, 2014). More comprehensive taxon based databases include the pioneering FishBase (Froese and Pauly 2015).

So far, 1000 publications have cited OBIS and on average 10 more each month (e.g., Basher et al. 2014a, b; Saeedi and Costello 2012; Costello et al. 2015a). OBIS directly contributes to several international activities, such as the UN Convention on Biological Diversity (for the identification of Ecologically or Biologically Significant Areas), the UN Food and Agriculture Organization (for the identification of Vulnerable Marine Ecosystems), the UN World Ocean Assessment, and the Global Environment Fund Transboundary Water Assessment. The Global Biodiversity Information Facility (GBIF) and OBIS use the same data standards and data sharing protocol (i.e., GBIF's Integrated Publishing Toolkit). GBIF contains all OBIS and additional marine data (e.g., Costello et al. 2013d). Most data in OBIS are available from the north-west and north-east Atlantic, South Africa and New Zealand, and some other locations (Fig. 6.2). The potential of data published through OBIS for time-series analysis was highlighted in a recent global scale analysis (Dornelas et al. 2014).

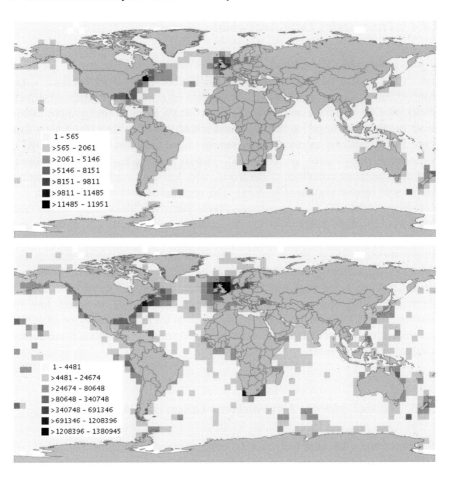

Fig. 6.2 A global map of the number of sampling days (*upper panel*) and sampling records (*lower panel*) in OBIS (downloaded October 2014) in 5-degree latitude longitude cells

An important development that will aid time series analysis, ecological niche modelling and climate change studies is currently underway as part of a two-year IODE project called 'Expanding OBIS with environmental data' (OBIS-ENV-DATA), which started in March 2015. The project is working on a solution to retain data in biological datasets that hold more than just species occurrence data, such as providing environmental and ecological context and data. The new approach will be based on the new Darwin Event Core and a modified 'MeasurementorFact' extension. The major change is that it will bring OBIS from a purely species occurrence database to one that can handle hierarchical sampling event structure with additional environmental and biometric measurements as well as details on the nature of the observations, measurements, and data collection methods, including equipment, data processing and sampling efforts.

6.4.4 Time-Series Data Availability

At present there are 20 monitoring programmes that have targeted species for more than five years that have been entered in OBIS. The focus of these efforts is on economically valuable and charismatic species (e.g., Antarctic krill, American lobster, marine mammals and seabirds). By contrast there are many more monitoring programmes targeting marine communities that have data for at least five years; 216 community monitoring programmes have uploaded their data to OBIS. When these programmes are combined, 16,616 stations have been monitored, encompassing most coastlines of the world, with less data available in developing countries or remote regions (Fig. 6.3a). The accumulation of time-series data has been exponential (Fig. 6.3b), reflecting both increasing monitoring efforts and global coordination. There may be an increasing willingness of scientists and institutions to share their data, with programmes such as the European Groundfish Survey showing up as being an important source of biodiversity data in the mid-1990s on a global scale (Fig. 6.3b). Even so, there are relatively fewer new stations that are being added to OBIS in comparison to the number of stations where monitoring surveys have ceased (Fig. 6.3b), leading to a net loss of time-series from OBIS in this decade. Explanations for this trend may be delays in data deposition, and/or perhaps the scope of specific monitoring efforts is increasing in extent and coordination.

6.4.5 Global Marine Environment Datasets (GMED)

GMED is a compilation of more than 60 publicly available climatic, biological and geophysical environmental layers featuring present, past and future environmental conditions (Basher et al. 2015). Marine biologists increasingly utilise geo-spatial techniques with modelling algorithms to visualise and predict species biodiversity at a global scale. Marine environmental datasets available for species distribution modelling (SDM) have different spatial resolutions and are frequently provided in assorted file formats. This makes data assembly one of the most time-consuming parts of any study using multiple environmental layers for biogeography visualisation or SDM applications. GMED covers the widest available range of environmental layers from in situ measured, remote-sensed, and modelled datasets for a broad range of quantitative environmental variables from the surface to the deepest part of the ocean. It has a uniform spatial extent, high-resolution land mask (to eliminate land areas in the marine regions), and high spatial resolution (5 arc-minute, ca. 9.2 km near equator). The free online availability of GMED enables rapid map overlay of species of interest (e.g., endangered or invasive) against different environmental conditions of the past, present and the future, and expedites mapping distribution ranges of species using popular SDM algorithms (e.g., Basher et al. 2014a, 2015; Basher and Costello 2016).

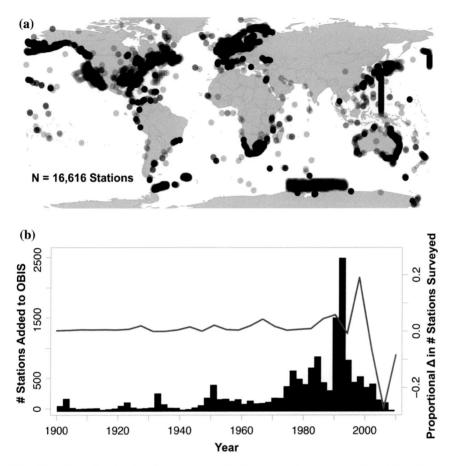

Fig. 6.3 a Map of station locations where monitoring surveys have been conducted for at least five years presently held in OBIS. **b** The *left axis* illustrates the number of stations where time-series data has been collected versus the year of the first survey (*black* histogram). Overall there has been an increase in monitoring. However, since the start of this century there has been a relative decrease in the number of stations being added to OBIS, evidenced by (*right axis*) the proportional difference in the number of new stations being added to OBIS versus those reaching completion. The *blue line* indicates where more monitoring stations were gained than lost from OBIS in a given year, while the *red line* indicates a loss

6.5 Data Analysis

Although marine biodiversity data analysis requires its own taxonomic, geographic and environmental information context, such as provided by WoRMS, Marine-Regions, OBIS, and GMED, the methods of data analysis are similar to biodiversity in other environments. The data are categorical (i.e., species, habitats, biotopes), numerical (e.g., species abundance, cover, biomass), and cartographic. Thus metrics of 'biodiversity' include species richness and abundance, phylogenetic structure (e.g., taxonomic distinctness; Warwick and Clarke 1998), indicator species, habitat and/or biotope richness in an area (Costello 2001). Data may be presented on maps, graphs, tables and as matrices (e.g., Figs. 6.1 and 6.2). Numerous software tools are available for this analysis, including PRIMER-E (www.primer-e.com), PAST (http://folk.uio.no/ohammer/past), MODESTR (www.ipez.es/ModestR), SAGA (System for Automated Geoscientific Analyses; www.saga-gis.org) and DIVA-GIS (www.diva-gis.org). The open-source software R has the benefit that the analytical process is documented and can be published to aid reproducibility of the analyses.

The massive size of modern datasets, such as in OBIS and GMED, can lead to a new set of difficulties in analysis and interpretation. These difficulties include processing times that can exceed the capabilities of extant computers, propagation of undetected errors, unfamiliarity with analytical assumptions (e.g., spatial auto-correlation), and difficulties in visualisation (Edgar et al. 2016). Fortunately, big-data techniques applied in other fields, such as high-performance and parallel computing, are helping to solve many of these problems. In addition packages to overcome significant challenges in compiling large datasets and maintaining these data through time are being improved. For example, the R package 'taxize' (Chamberlain and Szocs 2013), which relies on accessing freely available and accurate information on species taxonomy, including from WoRMS. This emphasises the benefits of scientists and institutes publishing monitoring data in order to advance our understanding of biodiversity change.

6.6 Discussion

Global marine biological databases are well-established for quality assurance of species nomenclature and associated information (WoRMS) and distribution data (OBIS) (Costello and Wieczorek 2014; Costello et al. 2015b). The coverage and quality of global marine environmental layers improves each year through a combination of remotely sensed, in situ, and modelling data. These layers and maps of marine regions are also freely available online at GMED and marineregions.org. Species trait information is being added to WoRMS, and more sample information can be added to OBIS so users can select datasets suitable for their purposes. The mapping of available data in OBIS shows how more sampling has been conducted in northern hemisphere and coastal environments compared to open-ocean, deep-sea and developing countries (Fig. 6.3a). However, because neither biodiversity nor human impacts are homogenously distributed, neither should it be expected that global sampling programmes will be. Sampling of particular guilds of biodiversity should thus be stratified to represent its spatial variation.

A major obstacle to engaging more scientists and citizens in recording marine biodiversity is the availability of guides to the identification of species. Generally, these are only widely available for vertebrates (Costello et al. 2006, 2015b). To identify invertebrates often requires numerous papers to be obtained, sometimes in different languages. The most useful publications are reviews of the taxonomy of particular taxa in a region that include images, drawings and keys that synthesise information on many species (Costello et al. 2013a, b, 2014a, b). The best long-term solution would be an online, pictorial, guide to all marine species accessible to people in several languages and scripts (Costello et al. 2015b).

All methods have their biases and this needs to be recognised in data analysis rather than assume a conventional method is representative of all biodiversity. In fact, it may be that pooling different sampling processes to gain insights into different aspects of biodiversity will create the most comprehensive understanding of how biodiversity is changing in the ocean. Methods must be selected that are 'best fit for the purpose' and limitations imposed by costs and environmental conditions should be considered in the interpretation of the samples obtained. Standardised methods have the advantage of apparent comparability between study locations and over time. However, this assumes the behaviour of animals is the same between species, and even within a species between locations and over time. This is not necessarily the case. Being ectothermal, fish appetite and activity is strongly temperature dependent (e.g., Darwall et al. 1993; Costello et al. 1995). Thus seasonal changes in the catch of fish and other mobile species may not reflect fish abundance or changing distribution, but rather their activity. Animal behaviour also needs to be considered. For example, fish are wary of people in places they are fished, especially spear-fished. However, in marine reserves they lose this fear and can be approached closely (Costello 2014). Where mammals, birds, fish and other animals may be fed, they become attracted to people. This mirrors the behaviour of animals on land. Thus not only do the physical features of sampling methods need to be considered in terms of bias, so do the behavioural responses of animals.

More recently developed methods, such as using photography, hydrophones, and tagging, avoid killing the species of interest. Artificial substrata, light-traps, hook and line, and traps can avoid killing unwanted by-catch species. However, most netting and trawling methods result in by-catch, and seabed dredging and trawling also damage habitat. It seems likely that scientific sampling will come under increasing ethical pressure to minimise habitat damage, by-catch and stress to species, especially in nature conservation areas and where species are threatened. Thus new in situ observation methods such as still and video image capture, sea-floor observatories, and sensors, are likely to become more important because they cause less disturbance of biodiversity.

In addition to the CPR, RLS and GEOHAB programmes reviewed here, new networking initiatives, marine biodiversity observation networks (mBON) in the USA (Muller-Karger et al. 2014), marine station networks and related organisations (Costello et al. 2015c), and groups of scientists interested in the biological and ecological effects of climate change, may establish globally coordinated marine biodiversity monitoring programmes. In addition, several international ocean observing systems, initially focused on the collection of physical and chemical ocean data, are now including biological data as well. These are comprised of the Australian led Integrated Marine Observing Systems (IMOS; www.imos.org.au) and the Southern Ocean Observing System (SOOS; www.soos.au), and NOAA's

Integrated Ocean Observing System (IOOS; www.ioos.noaa.gov). Some international efforts have a regional focus. For example, the Circumpolar Biodiversity Monitoring Program (CBMP), under the auspices of the Arctic Council, has an Arctic Marine Biodiversity Monitoring Plan (www.caff.is/marine).

Dornelas et al. (2014) compiled the first global time-series data base for analysis of trends in marine biodiversity. As described earlier, biodiversity data are available for many taxa and regions of the world and the challenge remains to access, compile and curate these data. A major obstacle is therefore not only the difficulty in maintaining funding for monitoring or data synthesis efforts, but fostering motivation for institutes and scientists to publish their data and overcoming communication and cultural differences. Building collaborative networks may be one means to begin to surmount these challenges to collate data across scientists, institutions, and data repositories. While efforts to collate the data that has been collected by the global monitoring community is certainly the best hope for generating historical knowledge, purpose-built global biodiversity platforms are fundamental for ensuring the capacity to track biodiversity change into the future. For example, MarineGEO (Duffy 2014) is establishing observatories where multiple components of biodiversity, including benthic and pelagic communities and food webs, will be monitored using globally standardised methods and experimentation, including artificial substrata (e.g., ARMS). Associated initiatives focus on global studies on seagrass (http://zenscience.org/about-zen; Reynolds et al. 2014) and kelp (www.kelpecosystems.org) habitats. Standard methods for these habitats have been published (e.g., Edgar et al. 2001; Davies et al. 2001). Such projects may utilise the Zooniverse platform for citizen science crowd sourcing (www.zooniverse.org).

Global sampling of surface water marine microbes is also underway utilising genomic methods, including synchronised sampling of hundreds of stations on 'Ocean Sampling Day' (www.microb3.eu/osd) (e.g., Davies et al. 2012a, b; Kopf et al. 2015). These and related research into molecular indicators may fill gaps that complement more conventional metrics of biodiversity (Leray and Knowlton 2015). Although there are issues to be resolved in the interpretation of DNA found in the environment (eDNA), including contamination, accuracy of matching results to species, and uncertainty about live versus dead material, it may prove invaluable in detecting rare and/or microscopic species that are otherwise hard to sample (Thomsen and Willerslev 2015).

There are two established and several emerging globally coordinated marine biodiversity monitoring programmes, covering surface plankton (CPR), mobile rocky and coral reef fauna (RLS), seagrass and kelp habitats, and pelagic microbes. There are similar sampling methods used internationally for other guilds of species; including mammals, whale sharks and birds; small fish and crustaceans in fishery trawls; macro-invertebrate infauna of coastal sediments; and sessile and sedentary biota on rocky seashores. For example, programmes such as the ICES North Sea Benthos Survey (e.g., Duineveld et al. 1991; Basford et al. 1993) and NaGISA (Benedetti-Cecchi et al. 2010; Cruz-Motta et al. 2010; Konar et al. 2010; Pohle et al. 2011; Miloslavich et al. 2013) could be continued and expanded internationally. NaGISA was one of several projects within the decade-long Census of Marine Life, the largest global collaboration in marine biology covering coastal to deep-sea, and polar to tropical environments, and which established OBIS (O'Dor et al. 2012). Thus opportunities exist to design globally standardised programmes for these ecological guilds that would be comparable with historic data. For example, the IOC–UNESCO endorsed IndiSeas (www.indiseas.org) has begun to provide indicators of biodiversity (including ecosystem health) related to fisheries and environment.

Gaps in time-series may be partly filled by using microfossils from sediment cores and specimen collections in museums, and also by revisiting places sampled in the past without continuous time-series. In addition, video cameras (baited and unbaited) are widely used for recording scavenging megafauna from coastal to deep-sea habitats (e.g., Costello et al. 2005). Gaps in these programmes include the species rich epi-benthic crustaceans and molluscs which together comprise one quarter of all marine species (Appeltans et al. 2012). However, the use of artificial substrata such as ARMS and light-traps may be able to fill this gap. Additional guilds that could be considered for monitoring include sediment meiofauna and parasites.

A common concern in launching global initiatives is both the start-up and long-term funding (Costello et al. 2014c). It is notable that the CPR, RLS, WoRMS and FishBase established their own legal organisations to 'own' their initiatives, even though they are largely funded by government and hosted by particular institutes. This community ownership may address issues of financial liability of individuals and their institutions, ownership of intellectual property, and perceptions of who benefits from the research. The establishment of global programmes must consider these and other issues so as to maximise the likelihood of support from individual scientists, host institutions and governments in the long term (Costello et al. 2014c).

References

Andréfouët, S., Costello, M. J., Faith, D. P., Ferrier, S., Geller, G. N., Höft, R., et al. (2008a). *The GEO biodiversity observation network concept document*. Geneva, Switzerland: GEO—Group on Earth Observations.

Andréfouët, S., Costello, M. J., Rast, M., & Sathyendranath, S. (2008b). Earth observations for marine and coastal biodiversity. *Remote Sensing of Environment, 112*, 3297–3299.

Andréfouët, S., Muller-Karger, F. E., Hochberg, E. J., Hu, C., & Carder, K. L. (2001). Change detection in shallow coral reef environments using Landsat 7 ETM+ data. *Remote Sensing of Environment, 78*, 150–162.

Anonymous. (2012). *Coastal and marine ecological classification standard*. Chalreston SC: Federal Geographic Data Committee, FGDC-STD-018-2012. www.csc.noaa.gov/digitalcoast/_/pdf/CMECS_Version%20_4_Final_for_FGDC.pdf Cited 8 February 2016.

Anonymous. (2014). *EUNIS habitat classification*. Copenhagen: European Environment Agency. http://www.eea.europa.eu/themes/biodiversity/eunis/eunis-habitat-classification   Cited   8 February 2016.

American Public Health Association (APHA), American Water Works Association, and Water Environment Federation. (2007) *Standard methods for the examination of water and wastewater*. American Public Health Association, American Water Works Association, and Water Environment Federation, Washington, DC., USA.

Appeltans, W., Ahyong, S.T., Anderson, G., Angel, M. V., Artois, T., & Bailly, N. et al. (2012). The magnitude of global marine species diversity. *Current Biology 22*, 2189–2202.

Ash, N., Jürgens, N., Leadley, P., Alkemade, R., Araújo, M. B., & Asner, G. P. (2009). *bioDISCOVERY: Assessing, monitoring and predicting biodiversity*. DIVERSITAS Report No. 7.

Atalah, J., Costello, M. J., & Anderson, M. (2007a). Temporal variability and intensity of grazing: A mesocosm experiment. *Marine Ecology Progress Series, 341*, 15–24.

Atalah, J., Otto, S., Anderson, M., Costello, M. J., Lenz, M., & Wahl, M. (2007b). Temporal variance of disturbance did not affect diversity and structure of a marine fouling community in north-eastern New Zealand. *Marine Biology, 153*, 199–211.

Atkinson, L. J., Leslie, R. W., Field, J. G., & Jarre, A. (2011). Changes in demersal fish assemblages on the west coast of South Africa, 1986–2009. *African Journal of Marine Science, 33*, 157–170.

Babin, M., Cullen, J., & Roesler, C. (Eds.). (2008). *Real time observations systems for marine ecosystem dynamics and harmful algal blooms: Theory, instrumentation and modelling*. Paris, France: UNESCO Publishing.

Barrett, N. S., Buxton, C. D., & Edgar, G. J. (2009). Changes in invertebrate and macroalgal populations in Tasmanian marine reserves in the decade following protection. *Journal of Experimental Marine Biology and Ecology, 370*, 104–119.

Basford, D. J., Eleftheriou, A., Davies, I. M., Irion, G., & Soltwedel, T. (1993). The ICES North Sea benthos survey: The sedimentary environment. *ICES Journal of Marine Science: Journal du Conseil, 50*, 71–80.

Basher, Z. & Costello, M. J. (2016). The past, present and future distribution of a deep-sea shrimp in the Southern Ocean. *PeerJ, 4*, e1713 https://doi.org/10.7717/peerj.1713

Basher, Z., Bowden, D. A., & Costello, M. J. (2014a). Diversity and distribution of deep-sea shrimps in the Ross Sea region of Antarctica. *PLoS ONE, 9*, e103195.

Basher, Z. & Costello, M. J. (2014b). Chapter 5.22: Shrimps (Crustacea: Decapoda). In C. De Broyer, P. Koubbi, H. J. Griffiths, B. Raymond, C. d'Udekem d'Acoz, A. Van de Putte, B. Danis, B. David, S. Grant, J. Gutt, C. Held, G. Hosie, F. Huettemann, A. Post, Y. Ropert-Coudert (Eds.), *Biogeographic Atlas of the Southern Ocean* (pp. 190–194). Cambridge, UK: Scientific Committee on Antarctic Research.

Basher, Z., Bowden, D. A. & Costello, M. J. (2015). *Global marine environment dataset (GMED)*. World Wide Web electronic publication. Version 1.0 (Rev.01.2014). http://gmed.auckland.ac.nz Cited 4 February 2015.

Benedetti-Cecchi, L., Iken, K., Konar, B., Cruz-Motta, J., Knowlton, A., et al. (2010). Spatial relationships between polychaete assemblages and environmental variables over broad geographical scales. *PLoS ONE, 5*, e12946.

Benoît, H. P., & Swain, D. P. (2003). Accounting for length-and depth-dependent diel variation in catchability of fish and invertebrates in an annual bottom-trawl survey. *ICES Journal of Marine Science, 60*, 1298–1317.

Bianchi, G., Gislason, H., Graham, K., Hill, L., Jin, X., Koranteng, K., et al. (2000). Impact of fishing on size composition and diversity of demersal fish communities. *ICES Journal of Marine Science: Journal du Conseil, 57*, 558–571.

Boero, F., Dupont, S., & Thorndyke, M. (2015). Make new friends, but keep the old: Towards a transdisciplinary and balanced strategy to evaluate good environmental status. *Journal of the Marine Biological Association of the United Kingdom, 95*, 1069–1070.

Bork, P., Bowler, C., de Vargas, C., Gorsky, G., Karsenti, E., & Wincker, P. (2015). Tara Oceans studies plankton at planetary scale. *Science, 348*, 873.

Boxshall, G., Mees, J., Costello, M. J., Hernandez, F., Gofas, S., & Hoeksema, B. W. (2015). *World register of marine species*. Available via VLIZ. http://www.marinespecies.org Cited 23 December 2015.

Bruno, J. F., & Selig, E. R. (2007). Regional decline of coral cover in the indo-pacific: Timing, extent, and subregional comparisons. *PLoS ONE, 2*, e711.

Buttigieg, P. L., & Ramette, A. (2014). A guide to statistical analysis in microbial ecology: A community-focused, living review of multivariate data analyses. *FEMS Microbiology Ecology, 90*, 543–550.

Chamberlain, S. & Szocs, E. (2013). Taxize—taxonomic search and retrieval in R. *F1000Research 2*, 191.

Claus, S., De Hauwere, N., Vanhoorne, B., Deckers, P., Souza Dias, F., Hernandez, F., et al. (2014). Marine regions: Towards a global standard for georeferenced marine names and boundaries. *Marine Geodesy, 37*, 99–125.

Connor, D. W., Allen, J. H., Golding, N., Howell, K. L., Lieberknecht, L. M., & Northen, K. O. et al. (2004). *The marine habitat classification for Britain and Ireland version 04.05*. Peterborough, UK: Joint Nature Conservation Committee (JNCC). ISBN 1 861 07561 8 (internet version).

Connor, D. W., Gilliland, P. M., Golding, N., Robinson, P., Todd, D., & Verling, E. (2006). *UKSeaMap: The mapping of seabed and water column features of UK seas*. Peterborough, UK: Joint Nature Conservation Committee.

Costello, M. J. (1992). Abundance and spatial overlap of gobies in Lough Hyne, Ireland. *Environmental Biology of Fishes, 33*, 239–248.

Costello, M. J. (2000). Developing species information systems: the European register of marine species. *Oceanography, 13*(3), 48–55.

Costello, M. J. (2001). To know, research, manage, and conserve marine biodiversity. *Océanis, 24*(4), 25–49.

Costello, M. J. (2009). Distinguishing marine habitat classification concepts for ecological data management. *Marine Ecology Progress Series, 397*, 253–268.

Costello, M. J. (2014). Long live marine reserves: A review of experiences and benefits. *Biological Conservation, 176*, 289–296.

Costello, M. J. (2015). Biodiversity: The known, unknown and rates of extinction. *Current Biology, 25*(9), 368–371.

Costello, M. J., & Thrush, S. F. (1991). Colonization of artificial substrata as a multi-species bioassay of marine environmental quality. In D. W. Jeffrey & B. Madden (Eds.), *Bioindicators and environmental management* (pp. 401–418). London, UK: Academic Press.

Costello, M. J., & Emblow, C. (2005). A classification of inshore marine biotopes. In J. G. Wilson (Ed.), *The intertidal ecosystem: The value of Ireland's shores* (pp. 25–35). Dublin, Ireland: Royal Irish Academy.

Costello, M. J., & Vanden Berghe, E. (2006). "Ocean biodiversity informatics" enabling a new era in marine biology research and management. *Marine Ecology Progress Series, 316*, 203–214.

Costello, M. J., & Baker, C. S. (2011). Who eats sea meat? Expanding human consumption of marine mammals. *Biological Conservation, 144*, 2745–2746.

Costello, M. J., & Myers, A. A. (1996). Turnover of transient species as a contributor to the richness of a stable amphipod (Crustacea) fauna in a sea inlet. *Journal of Experimental Marine Biology and Ecology, 202*, 49–62.

Costello, M. J., & Wieczorek, J. (2014). Best practice for biodiversity data management and publication. *Biological Conservation, 173*, 68–73.

Costello, M. J., Darwall, W. R. & Lysaght S. (1995). Activity patterns of north European wrasse (Pisces, Labridae) and precision of diver survey techniques. In A. Eleftheriou, A. D. Ansell, C. J. Smith (Eds.), *Biology and ecology of shallow coastal waters. Proceedings of the 28th European Marine Biology Symposium*, IMBC, Hersonissos, Crete 1993 (pp. 343–350). Fredensborg, Denmark: Olsen and Olsen Publ..

Costello, M. J., Emblow, C. & White, R. (Eds.). (2001). European register of marine species. A check-list of marine species in Europe and a bibliography of guides to their identification. *Patrimoines Naturels 50*, 1–463.

Costello, M. J., Pohle G., & Martin A. (2004). *Evaluating biodiversity in marine environmental assessments.* Research and Development Monograph Series 2001, Canadian Environmental Assessment Agency, Ottawa, Canada. http://www.collectionscanada.gc.ca/webarchives/20071213100057/www.ceaa.gc.ca/015/001/019/print-version_e.htm

Costello, M. J., McCrea, M., Freiwald, A., Lundälv, T., Jonsson, L., Bett, B. J., et al. (2005). Role of cold-water *Lophelia pertusa* coral reefs as fish habitat in the NE Atlantic. In A. Freiwald & J. M. Roberts (Eds.), *Cold-water corals and ecosystems* (pp. 771–805). Berlin Heidelberg, Germany: Springer.

Costello, M. J., Emblow, C. S., Bouchet, P., & Legakis, A. (2006). European marine biodiversity inventory and taxonomic resources: State of the art and gaps in knowledge. *Marine Ecology Progress Series, 316*, 257–268.

Costello M. J., Stocks K., Zhang Y., Grassle J. F. & Fautin D. G. (2007). About the Ocean biogeographic information system. Retrieved from http://hdl.handle.net/2292/236

Costello, M. J., Cheung, A., & De Hauwere, N. (2010a). Topography statistics for the surface and seabed area, volume, depth and slope, of the world's seas, oceans and countries. *Environmental Science and Technology, 44*, 8821–8828.

Costello, M. J., Coll, M., Danovaro, R., Halpin, P., Ojaveer, H., & Miloslavich, P. (2010b). A census of marine biodiversity knowledge, resources and future challenges. *PLoS ONE, 8*, e12110.

Costello, M. J., May, R. M., & Stork, N. E. (2013a). Can we name Earth's species before they go extinct? *Science, 339*, 413–416.

Costello, M. J., May, R. M., & Stork, N. E. (2013b). Response to comments on "can we name Earth's species before they go extinct?". *Science, 341*, 237.

Costello, M. J., Bouchet, P., Boxshall, G., Fauchald, K., Gordon, D. P., Hoeksema, B. W., et al. (2013c). Global coordination and standardisation in marine biodiversity through the World Register of Marine Species (WoRMS) and related databases. *PLoS ONE, 8*(1), e51629.

Costello, M. J., Michener, W. K., Gahegan, M., Zhang, Z.-Q., & Bourne, P. (2013d). Data should be published, cited and peer-reviewed. *Trends in Ecology & Evolution, 28*(8), 454–461.

Costello, M. J., Appeltans, W., Bailly, N., Berendsohn, W. G., de Jong, Y., Edwards, M., et al. (2014a). Strategies for the sustainability of online open-access biodiversity databases. *Biological Conservation, 173*, 155–165.

Costello, M. J., Houlding, B., & Wilson, S. (2014b). As in other taxa, relatively fewer beetles are being described by an increasing number of authors: Response to Löbl & Leschen. *Systematic Entomology, 39*, 395–399.

Costello, M. J., Houlding, B. & Joppa, L. (2014c). Further evidence of more taxonomists discovering new species, and that most species have been named: response to Bebber *et al.* (2014). *New Phytologist 202*, 739–740.

Costello, M. J., Claus, S., Dekeyzer, S., Vandepitte, L., Tuama, É. Ó., & Lear, D. et al. (2015a). Biological and ecological traits of marine species. *PeerJ 3*, e1201.

Costello, M. J., Vanhoorne, B., & Appeltans, W. (2015b). Conservation of biodiversity through taxonomy, data publication and collaborative infrastructures. *Conservation Biology, 29*(4), 1094–1099.

Costello, M. J., Archambault, P., Chavanich, S., Miloslavich, P., Paterson, D. M., Phang, S.-W., et al. (2015c). Organizing, supporting and linking the world marine biodiversity research community. *Journal of the Marine Biological Association of the United Kingdom, 95*(3), 431–433. doi:10.1017/S0025315414001969

Cruz-Motta, J. J., Miloslavich, P., Palomo, G., Iken, K., Konar, B., Pohle, G., et al. (2010). Patterns of spatial variation of assemblages associated with intertidal rocky shores: A global perspective. *PLoS ONE, 12,* e14354.

Daan, N., Gislason, H., Pope, J. G., & Rice, J. C. (2005). Changes in the North Sea fish community: Evidence of indirect effects of fishing? *ICES Journal of Marine Science, 62*(2), 177–188.

Darwall, W. R. T., Costello, M. J., Donnelly, R., & Lysaght, S. (1993). Implications of life history strategies for a new wrasse fishery. *Journal of Fish Biology, 41B,* 111–123.

Davies, J., Baxter, J., Bradley, M., Connor, D., Khan, J., & Murray, E. et al. (2001), *Marine monitoring handbook* (405 p.). Peterborough: Joint Nature Conservation Committee.

Davies, N., Field, D., & Genomic Observatories Network. (2012a). Sequencing data: A genomic network to monitor Earth. *Nature, 481*(7380), 145.

Davies, N., Meyer, C., Gilbert, J. A., Amaral-Zettler, L., Deck, J., Bicak, M., et al. (2012b). A call for an international network of genomic observatories (GOs). *GigaScience, 1*(1), 5.

De Monte, S., Soccodato, A., Alvain, S., & d'Ovidio, F. (2013). Can we detect oceanic biodiversity hotspots from space? *ISME Journal, 7,* 2054–2056.

Doherty, P. J. (1987). Light-traps: Selective but useful devices for quantifying the distributions and abundances of larval fishes. *Bulletin of Marine Science, 41*(2), 423–431.

Dornelas, M., Gotelli, N. J., McGill, B., Shimadzu, H., Moyes, F., Sievers, C., et al. (2014). Assemblage time series reveal biodiversity change but not systematic loss. *Science, 344*(6181), 296–299.

Duffy, J. E. (2014). Sustaining coastal resilience. *Pan-European Networks: Science and Technology, 12,* 1–4.

Duineveld, G. C. A., Künitzer, A., Niermann, U., De Wilde, P. A. W. J., & Gray, J. S. (1991). The macrobenthos of the North Sea. *Netherlands Journal of Sea Research, 28*(1–2), 53–65.

Edgar, G. J., & Barrett, N. S. (1999). Effects of the declaration of marine reserves on Tasmanian reef fishes, invertebrates and plants. *Journal of Experimental Marine Biology and Ecology, 242,* 107–144.

Edgar, G. J., & Stuart-Smith, R. D. (2009). Ecological effects of marine protected areas on rocky reef communities: a continental-scale analysis. *Marine Ecology Progress Series, 388,* 51–62.

Edgar, G. J., & Stuart-Smith, R. D. (2014). Systematic global assessment of reef fish communities by the Reef Life Survey program. *Scientific Data, 1*(140007), 1–8.

Edgar, G. J., Orth, R. J., & Mukai, H. (2001). Fish, crabs, shrimps and other large mobile epibenthos: Measurement methods for their biomass and abundance in seagrass. In F. T. Short & R. G. Coles (Eds.), *Global seagrass research methods* (pp. 255–270). Amsterdam, The Netherlands: Elsevier.

Edgar, G. J., Stuart-Smith, R. D., Willis, T. J., Kininmonth, S., Banks, S., Barrettk, N. S., et al. (2014). Global conservation outcomes depend on marine protected areas with five key features. *Nature, 506,* 216–220.

Edgar, G. J., Bates, A. E., Bird, T. J., Jones, A. H., Kininmonth, S., Stuart-Smith, R. D., et al. (2016). New approaches to marine conservation through the scaling up of ecological data. *Annual Review of Marine Science, 8,* 435–461.

Edwards, M., Beaugrand, G., Hays, G. C., Koslow, J. A., & Richardson, A. J. (2010). Multi-decadal oceanic ecological datasets and their application in marine policy and management. *Trends in Ecology & Evolution, 25,* 602–610.

Edwards, M., Helaouet, P., Johns, D. G., Batten, S., Beaugrand, G., & Chiba, S. et al. (2012). Global marine ecological status report: Results from the global CPR survey 2010/2011. *SAHFOS Technical Report* 9, 1–40. Plymouth, U.K.

Eleftheriou, A. (Ed.). (2013). *Methods for the study of marine benthos.* Chichester, UK: Wiley-Blackwell.

Elliott, M., & Hemingway, K. (Eds.). (2002). *Fishes in estuaries.* Oxford, UK: Blackwell Science.

English, S. S., Wilkinson, C. C. & Baker, V. V. (1994). *Survey manual for tropical marine resources*. Townsville, Australia: Australian Institute of Marine Science (AIMS).

Facon, M., Pinault, M., Obura, D., Pioch, S., Pothin, K., Bigot, L., et al. (2016). A comparative study of the accuracy and effectiveness of line and point intercept transect methods for coral reef monitoring in the southwestern Indian Ocean islands. *Ecological Indicators, 60*, 1045–1055.

Francisco-Ramos, V., & Arias-González, J. E. (2013). Additive partitioning of coral reef fish diversity across hierarchical spatial scales throughout the Caribbean. *PLoS ONE, 8*(10), e78761.

Fraser, H. M., Greenstreet, S. P. R., & Piet, G. J. (2007). Taking account of catchability in groundfish survey trawls: Implications for estimating demersal fish biomass. *ICES Journal of Marine Science, 64*, 1800–1819.

Fraser, H. M., Greenstreet, S. P. R., Fryer, R. J., & Piet, G. J. (2008). Mapping spatial variation in demersal fish species diversity and composition in the North Sea: accounting for species- and size-related catchability in survey trawls. *ICES Journal of Marine Science, 65*, 531–538.

Froese, R. & Pauly, D. (2015). *Fishbase*. World Wide Web database. Accessed at http://www.fishbase.org on 19th February 2015.

GBO. (2014). Global biodiversity outlook 4. Secretariat of the convention on biological diversity. Montréal. 155 p. https://www.cbd.int/GBO4/

Giongo, A., Crabb, D. B., Davis-Richardson, A. G., Chauliac, D., Mobberley, J. M., Gano, K. A., et al. (2010). PANGEA: Pipeline for analysis of next generation amplicons. *The ISME journal, 4*(7), 852–861.

Gotelli, N. J., & Colwell, R. K. (2001). Quantifying biodiversity: Procedures and pitfalls in the measurement and comparison of species richness. *Ecology Letters, 4*(4), 379–391.

Haegeman, B., Hamelin, J., Moriarty, J., Neal, P., Dushoff, J., & Weitz, J. S. (2013). Robust estimation of microbial diversity in theory and in practice. *The ISME Journal, 7*, 1092–1101.

Hallegraeff, G. M., Anderson, D. M. & Cembella, A. D. (Eds.). (2003). *Manual on harmful marine microalgae* (794 p.). Paris: UNESCO Publishing.

Harris, S. A., Shears, N. T., & Radford, C. A. (2015). Ecoacoustic indices as proxies for biodiversity on temperate reefs. *Methods in Ecology and Evolution*.

Hiscock, K. (2014). *Marine biodiversity conservation: A practical approach*. Abingdon: Routledge. 289 pp.

Hodgson, G. (1999). A global assessment of human effects on coral reefs. *Marine Pollution Bulletin, 38*(5), 345–355.

Holmes, J. M. C., & O'Connor, J. P. (1988). A portable light-trap for collecting marine crustaceans. *Journal of the Marine Biological Association of the United Kingdom, 68*(2), 235–238.

Hungerford, H. B., Spangler, P. J. & Walker, N. A. (1955). Subaquatic light traps for insects and other animal organisms. *Transactions of the Kansas Academy of Science (1903–) 58*(3), 387–407.

Hunt, G., Cronin, T. M., & Roy, K. (2005). Species–energy relationship in the deep sea: A test using the quaternary fossil record. *Ecology Letters, 8*, 739–747.

Inniss, L., Simcock, A., Ajawin, A. Y., Alcala, A. C., Bernal, P., & Calumpong, H. P. et al. (2016). The first global integrated marine assessment. United Nations. Accessed at on 5th February 2016 http://www.un.org/Depts/los/global_reporting/WOA_RegProcess.htm

IODE Steering Group for OBIS (SG-OBIS). (2013). Reports of meetings of experts and equivalent bodies. Third session, IOC Project Office for IODE, 4–6 December 2013, IOC/IODE-SG-OBIS-III/3. UNESCO, Oostende. 40 p.

Jackson, J., Donovan, M., Cramer, K., & Lam, V. (2014). *Status and trends of Caribbean coral reefs: 1970–2012* (304 p.). Washington, D.C: Global Coral Reef Monitoring Network.

Karlson B., Cusack C. & Bresnan, E. (Eds.). (2010). *Microscopic and molecular methods for quantitative phytoplankton analysis*. Paris: Intergovernmental Oceanographic Commission of UNESCO. 110 p.

Kingsford, M., & Battershill, C. (Eds.). (1998). *Studying temperate marine environments: a handbook for ecologists* (355 p.). Canterbury University Press.

Kohler, K. E., & Gill, S. M. (2006). Coral point count with excel extensions (CPCe): A Visual Basic program for the determination of coral and substrate coverage using random point count methodology. *Computers & Geosciences, 32*, 1259–1269.

Konar, B., Iken, K., Cruz-Motta, J. J., Benedetti-Cecchi, L., Knowlton, A., Pohle, G., et al. (2010). Current patterns of macroalgal diversity and biomass in northern hemisphere rocky shores. *PLoS ONE, 5*(10), e13195.

Kopf, A., Bicak, M., Kottmann, R., Schnetzer, J., Kostadinov, I., Lehmann, K., et al. (2015). The ocean sampling day consortium. *GigaScience, 4*(1), 1–5. doi:10.1186/s13742-015-0066-5

Kotwicki, S., De Robertis, A., von Szalay, P., & Towler, R. (2009). The effect of light intensity on the availability of walleye pollock (*Theragra chalcogramma)* to bottom trawl and acoustic surveys. *Canadian Journal of Fisheries and Aquatic Sciences, 66*(6), 983–994.

Leleu, K., Remy-Zephir, B., Grace, R., & Costello, M. J. (2012). Mapping habitat change after 30 years in a marine reserve shows how fishing can alter ecosystem structure. *Biological Conservation, 155*, 193–201.

Leray, M., & Knowlton, N. (2015). DNA barcoding and metabarcoding of standardized samples reveal patterns of marine benthic diversity. *Proceedings of the National Academy of Sciences*, 201424997.

Leujak, W., & Ormond, R. F. G. (2007). Comparative accuracy and efficiency of six coral community survey methods. *Journal of Experimental Marine Biology and Ecology, 351*(1), 168–187.

McGeoch, A. M., Genovesi, P., Bellingham, P. J., Costello, M. J., McGrannachan, C., & Sheppard, A. (2016). Prioritizing species, pathways, and sites to achieve conservation targets for biological invasion. *Biological Invasions, 18*(2), 299–314.

Meekan, M. G., Wilson, S. G., Halford, A., & Retzel, A. (2001). A comparison of catches of fishes and invertebrates by two light trap designs, in tropical NW Australia. *Marine Biology, 139*(2), 373–381.

Miller, T. J. (2013). A comparison of hierarchical models for relative catch efficiency based on paired-gear data for US Northwest Atlantic fish stocks. *Canadian Journal of Fisheries and Aquatic Sciences, 70*(9), 1306–1316.

Miloslavich, P., Cruz-Motta, J. J., Klein, E., Iken, K., Weinberger, V., Konar, B., et al. (2013). Large-scale spatial distribution patterns of gastropod assemblages in rocky shores. *PLoS ONE, 8*(8), e71396. doi:10.1371/journal.pone.0071396

Moestrup, Ø., Akselmann, R., Fraga, S., Hansen, G., Hoppenrath, M., & Iwataki, M., et al. (Eds.). (2009 onwards). IOC-UNESCO taxonomic reference list of harmful micro algae. Accessed at on 18 January 2015 http://www.marinespecies.org/hab/

Muller-Karger, F. E., Kavanaugh, M. T., Montes, E., Balch, W. M., Breitbart, M., Chavez, F. P., et al. (2014). A framework for a marine biodiversity observing network within changing continental shelf seascapes. *Oceanography, 27*(2), 18–23.

Neilson, B., & Costello, M. J. (1999). The relative lengths of seashore substrata around the coastline of Ireland as determined by digital methods in a geographical information system. *Estuarine, and Coastal Shelf Sciences, 49*, 501–508.

O'Dor, R., Boustany, A. M., Chittenden, C. M., Costello, M. J., Moustahfid, H., Payne, J., et al. (2012). A census of fishes and everything they eat: How the census of marine life advanced fisheries science. *Fisheries, 37*(9), 398–409.

Øresland, V. (2007). Description of the IMR standard light trap and the vertical distribution of some decapod larvae (*Homarus* and *Nephrops)*. *Western Indian Ocean Journal of Marine Science, 6*(2), 225–231.

Pagad S., Hayes K, Katsanevakis S., & Costello M. J. (2015). World Register of Introduced Marine Species (WRIMS). Accessed at on 12 May 2015 http://www.marinespecies.org/introduced

Pereira, H. M., Ferrier, S., Walters, M., Geller, G. N., Jongman, R. H. G., Scholes, R. J., et al. (2013). Essential biodiversity variables. *Science, 339*, 277–278.

Perry, A. L., Low, P. J., Ellis, J. R., & Reynolds, J. D. (2005). Climate change and distribution shifts in marine fishes. *Science, 308*(5730), 1912–1915.

Pohle, G., Iken, K., Clarke, K. R., Trott, T., Konar, B., Cruz-Motta, J. J., et al. (2011). Aspects of benthic decapod diversity and distribution from rocky nearshore habitat at geographically widely dispersed sites. *PLoS ONE, 6*(4), e18606.

Queirolo, D., Gaete, E., Montenegro, I., Soriguer, M. C., & Erzini, K. (2012). Behaviour of fish by-catch in the mouth of a crustacean trawl. *Journal of Fish Biology, 80*(7), 2517–2527.

Reguera, B., Alonso, R., Moreira, A. & Méndez, S. 2011. Guía para el diseño y puesta en marcha de un plan de seguimiento de microalgas productoras de toxinas. *IOC Manuals and Guides*, no. 59 – (IOC/2011/MG/59). COI de UNESCO y OIEA, Paris. 65 p.

Remy-Zephir, B., Leleu, K., Grace, R. & Costello, M.J. (2012). Geographical information system (GIS) files of seabed habitats and biotope maps from Leigh marine reserve in 1977 and 2006. Accessed at Dryad data archive http://dx.doi.org/10.5061/dryad.6vr28

Reynolds, P. L., Richardson, J. P., & Duffy, J. E. (2014). Field experimental evidence that grazers mediate transition between microalgal and seagrass dominance. *Limnology and Oceanography, 59*(3), 1053–1064.

Saeedi, H., & Costello, M. J. (2012). Aspects of global distribution of six marine bivalve mollusc families. In F. da Costa (Ed.), *Clam fisheries and aquaculture* (pp. 27–44). New York: Nova Science Publishers Inc.

Santhanam, R. & Srinivasan, A. (1994). *A manual of marine zooplankton*. Oxford and IBH Pub. Co.

Shackell, N. L., & Frank, K. T. (2003). Marine fish diversity on the Scotian Shelf, Canada. *Aquatic Conservation: Marine and Freshwater Ecosystems, 13*(4), 305–321.

Sistiaga, M., Herrmann, B., Grimaldo, E., Larsen, R. B., & Tatone, I. (2015). Effect of lifting the sweeps on bottom trawling catch efficiency: A study based on the Northeast arctic cod (*Gadus morhua*) trawl fishery. *Fisheries Research, 167*, 164–173.

Stuart-Smith, R. D., Bates, A. E., Lefcheck, J. S., Duffy, J. E., Baker, S. C., Thomson, R. J., et al. (2013). Integrating abundance and functional traits reveals new global hotspots of fish diversity. *Nature, 501*(7468), 539–542.

Stuart-Smith, R. D., Edgar, G. J., Barrett, N. S., Kininmonth, S. J., & Bates, A. E. (2015). Thermal biases and vulnerability to warming in the world's marine fauna. *Nature, 528*, 88–92.

Sudia, W. D., & Chamberlain, R. W. (1988). Battery-operated light trap, an improved model. *Journal of the American Mosquito Control Association, 4*(4), 536–538.

Szentkiralyi, F. (2002). Fifty year long insect survey in Hungary: T. Jermy's contributions to light-trapping. *Acta Zoologica Academiae Scientiarum Hungaricae, 1*, 85–105.

Tait, R. V. & Dipper, F. (1998). *Elements of marine ecology* (462 p.). Oxford: Butterworth-Heinemann.

Thomsen, P. F., & Willerslev, E. (2015). Environmental DNA—an emerging tool in conservation for monitoring past and present biodiversity. *Biological Conservation, 183*, 4–18.

UNESCO. (2012). *A framework for ocean observing. By the task team for an integrated framework for sustained ocean observing*. UNESCO, IOC/INF-1284. doi:10.5270/OceanObs09-FOO

Vandepitte, L., Hernandez, F., Claus, S., Vanhoorne, B., De Hauwere, N., Deneudt, K., et al. (2011). Analysing the content of the European ocean biogeographic information system (EurOBIS): Available data, limitations, prospects and a look at the future. *Hydrobiologia, 667*, 1–14.

Vandepitte, L., Bosch, S., Tyberghein, L., Waumans, F., Vanhoorne, B., & Hernandez, F. et al. (2015a). Fishing for data and sorting the catch: Assessing the data quality, completeness and fitness for use of data in marine biogeographic databases. *Database 2015*.

Vandepitte, L., Vanhoorne, B., Decock, W., Dekeyzer, S., Trias Verbeeck, A., Bovit, L., et al. (2015b). How Aphia—The platform behind several online and taxonomically oriented databases—can serve both the taxonomic community and the field of biodiversity informatics. *Journal of Marine Science and Engineering, 3*(4), 1448–1473.

Venter, J. C., Remington, K., Heidelberg, J. F., Halpern, A. L., Rusch, D., Eisen, J. A., et al. (2004). Environmental genome shotgun sequencing of the Sargasso Sea. *Science, 304*(5667), 66–74.

Wahl, M., Link, H., Alexandridis, N., Thomason, J. C., Cifuentes, M., Costello, M. J., et al. (2011). Re-structuring of marine communities exposed to environmental change: A global study on the interactive effects of species and functional richness. *PLoS ONE, 6*(5), e19514.

Warwick, R. M., & Clarke, K. R. (1998). Taxonomic distinctness and environmental assessment. *Journal of Applied Ecology, 35*(4), 532–543.

Webb, T. J., & Mindel, B. L. (2015). Global patterns of extinction risk in marine and non-marine systems. *Current Biology, 25*(4), 506–511.

Weinberg, K. L., & Kotwicki, S. (2008). Factors influencing net width and sea floor contact of a survey bottom trawl. *Fisheries Research, 93*(3), 265–279.

Wilkinson, C., & Hill, J. (2004). *Methods for ecological monitoring of coral reefs.* Townsville, Australia: Australian Institute of Marine Science.

Winger, P. D., Eayrs, S., & Glass, C. W. (2010). Fish behaviour near bottom trawls. In P. He (Ed.), *Behaviour of marine fishes: Capture processes and conservation challenges* (pp. 67–103). Oxford, UK: Wiley-Blackwell.

Yasuhara, M., & Cronin, T. M. (2008). Climatic influences on deep-sea ostracode (Crustacea) diversity for the last three million years. *Ecology, 89,* S52–S65.

Yasuhara, M., Cronin, T. M., deMenocal, P. B., Okahashi, H., & Linsley, B. K. (2008). Abrupt climate change and collapse of deep-sea ecosystems. *Proceedings of the National Academy of Sciences of the United States of America, 105,* 1556–1560.

Yasuhara, M., Hunt, G., Cronin, T. M., & Okahashi, H. (2009). Temporal latitudinal-gradient dynamics and tropical instability of deep-sea species diversity. *Proceedings of the National Academy of Sciences of the United States of America, 106,* 21717–21720.

Yasuhara, M., Okahashi, H., Cronin, T. M., Rasmussen, T. L., & Hunt, G. (2014). Deep-sea biodiversity response to deglacial and Holocene abrupt climate changes in the North Atlantic Ocean. *Global Ecology and Biogeography, 23,* 957–967.

Yasuhara, M., Tittensor, D. P., Hillebrand, H., & Worm, B. (2015). Combining marine macroecology and palaeoecology in understanding biodiversity: Microfossils as a model. *Biological Reviews.* doi:10.1111/brv.12223.

Chapter 7
Observations of Inland Water Biodiversity: Progress, Needs and Priorities

Eren Turak, David Dudgeon, Ian J. Harrison, Jörg Freyhof,
Aaike De Wever, Carmen Revenga, Jaime Garcia-Moreno,
Robin Abell, Joseph M. Culp, Jennifer Lento, Brice Mora,
Lammert Hilarides and Stephan Flink

Abstract This chapter aims to assist biodiversity observation networks across the world in coordinating comprehensive freshwater biodiversity observations at national, regional or continental scales. We highlight special considerations for freshwater biodiversity and methods and tools available for monitoring. We also discuss options for storing, accessing, evaluating and reporting freshwater

E. Turak
Office of Environment and Heritage, Level 18, 59-61 Goulburn Street,
Sydney South, PO Box A290, Sydney, NSW 1232, Australia

E. Turak (✉)
Australian Museum, 6 College Street, Sydney, NSW 2000, Australia
e-mail: eren.turak@environment.nsw.gov.au

D. Dudgeon
School of Biological Sciences, The University of Hong Kong,
Pokfulam, Hong Kong, SAR, China
e-mail: ddudgeon@hku.hk

I.J. Harrison
Center for Environment and Peace Conservation International,
2011 Crystal Drive, Suite 500, Arlington, VA 22202, USA
e-mail: iharrison@conservation.org

J. Freyhof
German Centre for Integrative Biodiversity Research (IDiv) Halle-Jena-Leipzig,
Deutscher Platz 5a, Leipzig 04103, Germany
e-mail: freyhof@igb-berlin.de

A. De Wever
OD Natural Environment, Royal Belgian Institute of Natural Sciences,
Vautierstraat 29, 1000 Brussels, Belgium
e-mail: aaike.dewever@naturalsciences.be

© The Author(s) 2017
M. Walters and R.J. Scholes (eds.), *The GEO Handbook on Biodiversity
Observation Networks*, DOI 10.1007/978-3-319-27288-7_7

biodiversity data and for ensuring their use in making decisions about the conservation and sustainable management of freshwater biodiversity and provision of ecosystem services.

Keywords Freshwater · Biodiversity · Monitoring · Essential biodiversity variables · Methods · Observations

7.1   Freshwater Biodiversity Observation

This chapter aims to assist biodiversity observation networks across the world in coordinating comprehensive freshwater biodiversity observations at national, regional or continental scales. We highlight special considerations for freshwater biodiversity and methods and tools available for monitoring. We also discuss options for storing, accessing, evaluating and reporting freshwater biodiversity data and for ensuring their use in making decisions about the conservation and sustainable management of freshwater biodiversity and provision of ecosystem services.

C. Revenga · R. Abell
The Nature Conservancy, 4245 N. Fairfax Drive, Suite 100, Arlington,
VA 22203-1606, USA
e-mail: crevenga@tnc.org

R. Abell
e-mail: robin.abell@tnc.org

J. Garcia-Moreno
ESiLi, Het Haam 16, 6846, KW Arnhem, The Netherlands
e-mail: jgarciamoreno@esili.net

J.M. Culp · J. Lento
Canadian Rivers Institute and Department of Biology, University of New Brunswick,
Fredericton, New Brunswick E3B 6E1, Canada
e-mail: jlento@gmail.com

B. Mora
GOFC-GOLD LC, PO Box 47, 6700, AA Wageningen, The Netherlands
e-mail: brice.mora@wur.nl

L. Hilarides · S. Flink
Wetlands International, PO Box 471, 6700, AL Wageningen, The Netherlands
e-mail: Lammert.Hilarides@wetlands.org

S. Flink
e-mail: Stephan.Flink@wetlands.org

7.1.1 What Is Freshwater Biodiversity?

Freshwater biodiversity is the diversity of life in inland (non-marine) waters. It includes both species that accomplish all, or parts of their lifecycles in or on water (i.e., 'real' aquatic species) and 'water-dependent' or 'paraquatic' species such as amphibians and water birds, which depend on inland water habitats during at least parts of their lives. The domain of freshwater biodiversity is defined by the extent of inland water ecosystems, which may be categorised as follows: (1) flowing waters (rivers and streams); (2) lacustrine wetlands (lakes, ponds, etc.); (3) palustrine wetlands (swamps, marshes, fens, bogs); and (4) groundwater systems (e.g., karstic systems, aquifers). Some of these inland waters are not best described as 'fresh', in particular, many lakes and aquifers contain high levels of dissolved salts. Nevertheless, it is more appropriate to consider the biodiversity of these systems as freshwater rather than as part of the terrestrial (Chap. 2) or marine realms (Chap. 6).

7.1.2 The Need for Special Attention to Freshwater Biodiversity Observations

Several lines of evidence suggest that rates of decline in freshwater biodiversity have been greater during the last few decades than that of their marine and terrestrial counterparts (Collen et al. 2014; Garcia-Moreno et al. 2014). Monitored populations of freshwater vertebrate species have declined by an average of 76 % over the past 40 years, compared to an average of 52 % decline of all vertebrate populations (McLeland et al. 2014). A panoply of direct and indirect threats affect freshwater species and their habitats (Strayer and Dudgeon 2010). For example, one estimate based largely on global models reports that approximately 65 % of global river discharge—and by extension the aquatic biodiversity supported by these rivers —is under considerable pressure from human activities (Vörösmarty et al. 2010).

The intensity of threats to freshwater species is likely to increase as a result of climate change. Higher temperatures and changed precipitation patterns combined with greater frequency of floods and droughts could result in the loss of freshwater species from their last refuges including from locations currently relatively free from anthropogenic threats or stressors. The reduction and degradation of suitable habitats, the difficulties of dispersal through aquatic environments, and the lack of corridors that link freshwater fragments will make it difficult for fully-aquatic species to move into new, more suitable areas following climate change. Conversely, certain invasive species will be able to expand their ranges, putting greater pressure on resident species and accelerate local extinctions (Strayer and Dudgeon 2010). In addition, climate change is creating concerns about water security that could precipitate management decisions that further degrade freshwater ecosystems (Poff et al. 2015).

Curtailing biodiversity declines and securing freshwater ecosystem services will require local and regional actions specific to these systems at appropriate scales, even when the systems cross national boundaries. Many if not most management and conservation interventions will rightly target freshwater ecosystems rather than species, yet the design of those interventions and the evaluation of their impact on achieving biodiversity goals will require information on multiple dimensions of freshwater biodiversity (i.e., genes, species, populations, communities, and ecosystem structure and function). Monitoring programs, using both traditional and recent, high-technology methods, that take into consideration the special features and structural organisation of inland waters can generate that information.

7.1.3   Freshwater Biodiversity Observations and Global Targets

It is widely agreed that goals set by parties to the Convention of Biological Diversity (CBD) to reduce the rate of biodiversity loss by 2010 were not met (Butchart et al. 2010). It would be hard to know if they were, since an evaluation of the state of freshwater biodiversity monitoring networks (Revenga et al. 2005) had earlier identified major shortfalls and gaps in monitoring capacity. One important finding was that existing data on freshwater species and populations were not readily accessible or harmonised in a way that they could be used to inform management decisions (Revenga et al. 2005). Freshwater fishes and water birds were by far the best studied groups, although there were considerable regional differences in completeness of data coverage. By contrast, aquatic plants, freshwater insects, molluscs and crustaceans were poorly known or not assessed in most regions and especially in the tropics (Balian et al. 2008a, b). Nonetheless, even in 2005 there were some well-established regional and continental assessments of freshwater biota (Revenga et al. 2005).

More recently, a 2011 evaluation of the Adequacy of Biodiversity Observation Systems to Support the CBD 2020 Targets (GEO BON 2011) showed that some progress has been made to address the gaps identified by Revenga et al. (2005). A global system of freshwater ecoregions has been completed (Abell et al. 2008), a global database of stream and networks at high spatial resolution has been developed (Lehner et al. 2008), large systematic biomonitoring programs have been established (e.g., CSIR 2007; Hatton-Ellis 2008; Davies et al. 2010; USEPA 2013), and additional regional assessments of freshwater species have been completed (Freyhof and Brooks 2011; Darwall et al. 2011). To address the past under-representation of biodiversity targets in the Millennium Development Goals, their sequel, the Sustainable Development Goals, now more explicitly include targets that are based on the CBD 2020 targets. Importantly, there have also been improvements in access to freshwater biodiversity data which we describe below.

7.1.4   Access and Management of Freshwater
Biodiversity Data

Ready access to freshwater biodiversity data and information from all parts of the
world is fundamental for the success of freshwater biodiversity observation programs and systems at global, national, regional or local scales. There has been
significant progress in this regard during recent years. For example, the EU-funded
BioFresh project (http://project.freshwaterbiodiversity.eu), which ran from 2010
until 2014, started building a global platform for freshwater biodiversity data. After
the termination of the project, four partner institutes committed to continue the
development of this on-line resource through the Freshwater Information Platform
(http://www.freshwaterplatform.eu) Major components of this platform include the
freshwater metadata journal and meta-database, the freshwater biodiversity data
portal, the Global Freshwater Biodiversity Atlas (see Box 7.1) and the freshwater
blog. The Freshwater Information Platform is an open body and additional global or
continental organisations are welcome to join.

> **Box 7.1. Global Freshwater Biodiversity Atlas**
>
> The Global Freshwater Biodiversity Atlas (http://atlas.freshwaterbiodiversity.eu)
> is a global collection of maps to showcase information on freshwater biodiversity
> and freshwater ecosystems, and includes background data such as freshwater
> resources, stressors and drivers of biodiversity and ecosystem change. It is a
> product of collaboration by numerous organisations, initiatives, scientists and
> projects active in the freshwater biodiversity community. This online information
> source aims to raise awareness about freshwater biodiversity from multiple
> perspectives (Fig. 7.1).
>
> As its name suggests, the Atlas includes a collection of published and
> open-access freshwater biodiversity maps as well as maps developed by
> different organisations from open-access data. The dynamic maps are
> accompanied by short articles explaining the maps, including background
> information and links to publications and data sources related to the specific
> maps. Contact points of the sources of maps are also provided to ease the
> access to data and additional information by users.
>
> The atlas provides stakeholders at the science-policy interface, the public
> and scientists interested in future conservation and sustainable management,
> with comprehensive information about freshwater biodiversity and its drivers
> and stressors. It allows those working in freshwater biodiversity to feature
> their results and make their research outputs visible to the broader
> community.

Despite such initiatives, much freshwater biodiversity data remain difficult to
access. There is a large number of smaller datasets or individual observations of
occurrence data that are not integrated into public repositories even though these

Fig. 7.1 Examples of maps in the Global Freshwater Biodiversity Atlas (*Source* http://atlas. freshwaterbiodiversity.eu)

data may have been used in scientific papers. Together with editors of leading freshwater journals, BioFresh led a call to make such data available in a standardised format (De Wever et al. 2012), but this has had limited impact so far. Adoption of data publishing practices as part of a mandatory archiving policy may well be required to effect changes in data management practices. In that respect institutes, research groups or individuals could relatively easily set up a data publishing infrastructure by making use of the GBIF Integrated Publishing Toolkit (IPT; http://www.gbif.org/ipt). This could allow the automation of the data publishing process while allowing authors to retain full control of that data. BioFresh or national GBIF nodes (see http://www.gbif.org/participation/list for a list of participants and associated nodes) are able to provide assistance in setting up such a system and often also have a central publishing infrastructure for those who do not have easy access to a server to run the IPT (e.g., http://data.freshwaterbiodiversity. eu/ipt/ for BioFresh). For datasets under construction or that cannot (yet) be released for particular reasons, we recommend documenting their existence in the freshwater metadatabase (see http://data.freshwaterbiodiversity.eu/metadb/bf_mdb_ help.php).

7.1.5 Improving Our Ability to Track Changes Through Freshwater Biodiversity Observations

Establishing baseline measures for the conservation status of the Earth's freshwater biodiversity remains an urgent challenge. This baseline is an essential first step for tracking changes in relation to the CBD 2020 targets. Considering the challenges of assessing the status of a sufficiently large proportion of freshwater species, Revenga et al. (2005) suggested beginning with a baseline assessment of the extent and conditions of freshwater habitats. Despite the expansion of monitoring programs focussed on river and lake conditions and the improvement in remote sensing technology for tracking wetland extent, a global assessment of the condition and extent of freshwater ecosystems is yet to be completed. A global assessment of threats to human water security and river biodiversity, based mostly on drainage-basin or in-stream indicators, was completed in 2010 (Vörösmarty et al. 2010), providing a coarse picture of the likely extent of imperilment of freshwater habitats.

Biodiversity observation networks can contribute to addressing these challenges by helping to coordinate data collection across large areas. A good example of such harmonisation is the Arctic Freshwater Biodiversity Monitoring Plan (Box 7.2) which details the need for coordinated assessment of Arctic freshwaters, including ponds, lakes and rivers as well as their tributaries and associated wetlands, and provides a framework for improving monitoring efforts in the Arctic region (Culp et al. 2012a, b). The plan represents an agreement among the Arctic nations on the approach to be taken to monitor and assess freshwater biodiversity across the pan-Arctic region. This program is coordinating the efforts of the Arctic countries as they inventory and collect freshwater monitoring data with the goal of producing the first status and trends assessment of Arctic freshwater biodiversity, which is planned for completion in 2017. The initial assessment will evaluate spatial and temporal trends from contemporary and historical time periods, where data allow, which means that by the end of this decade there should be sufficient time-series data to report on changes towards the 2020 CBD targets for the Arctic region. Furthermore, planned periodic re-assessments will continue to inform management decisions beyond 2020. In many other regions of the world there are comparable programs (albeit mostly at much smaller spatial scales) involving the collection of freshwater biodiversity data in a standardised way at least for each individual site and often for a group of sites.

One recent example is the Delaware River Watershed Initiative, a collaborative effort of over 50 organisations working across the Northeast U.S.A.'s 36,570 km^2 Delaware River Basin. The initiative has at its core the implementation of standardised monitoring protocols to assess its impact on water quality (see www.ansp.org/drwi). Although freshwater species and population data are not being collected in the service of assessing biodiversity per se, the data are being housed in an open-access database and may prove useful for evaluating species trends in the basin over time.

In general, the data collection protocols of such basin-scale efforts are tailored to the specific goals of individual programs or research efforts, creating challenges for directly combining the primary data for global or regional assessments. It may, however, be possible to use these primary data to quantify essential biodiversity variables representing main components of freshwater biodiversity (e.g., genetic diversity, community composition, ecosystem function; Pereira et al. 2013).

Box 7.2. The Arctic Freshwater Biodiversity Monitoring Plan: Circumpolar Biodiversity Monitoring Program

The Arctic Freshwater Biodiversity Monitoring Plan (CBMP-Freshwater Plan) details the rationale and framework for improvements in Arctic freshwater monitoring, including ponds, lakes, rivers, their tributaries and associated wetlands. The framework facilitates circumpolar assessments by providing Arctic countries with a structure and a set of guidelines for initiating and developing monitoring activities that employ common approaches and indicators. The CBMP-Freshwater Plan is part of the Circumpolar Biodiversity Monitoring Program (CBMP) of the Conservation of Arctic Flora and Fauna (CAFF) that is working with partners to harmonise and enhance long-term Arctic biodiversity monitoring efforts. A major goal is to facilitate detection and communication of environmental and biological change in the Arctic, and stimulate societal responses to significant trends and pressures (Fig. 7.2).

The CBMP-Freshwater Plan resulted from the collaboration of the CBMP Freshwater Expert Monitoring Group (represented by Canada, Sweden, Denmark, Finland, Iceland, Norway, Russia, and USA) and additional international freshwater scientists with a broad range of expertise. Contributors assessed the spatial and temporal coverage of available monitoring data and identified important elements, including environmental stressors, indicators, and Focal Ecosystem Components (FECs) to be incorporated into the pan-Arctic Freshwater Plan. FECs are biotic or abiotic elements, such as taxa or key abiotic processes, which are ecologically pivotal, charismatic or sensitive to changes in biodiversity). The mechanistic link between an environmental or anthropogenic stressor and the FECs was identified through 'impact hypotheses', i.e. predictive statements that outline the potential ways in which selected stressors might impact the structure or function of FECs. Preliminary information on the spatial and temporal coverage of available freshwater monitoring data for FECs was summarised to identify high-quality data sets that will form the basis for the first report on status and trends in freshwater biodiversity in the Arctic, which is planned for completion in 2017. This report will evaluate trends in existing data and identify gaps in monitoring efforts and scientific knowledge of Arctic freshwaters. It will also provide recommendations and guidance for more effective monitoring activities that are coordinated and stressor-targeted. By establishing common approaches for monitoring and assessment, the CBMP-Freshwater Plan and the first status and trends report are intended to improve our ability to detect changes to biodiversity and evaluate stressor impacts on a circumpolar scale, thus facilitating more effective management of these systems.

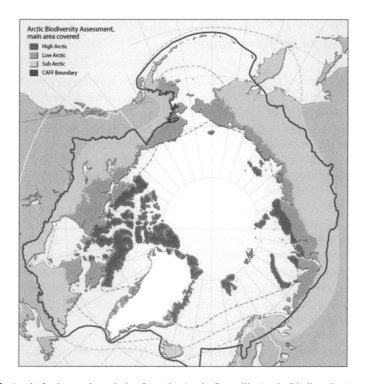

Fig. 7.2 Arctic freshwater boundaries from the Arctic Council's Arctic Biodiversity Assessment developed by CAFF, showing the three sub-regions of the Arctic that are the focus of the CBMP-Freshwater Plan, namely the *high* (*dark purple*), *low* (*purple*) and sub-Arctic (*light purple*), and the CAFF boundary (*grey line*) (*Source* Culp et al. 2012a)

7.2 Observations on Components of Freshwater Biodiversity

Biological monitoring of fresh or inland waters is developing rapidly. There is a diverse array of methods to assess many components of freshwater biodiversity (http://nepis.epa.gov/Exe/ZyPURL.cgi?Dockey=P100AVOF.TXT; see, for example, review by Friberg et al. 2011) and there are practical guides for setting up monitoring programs (e.g., Silk and Ciruna 2005). The priorities for measuring global freshwater biodiversity were identified by Turak et al. (2016) using and Essential Biodiversity Variables (EBV) Framework (Pereira et al. 2013). Here we present some of the important considerations specific to freshwater biodiversity monitoring described by Turak et al. (2016) together with additional information that would be useful for biodiversity observation networks. We have organised this information under the six broad classes of Essential Biodiversity Variables (EBVs; Pereira et al. 2013): i.e., genetic composition, species populations, species traits, community composition ecosystem structure, and ecosystem functioning. We stress, however, that some widely-used indicators for the condition of freshwater ecosystems (e.g., water quality variables) do not fit neatly into these categories.

7.2.1 The Spatial Context for Freshwater Biodiversity Observations

In situ observations of freshwater biodiversity provide information about species or biological communities at discrete locations within a freshwater body (e.g., a river section, a lake margin, or a portion of an aquifer). The use of these observations to infer the status of biodiversity across any large area at a given point in time requires aggregating disparate observations according to relationships between geography and the physical environment on the one hand and geography and freshwater biodiversity on the other. These relationships can also indicate how monitoring efforts can be distributed most efficiently across any given region. Two recent developments provide a foundation for formulating and applying such relationships at regional to global scales.

The first of these (as mentioned in Sect. 7.1.3) is the global biogeographic regionalisation of the world's freshwaters (Freshwater Ecoregions of the World or FEOW; www.feow.org; Abell et al. 2008). FEOW was developed based on freshwater biogeography, defined broadly to include the influences of phylogenetic history, palaeogeography, and ecology. FEOW development used fish species as proxies for the distinctiveness of biotic assemblages, with a few exceptions for extremely data-poor regions and inland seas, where some invertebrates and brackish-water fish were considered, respectively. FEOW offers a framework for development of broad-scale conservation strategies and represents a global-scale knowledge base with the potential for increasing freshwater biogeographic literacy, but it does not provide species occurrence data at a level of resolution that is especially useful for monitoring change over time (Abell et al. 2008).

The second important development is the availability of databases and tools such as HydroBASINS (Lehner and Grill 2013), the most accurate, globally consistent, digital catchment dataset currently available. It provides rapid access to reliable information about drainage basins, globally, at twelve levels of spatial resolution, and includes information on network connectivity. Such landscape units are probably better suited to mapping patterns of biodiversity across broad regions than the uniform, arbitrarily-scaled (typically square) grids used to map patterns of terrestrial or marine biodiversity. These drainage units also have great potential for planning freshwater conservation initiatives and identifying inland water areas for protection (e.g., Heiner et al. 2011).

7.2.2 Genetic Composition of Freshwater Biodiversity

Knowledge of the genetic composition and, especially, inter-population variability of freshwater species is of particular importance as river basins and lakes can be relatively isolated 'islands' separated from each other within a terrestrial or marine matrix that most freshwater animals cannot traverse. As a result, gene flow is limited and populations of the same species may vary considerably in their genetic composition. This variability has particular applications to the management of freshwater fisheries where loss of genetic variants may have major consequences for ecosystem service provision. Knowledge of inter-population genetic variability can also assist in deciding which populations should be priorities for conservation action, and may be important for assessing risks from invasive species. At present, most genetic data for freshwater species are accessible through GenBank (http://www.ncbi.nlm.nih.gov/genbank/) where fishes, amphibians, waterbirds and mammals are the best documented groups of freshwater organisms.

Recent advances in high-speed environmental DNA technology (see Taberlet et al. 2012; Goldberg et al. 2015 and references therein) offer great potential for assessing the presence of species and the genetic diversity of biological communities directly from their DNA fragments in the water. DNA extracted from water samples can be used to determine the genetic diversity of the community organisms that were present in that waterbody within up to two weeks before sample collection (see Thomsen et al. 2012). The molecular markers used are usually fragments of the mitochondrial CO1 gene (micro-barcodes), 16s, 18s or 18sV4 rDNA fragments.

Analysis of the mitochondrial CO1 gene is also being used in a global DNA barcoding initiative to catalogue the Earth's biota that already includes many freshwater fishes, amphibians and macroinvertebrates (Hebert et al. 2003; http://www.barcodeoflife.org). The CO1 gene was selected for barcoding because of its utility in species identification, but it also shows inter-population polymorphism and is used to identify genetic variants in commercial fish species (Ardura et al. 2011). Environmental DNA methods offer possibilities for monitoring metagenomes (i.e., genetic material recovered directly from environmental samples) of entire freshwater ecosystems, capturing both the variability among species and that among populations within species. It also offers new possibilities in freshwater biodiversity monitoring such as obtaining direct measures of the species diversity (though not, at present, species abundances) of individual water bodies including the diversity of microorganisms; enhancing the detection of cryptic, rare or endangered species without having to physically capture individuals; and early detection of invasive species at the expansion front. Nevertheless, this technology is still in its infancy; it would thus be pertinent to caution against over-reliance on it until issues around its sensitivity are resolved (Iversen et al. 2015).

7.2.3 Observations of Freshwater Species

The information available on the distribution, population sizes and population structure of freshwater species has greatly improved in recent years, allowing a general enhancement of regional, national, and global biodiversity observation networks. The Freshwater Animal Diversity Assessment (FADA; Balian et al. 2008a, b) provides an overview of genus- and species-level diversity of selected animal taxa groups and macrophytes of the Earth's inland waters. The raw data provided by the 163 experts who undertook the initial FADA is accessible through an online database (www.fada.biodiversity.be). Despite many obvious taxonomic and geographic gaps, and hence a need to collect more data (Balian et al. 2008b), FADA provides a much more detailed overview of freshwater biodiversity than had been available previously, and generates essential statistics such as the species richness of major organism groups. In particular, the disproportionate richness of global freshwaters is striking: the total number of freshwater animal species was estimated at 125,531 species, representing 9.5 % of 1,324,000 animal species described thus far. Insects make up the majority (60.4 %), while only 14.5 % are vertebrates. Furthermore, the 18,235 species of freshwater vertebrates represent 35 % of all vertebrates (about 52,000 species), despite the fact that inland waters occupy less than 1 % of the Earth's surface. Most of these vertebrates are fish (69 %), followed by amphibians (24 %). The total global number of fish species is presently estimated at 33,715 (Eschmeyer and Fong 2015, based on estimates from Reid et al. 2013). It is apparent that almost 50 % of all fish species inhabit fresh and brackish waters (i.e., 15,062 species, 12,470 of which are strictly freshwater). Freshwater habitats also support 73 % of amphibian species.

The Freshwater Biodiversity Unit of the International Union for the Conservation of Nature (IUCN) has been leading the development of a global assessment of the distribution and conservation status of freshwater organisms (Carrizo et al. 2013). These assessments bring together the most updated taxonomy and the extensive knowledge from thousands of regional experts. Assessments undertaken thus far have focused on fishes, molluscs (mainly unionid bivalves), decapods (crabs, crayfish and shrimps), Odonata (dragonflies and damselflies), and selected plant families (Carrizo et al. 2013). These taxonomic groups encompass a range of biogeographic distributions, habitat preferences and feeding habits, thereby offering a representative view of the ecology and conservation status of freshwater ecosystems. In addition, many of the assessed taxa are good indicators for environmental health in freshwater systems.

Importantly, the IUCN assessments of species are based upon the most comprehensive and accurate information available, involving collation of data on taxonomic status, ecology, distribution, spatial and temporal trends in abundance, as well as the threats they face, their use by humans and conservation measures in place to protect them. The integration of these data results in a classification of extinction risk according to IUCN Red List categories (Extinct, Extinct in the Wild, Critically Endangered, Endangered, Vulnerable, Near threatened, Least Concerned, Data Deficient). The species' ranges are mapped to HydroSHEDS (http://www.hydrosheds.org/; Lehner et al. 2008), but in the near future these data will be transferred to HydroBASINS (http://www.hydrosheds.org/page/hydrobasins), an updated version of HydroSHEDS that includes a coding system that captures the hierarchical spatial relationship among basins. All information on species included in the IUCN database is both widely available and freely accessible through the Red List of Threatened Species (http://www.iucnredlist.org/). Because these data are available at basin or sub-catchment units, they can be combined with information on population, land use and other types of data that are used for water resource management.

Modelling techniques that allow mapping of suitable habitats for individual species are increasingly being applied to freshwater species (e.g., Bush et al. 2014b). These models use species occurrence data together with digital data on environmental layers to help predict where species might occur, allowing targeted in situ observations or monitoring of species of particular interest or of conservation concern. If climatic variables are included among the environmental data, these models offer the potential to coarsely predict how species distributions may shift in response to global climate change (e.g., Bush et al. 2014a).

7.2.4   Observations of Freshwater Species Traits

Species traits have widely been used to characterise freshwater assemblages or communities, and may include aspects of morphology, function, physiology, behaviour, habitat use, reproduction and life history. Commonly documented traits include: trophic ecology (or functional feeding groups); oxygen or nutrient requirements; thermal range, or tolerance to pollutants, acidity, desiccation, turbidity, etc.; preference for particular substrates, flow regimes of microhabitats; locomotion or dispersal ability; body form; and life span, dormancy, and timing and frequency of breeding etc. Species-trait databases have been developed in some regions for certain taxa, most commonly fishes and macroinvertebrates (http://www.freshwaterecology.info/; http://eol.org/traitbank; http://www.epa.gov/ncea/global/traits/; Schmidt-Kloiber and Hering 2015), but plankton, diatoms and macrophytes are also represented. However, such trait-specific data are still lacking for many taxa and in most parts of the world, and fundamental facts about even the basic ecology of many common species are lacking, especially in the tropics.

7.2.5   Observations of the Composition of Freshwater Communities

Information on the composition of freshwater assemblages has been employed with some success to assess the condition of freshwater ecosystems, and statements about desirable composition of freshwater biota have been integrated into environmental legislation in countries in Europe and elsewhere (Friberg et al. 2011). The groups most widely used in examining the composition of biological communities in freshwater include macroinvertebrates, benthic algae, macrophytes, phytoplankton and fishes. Community composition metrics typically provide a quantitative measure of departure from reference conditions representing taxonomic completeness of the community (see Hawkins 2006). Reference conditions may be represented by relatively undisturbed reference sites or constructed using multiple lines of evidence, and may include the opinion of expert panels (Stoddard et al. 2006).

When the reference condition is represented by extant reference sites, environmental variables thought to be unaffected by human activities may be used to predict the probability of occurrence of a taxon at a site based on the site environmental characteristics. Taxa that have a high probability of occurring at a location are considered to be a natural component of the community at a site if the site's condition was equivalent to a reference site (i.e., unimpaired). Different metrics may then be used to quantify the difference between the predicted and observed community at any putatively impaired or impacted site. The simplest of these metrics is the number of taxa, which is in essence a measure of taxonomic completeness. Another widely-used metric is the ratio of average scores of pollution tolerance of the predicted and observed communities, based upon the combined pollution tolerance scores assigned to each taxon. Note that assignment of such tolerances typically requires good knowledge of the ecology of component species in the community and such information is frequently unavailable.

Widely used metrics of community composition assume that the detection of a species at a location is determined by the suitability of a habitat for colonisation by that species together with its ability to get there. Hence species interactions, for example, predation, competition, parasitism etc., are not incorporated into these assessments.

Despite being integrated into large biomonitoring programs, the data available on the composition of freshwater biological communities are biased and patchy, with no data being available for extensive areas of the Earth at any given point in time. This makes it difficult to determine temporal changes in biodiversity or to compare the status and trends in biodiversity among regions. However, the data that are available can be combined with spatially-continuous remotely-derived environmental layers to model community-level properties of biodiversity such as richness (alpha diversity) and compositional turnover (beta diversity) across large regions (see Ferrier 2011). The applications of these modelling approaches to regional and global biodiversity observations are discussed in Chap. 10.

7.2.6   Observations of the Structure of Freshwater Ecosystems

The persistence of freshwater species and communities is greatly influenced by the spatial arrangements of suitable habitats in the landscape and in particular, the presence and location of barriers to the movement of freshwater species, including those introduced by humans (e.g., dams). Observations of ecosystem structure for tracking changes in freshwater ecosystems include measuring changes in the extent of inland water habitats such as wetlands, lakes, rivers and aquifers. Remote sensing technologies for mapping the extent of wetlands and lakes is advancing rapidly (see Chap. 8). Smaller-scale habitat extent observations may encompass the extent of pools, riffles, and runs in streams, or the substratum (e.g., grain size) and flow characteristics in riffles; the area and depth of large pools in rivers, or the presence of gravel beds and channel braiding; and the connectivity of floodplains and backwaters with river channels. Such smaller-scale observations are particularly useful in mapping habitat needs of single species. For example, salmonid habitats in rivers can be mapped based on combined measurements of substratum grain size and water depth, and today's remote sensing capability can facilitate these observations (Carbonneau and Piégay 2012). In situ observations of physical and chemical characteristics of water are also an essential component of assessments of ecosystem structure in the context of monitoring freshwater biodiversity.

Advances in remote sensing technologies are increasingly enabling these observations on habitat structure to be made from space and depending on the ecosystem, with fewer in situ observations (see also Chap. 8). The advent of the Sentinel constellations (Sentinel-1 and -2 satellites in particular) as part of the European Copernicus Programme and the NASA Landsat Data Continuity Mission, will ensure continuous provision of Earth Observation data at high spatial resolution (10–30 m) and at higher time frequency (3–5 days combining Landsat 8 and Sentinel-2 satellites). The recent advent of time-series analysis algorithms combined with higher processing capabilities will enable monitoring of seasonal variations of habitat biophysical characteristics, and support potential development of early warning systems.

7.2.7 *Observations of Freshwater Ecosystem Functioning*

The use of indicators of ecosystem functioning, other than those that may be extrapolated from water-quality data, in monitoring or reporting on the condition of freshwater ecosystems, is rare. The relationship between biodiversity and ecosystem functioning is a growing research area but will need considerable further development before it will be possible to include measures of ecosystem functioning in freshwater biodiversity observations or link changes in biodiversity or ecosystem health to changes in functioning (see Dudgeon 2010). The attributes of ecosystem function that offer the greatest potential for monitoring changes in freshwater ecosystems include rates of organic matter processing (especially leaf litter breakdown in streams), primary production, rates of ecosystem metabolism at different scales (e.g. small patches of river sections), and aspects of secondary production such as fishery yields. Functional measures provide information not provided by measures of community composition. They more directly indicate changes in ecosystem services and can serve as early warning signs of sub-lethal effects that may lead to changes in community composition and abundance of species of conservation concern.

A complicating factor in decisions about whether to use functional attributes in biodiversity observations (and how to interpret them) is that it is not generally possible to predict how functioning changes with species loss. Some species may be 'redundant' so that their loss has little impact on overall functioning (e.g., a loss of a

single algal species may have negligible effect on overall algal production).
Conversely, the loss of certain species (e.g., keystone species such as the beaver,
Castor spp.) may have a large effect on functioning even if their loss is reflected (at
least initially) in minor changes in community composition. For most freshwater
species, and virtually all ecosystems, we are not yet in a position to predict the
magnitude of structural redundancy in relation to a given ecosystem function, or to
identify the role of individual species maintaining that function. Thus structure may
change and function remain unchanged (hence structure is a more sensitive indicator and needs to be closely monitored), or function may change before any
structural change has occurred (so function is more sensitive), or there may be no
consistent relationship between the two and so, ideally, both need to be monitored
(Dudgeon 2010). Further complexity arises from the possibility that function does
not respond linearly to changes in environmental conditions: leaf-litter breakdown
rates in streams can increase in response to nutrient enrichment until some critical
level when they begin to decline; primary productivity is likely also to show a
positive or hump-shaped response to nutrient enrichment. Accordingly, our ability
to predict the condition of biodiversity at a site from measurements of ecosystem
functioning alone may be limited, nor are such measures likely to be helpful when
we are concerned with assessing trends in the populations of rare species that may
well be so scarce as to have become functionally 'redundant.'

7.3 Use of Freshwater Biodiversity Data in Decision-Making

Efficient investment of resources in protecting freshwater species requires combining actions targeted at the level of ecosystems and landscapes and those that
target individual species of conservation concern. The efforts invested in freshwater
biodiversity observations and the evaluation of monitoring data must take into
account the need to achieve a balance between information needed on individual
species of concern with information on other components of biodiversity (such as
community composition; Box 7.3, Fig. 7.3a).

In prioritising species for monitoring or for repeated or long-term observations,
some of the major factors to consider are the level of threat (IUCN Red List status;
local classifications of species at risk or the relevant protected-species legislation);
regional freshwater conservation targets; community interest in iconic species or
those otherwise of particular concern to humans; and species that are essential as
sources of food or habitat for threatened species.

Actions that can address the threats to freshwater ecosystems across drainage
basins or in broader regions are especially important for conserving biodiversity,
but these actions must be prioritised so that resources are spent where greatest
benefits can be achieved. Freshwater conservation planning tools can help this
prioritisation (Box 7.3 Fig. 7.3b; also see Linke et al. 2011). Such tools require data

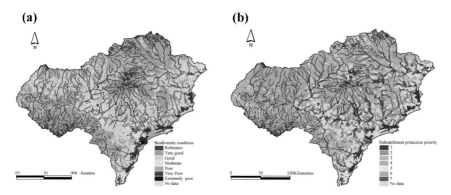

Fig. 7.3 Maps showing **a** patterns in river condition and **b** spatial priorities management actions aimed at protecting biodiversity, for the Hunter Catchment Management Region in south-eastern Australia (*Source* Turak et al. 2011)

on freshwater species or assemblages, as well as measures of environmental features that are intended to serve collectively as surrogates for all freshwater biodiversity. The success of actions at the drainage-basin scale is generally assessed through monitoring programs that use taxonomic composition of assemblages together with population trends of individual species.

Box 7.3. Multiple use of freshwater biodiversity monitoring data to support freshwater conservation

Biological monitoring programs in South-Eastern Australia have yielded extensive data on the composition of river macroinvertebrate and fish communities. These data were used in a variety of ways to support freshwater conservation in the region.

For example, occurrence records of macroinvertebrate families were used to develop predictive models that allowed quantitative scoring of river health at any given river site. These scores were then extrapolated using disturbance variables as predictors to generate digital layers of river condition (see Fig. 7.3a).

Another application of the data collected was in bottom-up biological classifications of rivers based on fish species records and macroinvertebrate family occurrences from relatively undisturbed reference sites. Digital layers representing these river classes together with the digital condition layers were used to generate maps representing spatial priorities for actions aimed at protecting river biodiversity (see Fig. 7.3b).

7.4 Future Directions for Freshwater Biodiversity Observations

Improved access to freshwater biodiversity data, refinement of frameworks for regional, national and continental monitoring programs, the widespread application of freshwater conservation planning tools and methods, and advances in remote-sensing technology have allowed the development of new programmes to enhance the freshwater components of national, regional and global biodiversity observation networks. Some notable examples are given below.

7.4.1 A Global Wetlands Observing System (GWOS)

In 2008 the Scientific and Technical Review Panel of the Ramsar Convention on Wetlands initiated establishment of a Global Wetlands Observing System (GWOS) to bring together available information on the status and values of wetlands and water in a way that can support policy processes and decision making at various geographic scales. It will describe extent and condition as well as change and trends over time of a variety of wetland types.

Although GWOS is still in a development phase, several thematic and regional pilot projects have been implemented already or are ongoing. As an example of a thematic project, the Japanese Aerospace Exploration Agency's Global Mangrove Watch (http://www.eorc.jaxa.jp/ALOS/en/kyoto/mangrovewatch.htm) aims to contribute to GWOS and support the Ramsar Convention. Examples of regional pilots have been the European Space Agency-sponsored GlobWetland and GlobWetland II projects that demonstrated the value of earth observation in mapping and monitoring of wetlands. The current GlobWetland Africa project will demonstrate this on a continental scale for Africa. The Mediterranean Wetlands Observatory (http://www.medwetlands-obs.org) serves as another regional pilot project and the Global Freshwater Biodiversity Atlas demonstrates some of the capabilities GWOS is expected to have when it is established as a global system.

As the first broad implementation of GWOS, the EU Horizon 2020-sponsored project 'Satellite-based Wetlands Observation Service' (http://swos-service.eu/) that started in 2015 will develop a monitoring and information service for wetlands tailored to specific policy needs on different levels. The project will bring together satellite observation data and validation datasets and will use citizen science to produce maps and metrics on wetlands and make available both these outputs as well as the toolkit required to produce them.

In the end GWOS will rely on NGOs, inter-governmental organisations, biodiversity observation networks, research institutions and government agencies for data, analyses and the development of tools. Biodiversity observation networks can contribute to GWOS as suppliers of freshwater biodiversity data. GWOS, in turn, can improve the utility of freshwater biodiversity observations by bringing together policy-relevant information and knowledge to support actions aimed at protecting freshwater biodiversity.

7.4.2   *Citizen Science in Freshwater Biodiversity Observations*

Recent advances in communication technology and the associated proliferation of citizen science protocols, web-services and phone apps, has opened up new opportunities for volunteers to collect and upload large volumes of biodiversity data, especially digital photographs (see Chap. 9). So far these methods have been or are being successfully applied to freshwater vertebrates only but there is potential to include other macroscopic taxa. Citizen scientists have been significant contributors to waterbird observations for over 50 years, with International Waterbird Census volunteers numbering over 10,000 and covering more than a hundred countries. Quality control is essential in citizen science and new technologies allow better quality control of these observations. The Global Amphibian BioBlitz (http://www.amphibians.org/citizen-science/) has helped to increase recorded observations and create awareness about amphibian declines. The recently launched Freshwater Fish BioBlitz (http://www.iucnffsg.org/ffsg-activities-2/global-freshwater-fish-bioblitz/) offers the possibility of capturing a vast number of observations made by recreational fishermen, aquarists and other fish enthusiasts.

Citizen science has the potential to make significant contributions to our knowledge about species distributions and their monitoring. Despite huge advances, problems with the geographic bias of observations towards developed countries remain. Improved public engagement in many countries of the world will be essential for the success of citizen science initiatives, starting with diversifying the language used for communications, as many people that could contribute observations do not speak English, which is the primary language used by these networks.

7.5   Conclusions

Recent developments in freshwater biodiversity observations indicate that there is potential for evaluating the condition of freshwater biodiversity by 2020 in 'real-time' or close to it. Despite the incompleteness of national and continental assessments, it now seems possible that we have sufficient tools for making periodic evaluations of freshwater biodiversity across large regions a realistic possibility by 2020. This alone will not ensure protection of freshwater biodiversity but it will provide evidence for the effectiveness of current management actions in conserving freshwater biodiversity. This evidence is essential for getting better results with existing resources and justifying claims for additional resources.

References

Abell, R., Thieme, M. L., Revenga, C., Bryer, M., Kottelat, M., Bogutskaya, N., et al. (2008). Freshwater ecoregions of the world: A new map of biogeographic units for freshwater biodiversity conservation. *BioScience, 58,* 403–414.

Ardura, A., Planes, S., & Garcia-Vazquez, E. (2011). Beyond biodiversity: Fish metagenomes. *PLoS ONE, 6,* e22592.

Balian, E. V., Lévêque, C., Segers, H., & Martens, K. (2008a). An introduction to the freshwater animal diversity assessment (FADA) project. *Hydrobiologia, 595,* 3–8.

Balian, E. V., Segers, H., Lévéque, C., & Martens, K. (2008b). The freshwater animal diversity assessment: An overview of the results. *Hydrobiologia, 595,* 627–637.

Bush, A. A., Nipperess, D. A., Duursma, D. E., Theischinger, G., Turak, E., & Hughes, L. (2014a). Continental-scale assessment of risk to the Australian Odonata from climate change. *PLoS ONE, 9,* e88958.

Bush, A. A., Nipperess, D. A., Theischinger, G., Turak, E., & Hughes, L. (2014b). Testing for taxonomic bias in the future diversity of Australian Odonata. *Diversity and Distributions, 20,* 1016–1028.

Butchart, S. H. M., Walpole, M., Collen, B., van Strien, A., Scharlemann, J. P. W., Almond, R. E. E., et al. (2010). Global biodiversity: Indicators of recent declines. *Science, 328,* 1164–1168.

Carbonneau, P. E., & Piégay, H. (Eds.). (2012). *Fluvial remote sensing for science and management.* Chichester, UK: Wiley.

Carrizo, S. F., Smith, K. G., & Darwall, W. R. T. (2013). Progress towards a global assessment of the status of freshwater fishes (Pisces) for the IUCN Red List: Application to conservation programmes in zoos and aquariums. *International Zoo Yearbook, 47,* 46–64.

Collen, B., Whitton, F., Dyer, E. E., Baillie, J. E. M., Cumberlidge, N., Darwall, W. R. T., et al. (2014). Global patterns of freshwater species diversity, threat and endemism. *Global Ecology and Biogeography, 23,* 40–51.

CSIR. (2007). *The South African River Health Programme.* CSIR Natural Resources and the Environment. http://www.csir.co.za/nre/water_resources/pdfs/factsheet_rhp.pdf. Accessed March 17, 2016.

Culp, J. M., Goedkoop, W., Lento, J., Christoffersen, K. S., Frenzel, S., Guðbergsson, G., et al. (2012a): *The Arctic Freshwater Biodiversity Monitoring Plan.* CAFF International Secretariat, CAFF Monitoring Series Report Nr. 7. CAFF International Secretariat. Akureyri, Iceland. http://www.caff.is/publications/doc_download/196-arctic-freshwater-biodiversity-monitoring-plan. Accessed March 17, 2016.

Culp, J. M., Lento, J., Goedkoop, W., Power, M., Rautio, M., Christoffersen, K. S., et al. (2012b). Developing a circumpolar monitoring framework for Arctic freshwater biodiversity. *Biodiversity, 13*, 215–227.

Darwall, W. R. T., Smith, K. G., Allen, D. J., Holland, R. A., Harrison, I. J., & Brooks, E. G. E. (Eds.). (2011). *The diversity of life in African freshwaters: Under water, under threat. An analysis of the status and distribution of freshwater species throughout mainland Africa.* Cambridge, United Kingdom and Gland, Switzerland: IUCN.

Davies, P. E., Harris, J. H., Hillman, T. J., & Walker, K. F. (2010). The sustainable rivers audit: assessing river ecosystem health in the Murray-Darling Basin. *Australian Journal of Marine and Freshwater Research, 61*, 764–777.

De Wever, A., Schmidt-Kloiber, A., Gessner, M. O., & Tockner, K. (2012). Freshwater journals unite to boost primary biodiversity data publication. *BioScience, 62*, 529–530.

Dudgeon, D. (2010). Prospects for sustaining freshwater biodiversity in the 21st century: Linking ecosystem structure and function. *Current Opinion in Environmental Sustainability, 2*, 422–430.

Eschmeyer, W. N. & Fong, J. D. (2015). Species by family/subfamily in the catalog of fishes. (http://researcharchive.calacademy.org/research/ichthyology/catalog/SpeciesByFamily.asp). Accessed December 24, 2015.

Ferrier, S. (2011). Extracting more value from biodiversity change observations through integrated modelling. *BioScience, 61*, 96–97.

Freyhof, J., & Brooks, E. G. E. (2011). *European red list of freshwater fishes.* Publications Office of the European Union, Luxembourg, Luxembourg, 61 pp. http://ec.europa.eu/environment/nature/conservation/species/redlist/downloads/European_freshwater_fishes.pdf.   Accessed March 17, 2016.

Friberg, N., Bonada, N., Bradley, D. C., Dunbar, M. J., Edwards, F. K., Grey, J., et al. (2011). Biomonitoring of human impacts in freshwater ecosystems: The good, the bad, and the ugly. *Advances in Ecological Research, 44*, 2–68.

GEO BON. (2011). *Adequacy of biodiversity observation systems to support the 2020 CBD targets.* Group on Earth Observations Biodiversity Observation Network Office, Pretoria. http://www.earthobservations.org/documents/cop/bi_geobon/2011_cbd_adequacy_report.pdf. Accessed March 17, 2016.

Goldberg, C. S., Strickler, K. M., & Pilliod, D. S. (2015). Moving environmental DNA methods from concept to practice for monitoring aquatic macroorganisms. *Biological Conservation, 183*, 1–3.

Hatton-Ellis, T. (2008). The hitchhiker's guide to the Water Framework Directive. *Aquatic Conservation: Marine and Freshwater Ecosystems, 18*, 111–116.

Hawkins, C. P. (2006). Quantifying biological integrity by taxonomic completeness: Evaluation of a potential indicator for use in regional and global-scale assessments. *Ecological Applications, 16*, 1277–1294.

Hebert, P., Cywinska, A., Ball, S., & deWaard, J. (2003). Biological identification through DNA barcodes. *Philosophical Transactions of the Royal Society B: Biological Sciences, 270*, 313–321.

Heiner, M., Higgins, J., Li, X. H., & Baker, B. (2011). Identifying freshwater conservation priorities in the Upper Yangtze River Basin. *Freshwater Biology, 56*, 89–105.

Iversen, L. L., Kielgast, J., & Sand-Jensen, K. (2015). Monitoring of animal abundance by environmental DNA—An increasingly obscure perspective: A reply to Klymus et al. 2015. *Biological Conservation, 192*, 479–480.

Lehner, B., Verdin, K., & Jarvis, A. (2008). New global hydrography derived from spaceborne elevation data. *EOS. Transactions of the American Geophysical Union, 89*, 93–94.

Lehner, B., & Grill, G. (2013). Global river hydrography and network routing: baseline data and new approaches to study the world's large river systems. *Hydrolic Process, 27*, 2171–2186.

Linke, S., Turak E., & Nel, J. (2011). Freshwater conservation planning: the case for systematic approaches. *Freshwater Biology, 56*, 6–20.

McLellan, R., Iyengar, L., Jeffries, B., & Oerlemans, N. (2014). *Living planet report 2014: species and spaces, people and places*. Gland, Switzerland: WWF International.

Pereira, H. M., Ferrier, S., Walters, M., Geller, G. N., Jongman, R. H. G., Scholes, R. J., et al. (2013). Essential Biodiversity Variables. *Science, 339*, 277–278.

Poff, N. L., Brown, C. M., Grantham, T. E., Matthews, J. H., Palmer, M. A., Spence, C. M., et al. (2015). Sustainable water management under future uncertainty with eco-engineering decision scaling. *Nature Climate Change*. Published online September 14, 2015.

Reid, G Mc G, Contreras MacBeath, T., & Csatádi, K. (2013). Global challenges in freshwater-fish conservation related to public aquariums and the aquarium industry. *International Zoo Yearbook, 47*, 6–45.

Revenga, C., Campbell, I., Abell, R., De Villiers, P., & Bryer, M. (2005). Prospects for monitoring freshwater ecosystems towards the 2010 targets. *Philosophical Transactions of the Royal Society B: Biological Sciences, 360*, 397–413.

Schmidt-Kloiber, A., & Hering, D. (2015). www.freshwaterecology.info—An online tool that unifies, standardises and codifies more than 20,000 European freshwater organisms and their ecological preferences. *Ecological Indicators, 53*, 271–282.

Silk, N., & Ciruna, K. (Eds.). (2005). *A practitioner's guide to freshwater biodiversity conservation. The Nature Conservancy*. Washington, USA: Island Press.

Strayer, D. L., & Dudgeon, D. (2010). Freshwater biodiversity conservation: Recent progress and future challenges. *Journal of the North American Benthological Society, 29*, 344–358.

Taberlet, P., Coissac, E., Hajibabaei, M., & Rieseberg, L. H. (2012). Environmental DNA. *Molecular Ecology, 21*, 1789–1793.

Thomsen, P., Kielgast, J. O. S., Iversen, L. L., Wiuf, C., Rasmussen, M., Gilbert, M. T. P., et al. (2012). Monitoring endangered freshwater biodiversity using environmental DNA. *Molecular Ecology, 21*, 2565–2573.

Turak, E., Ferrier, S., Barrett, T. O. M., Mesley, E., Drielsma, M., Manion, G., et al. (2011). Planning for the persistence of river biodiversity: Exploring alternative futures using process-based models. *Freshwater Biology, 56*, 39–56.

Turak, E., Harrison, I. J., Dudgeon, D., Abell, R., Bush, A., Darwall, D., et al. (2016). Essential Biodiversity Variables for measuring change in global freshwater biodiversity. *Biological Conservation*. http://dx.doi.org/10.1016/j.biocon.2016.09.005

USEPA. (2013). National rivers and streams assessment 2008–2009: a collaborative survey. EPA/841/D-13/001. Office of wetlands, oceans and watersheds and office of research and development, Washington, DC. http://nepis.epa.gov/Exe/ZyPURL.cgi?Dockey=P100AVOF.TXT

Vörösmarty, C. J., McIntyre, P. B., Gessner, M. O., Dudgeon, D., Prusevich, A., Green, P., et al. (2010). Global threats to human water security and river biodiversity. *Nature, 467*, 555–561.

Chapter 8
Remote Sensing for Biodiversity

Gary N. Geller, Patrick N. Halpin, Brian Helmuth, Erin L. Hestir,
Andrew Skidmore, Michael J. Abrams, Nancy Aguirre, Mary Blair,
Elizabeth Botha, Matthew Colloff, Terry Dawson, Janet Franklin,
Ned Horning, Craig James, William Magnusson, Maria J. Santos,
Steven R. Schill and Kristen Williams

Abstract Remote sensing (RS)—taking images or other measurements of Earth
from above—provides a unique perspective on what is happening on the Earth and
thus plays a special role in biodiversity and conservation applications. The periodic
repeat coverage of satellite-based RS is particularly useful for monitoring change
and so is essential for understanding trends, and also provides key input into

G.N. Geller (✉)
Group on Earth Observations Secretariat, Geneva, Switzerland
e-mail: ggeller@geosec.org

P.N. Halpin
Nicholas School of the Environment, Duke University, Durham, NC, USA
e-mail: phalpin@duke.edu

B. Helmuth
Northeastern University Marine Science Center, Nahant, MA, USA
e-mail: b.helmuth@neu.edu

E.L. Hestir
Department of Marine, Earth and Atmospheric Sciences, North Carolina State University,
Raleigh, NC, USA
e-mail: elhestir@ncsu.edu

A. Skidmore
Faculty ITC, University Twente, Enschede, The Netherlands
e-mail: skidmore@itc.nl

G.N. Geller · M.J. Abrams
NASA Jet Propulsion Laboratory, Pasadena, CA, USA
e-mail: michael.abrams@jpl.nasa.gov

N. Aguirre
Instituto de Investigación de Recursos Biológicos Alexander von Humboldt, Bogotá,
Colombia
e-mail: naguirre@humboldt.org.co

M. Blair
Center for Biodiversity and Conservation, American Museum of Natural History, New York,
NY, USA
e-mail: mblair1@amnh.org

M. Walters and R.J. Scholes (eds.), *The GEO Handbook on Biodiversity
Observation Networks*, DOI 10.1007/978-3-319-27288-7_8

187

assessments, international agreements, and conservation management. Historically, RS data have often been expensive and hard to use, but changes over the last decade have resulted in massive amounts of global data being available at no cost, as well as significant (if not yet complete) simplification of access and use. This chapter provides a baseline set of information about using RS for conservation applications in three realms: terrestrial, marine, and freshwater. After a brief overview of the mechanics of RS and how it can be applied, terrestrial systems are discussed, focusing first on ecosystems and then moving on to species and genes. Marine systems are discussed next in the context of habitat extent and condition and including key marine-specific challenges. This is followed by discussion of the special considerations of freshwater habitats such as rivers, focusing on freshwater ecosystems, species, and ecosystem services.

E. Botha
Division of Land and Water, Council for Scientific and Industrial Research, Canberra, Australia
e-mail: elizabeth.botha@csiro.au

M. Colloff · C. James · K. Williams
Council for Scientific and Industrial Research, Canberra, Australia
e-mail: matt.colloff@csiro.au

C. James
e-mail: craig.james@csiro.au

K. Williams
e-mail: kristen.williams@csiro.au

T. Dawson
School of the Environment, University of Dundee, Dundee, UK
e-mail: t.p.dawson@dundee.ac.uk

J. Franklin
School of Geographical Sciences and Urban Planning, Arizona State University, Tempe, AZ, USA
e-mail: janet.franklin@asu.edu

N. Horning
Center for Biodiversity and Conservation, American Museum of Natural History, New York, NY, USA
e-mail: horning@amnh.org

W. Magnusson
National Institute for Amazonian Research, Aleixo, Brazil
e-mail: bill@inpa.gov.br

M.J. Santos
Center for Spatial and Textual Analysis, Bill Lane Center for the American West, Stanford University, Stanford, CA, USA
e-mail: mjsantos@stanford.edu

S.R. Schill
The Nature Conservancy, Caribbean Program, Arlington, VA, USA
e-mail: sschill@tnc.org

Keywords Remote sensing · Earth observation · Satellite · Monitoring · Terrestrial · Marine · Freshwater · Ecosystem services

8.1 Remote Sensing

Every remotely sensed image of Earth can be considered a biological dataset. Each of these tells a story and a sequence tells the larger story of what is changing over time. Civilian satellite observations of Earth started over 40 years ago and provide an excellent historical record to help assess change. This chapter provides an overview of how remote sensing can be used for biodiversity and conservation applications, emphasizing change assessment. It focuses on satellite-based remote sensing because this provides global coverage with regular repeat cycles, sometimes providing a nearly daily view of the entire Earth, and is often available at no cost.

The potential for applying remote sensing (RS; sometimes referred to as Earth Observation, or EO, though this term is better used to refer to all kinds of observations, not just RS) for monitoring biodiversity and guiding conservation efforts has not been fully realised due to concerns about ease-of-use and cost. Historically, RS data have not always been easy to find or use because of specialised search and order systems, unfamiliar file formats, large file size, and the need for expensive and complex analysis tools. That is gradually changing with increasing implementation of standards, web delivery services, and the proliferation of free and low-cost analysis tools. Although data cost used to be a common prohibitive factor, it is no longer a big stumbling block for most users except where high resolution commercial images are needed.

8.1.1 How Remote Sensing Works

Remote sensing measures the energy that is reflected and emitted from the Earth's surface (for a good background on RS basics see https://www.nrcan.gc.ca/sites/ www.nrcan.gc.ca/files/earthsciences/pdf/resource/tutor/fundam/pdf/fundamentals_ e.pdf). Because the properties of materials commonly found on the surface (e.g., plants, soils, phyto-plankton-containing surface waters, ice bodies) are known, RS provides insight into the surface composition. There are also biodiversity-relevant situations which may not be directly observable with RS but which may be correlated with what *can* be observed. This allows remotely sensed observations to act as a "proxy" for surface activities if sufficient surface measurements are available to establish the link. For example, sea surface height can be measured and is correlated with upwelling and therefore with higher nutrient concentrations that affect the ecosystem in a variety of important ways.

Optical data such as that from the Landsat series of satellites and many others are a measure of the amount of light reflected from Earth's surface. Typically, the various wavelengths that are reflected are measured in separate "bands", each of which is stored as a separate image layer. Thus a typical "image" file contains a separate monochromatic image for each band. Creating a natural-looking colour image requires the user to combine red, green and blue bands. Other band combinations can also be used and these can highlight different components of interest in the image, often using wavelengths beyond what our eyes can see, particularly in the near-infrared region. Many optical sensors, including Landsat, have a "thermal" band that measures the long-wave infrared (thermal) radiation emitted from the Earth's surface, information particularly useful for estimating surface temperature.

The information available for understanding what is happening on the ground increases with the number of bands that a sensor has, but not in a directly proportional way. Thus typical "multi-spectral" sensors with 4 to 20 carefully selected and well-calibrated bands provide a great deal of information, and adding more bands can help with specific issues. "Hyperspectral" sensors can have more than 200 bands and can provide a wealth of information to help, for example, identify specific species. Processing such datasets requires special expertise and satellite-based hyperspectral sensors are not yet common. Other sensor types include radar and lidar which actively emit electromagnetic energy and measure the amount that is reflected—these sensors are useful for measuring surface height as well as tree canopy characteristics and surface roughness. Lidar is generally more precise than radar and ideal for measuring tree height. Radar is particularly useful where cloud cover is a problem (for instance, in the biodiversity-rich tropical rainforests) because it penetrates clouds. However, availability of lidar data is quite limited, and although radar data are more widely available it may be expensive and its use is less intuitive than the interpretation of optical images.

8.1.2   Combining Remote Sensing with in situ Observations

Remote sensing is generally most useful when combined with in situ observations, and these are usually required for calibration and for assessing RS accuracy. RS can provide excellent spatial and temporal coverage, for example, though its usefulness may be limited by pixel size which may be too coarse for some applications. On the other hand, in situ measurements are made at very fine spatial scales but tend to be sparse and infrequent, as well as difficult and relatively expensive to collect. Combining RS and in situ observations takes advantage of their complementary features.

8.1.3   Detecting Change

With the systematic coverage and long time-series provided by satellite observations, RS is particularly useful for detecting change. RS plays a major role in detecting and monitoring global- to local-scale processes that affect ecosystems, species, and ecosystem services, with effects on genes being an emerging field. RS data or its derivatives are an important input to models projecting future states and trends, which can provide an early warning of change and facilitate a timely response. Note, however, that *rates* of change may be more important than change in absolute conditions, making the frequency of repeats in a time-series, as well as data continuity, very important.

Several key factors require consideration to prevent the appearance of false positives (detection of change when none exists) or false negatives (failure to detect significant change). Because some degree of temporal and spatial variance is inherent in all ecosystems and species distributions and the physical variables that affect them, it is important to match the temporal and spatial scale of the environmental data with that of the ecosystem or species of interest. Such variation can occur over annual, seasonal, daily, or even hourly scales. Matching appropriate RS observations to ecological processes or species distributions often requires a multi-scale approach where one spatial and temporal scale provides information on a portion of an ecological process or species' life-history while other scales are required to observe another portion.

Next, the role of RS in monitoring Terrestrial, Marine, and Freshwater environments will be discussed. These are treated individually because they differ in many physical respects and RS is often applied in rather different ways.

8.2 Terrestrial

Remote sensing plays a major role in mapping and understanding terrestrial biodiversity. It is the basis of most land cover/land use maps, provides much of the environmental data used in species distribution modelling, can characterise ecosystem functioning, assists in ecosystem service assessment, and is beginning to be used in genetic analyses. Except in cases of direct observation of species, which generally require expensive high resolution images, biological RS data are usually combined with physical data such as elevation or climate (which in fact may be derived from RS data) and, increasingly, with socio-economic data.

8.2.1 Ecosystems

This section discusses the important role that RS can play in monitoring various aspects of ecosystems and the services they can provide.

8.2.1.1 Ecosystem Structure and Composition

Remote sensing is frequently used to generate maps of terrestrial ecosystems, which are often based on a map that delineates different vegetation types or land uses. Figure 8.1 is a simple example of an image that has been classified into five different types of cover based on how the spectral characteristics varied across the landscape (the UN Food and Agriculture Organization, FAO, has tools and information on class determination e.g., http://www.glcn.org, though note that other classification systems exist). One of the simplest and most common maps shows areas of forest and of non-forest, the latter often being a result of conversion to other uses. One point to remember is that it is essential that such maps be calibrated and validated with ground observations, otherwise the level of mis-classification can be very high, as well as unknown. Note that even with ground calibration such maps are often only about 80 % accurate, meaning that one pixel in five is classified incorrectly. Another point to remember is that spectrally similar vegetation types will be put in the same class. For example, tree plantations such as oil palm may be spectrally similar to native vegetation and both could be placed in the 'forest' class, and often are. Also note that classified maps use discrete categories, yet the actual landscape often varies continuously from one habitat to another, so the cut-off between classes can be somewhat arbitrary.

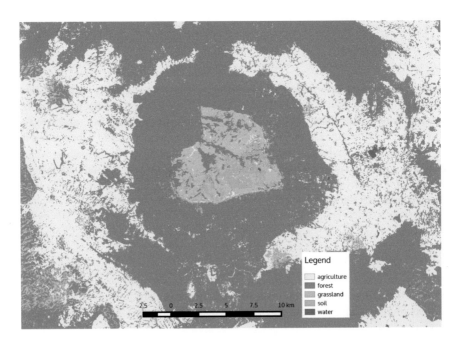

Fig. 8.1 Example of a simple classification with five classes. A national park in Thailand is in the centre of the image—a heart-shaped plateau surrounded by forest on the downslope, and then agriculture (*source* Classified Landsat image courtesy of Martin Wegmann)

In practice there are several classification techniques. The oldest is simply to visually interpret an image, drawing lines at the vegetation boundaries. Positioning these boundaries should combine what can be seen in the image with knowledge of what is known to occur on the ground, thus accurate interpretation requires that the analyst is familiar with the local vegetation. Software can also be used that automatically creates classes based only on spectral properties—the user can select the number of classes and what distinguishes them (i.e., how different two pixels must be before they are placed in different classes). This is called an "unsupervised" classification because once the number of classes and the difference thresholds are specified there is no human oversight of the process until the operator labels each class after the software finishes. A "supervised" classification takes a somewhat different approach. Here the user "trains" the software by selecting groups of pixels which are known to correspond to a particular class. A simple example might be to delineate areas of evergreen forest, deciduous forest, and agriculture based on a combination of visual interpretation and ground knowledge. For each of these three classes the software would then find all the other pixels in the image that had similar characteristics and identify them, allowing a map to be created.

Free software exists to do supervised and unsupervised classification, for example, https://www.orfeo-toolbox.org/ and http://www.dpi.inpe.br/spring/. One additional very useful tool is the Rapid Land Cover Mapper (http://lca.usgs.gov/lca/rlcm/), which provides a very simple way of visually mapping Land Use/Land Cover and change; it is free though requires ArcGIS ArcMap software. And increasingly, the open source R statistical software (http://www.r-project.org) is being used for image analysis, and many classification techniques and other geo-statistical models can be easily applied to images using existing user-supplied "packages".

8.2.1.2 Ecosystem Function

Ecosystem function can be thought of as the "work" that is done by an ecosystem. In other words, ecosystem function measures and monitors the energy dynamics as well as exchange of matter within an ecosystem, for example between the biota and the atmosphere, or within the biota. Examples of ecosystem functions include primary production, albedo, land surface temperature, evapotranspiration, as well as functional classifications such as Ecosystem Functional Types that characterise ecosystems based on similarities in energy dynamics or exchange of matter. To give a more concrete example, consider net primary productivity (NPP) which is an Essential Biodiversity Variable for the energy flow through ecosystems. NPP is the amount of biomass produced by an ecosystem within a defined period. Since plant biomass has a fairly constant carbon content, this can also be expressed as the amount of carbon assimilated by photosynthesis minus the carbon released by plant respiration.

8.2.1.3   Ecosystem Change

Ecosystem change usually refers to changes in land cover or land use, and identifying and quantifying it is particularly important. Land cover is a measurement of ecosystem state, and there is only one land cover at a point. Land use is what that land is used for, by people, and is typically defined by the calendar of activities which take place. There can be many land uses for any given land cover: for instance, a forest might simultaneously be used for the harvest of timber and for recreation and water provision. Land use is hard to detect from space, but it can be inferred from land cover and other remotely sensed attributes, and validated and enriched using ground-observed information. Evaluating the conversion of forest to non-forest is a common application for which images taken at two different times are compared (Fig. 8.2). The comparison can be done visually or using automated software; maps can then be made and the number of pixels that have changed calculated to provide a measure of forest loss. Recently, global maps of forest cover change at 30 m have become available at no cost, and these will be updated periodically (see http://earthenginepartners.appspot.com/science-2013-global-forest; and http://www.globalforestwatch.org/). Such maps may prove useful for users needing forest cover change information, however, local accuracy can vary significantly so users should be very careful to validate the information for their location. Beyond just forest conversion it is also possible to map how individual classes of vegetation or land use are changing, and to indicate what each class is changing into. This can be useful in trying to understand what the causes and consequences of change are.

Fig. 8.2 Monitoring forest loss. In this example from Rondonia, Brazil, images from 2000 and 2013 are compared and the difference, indicating loss of forest, is shown in *red* in the 3rd panel. A simple two class ("forest" and "clearing") supervised classification, using software such as that mentioned, can be used for such analyses. In this example the percentage of pixels classified as forest in 2000 was 65 %, while in 2013 it was 20 % (with 15 % not falling into either class, e.g., clouds and urban areas) (*source* ASTER images courtesy of NASA and Japan's METI; classification courtesy of Michael Abrams)

8.2.1.4 Ecosystem Services

Ecosystem services are comprehensively covered in Chap. 3; this section focuses on how terrestrial ecosystem services are measured and monitored remotely.

Ecosystem services are rarely if ever directly sensed with RS, rather, an ecosystem response is sensed as an indicator of status or change in a service (Fig. 8.3). Often one remotely sensed variable can be used to infer a range of ecosystem service changes through different model interpretations and interactions with other variables. For example, greenness measures such as the Normalized Difference Vegetation Index (NDVI) indicate plant vigour and potential productivity and can be used to indicate agricultural output (when the crop species is known), phenology, and CO_2 respiration. System status signals are related to ecosystem services through a number of pathways because biota are integrators of many physical and chemical factors in the environment. In other words the biophysical dimension of ecosystem services has seen the most application of RS when estimating provisioning and regulation. There have been few attempts to retrieve cultural services.

The process of inferring an ecosystem service from a remotely sensed ecosystem state often requires additional information and a considerable amount of modelling. For example, to estimate crop production output, information on phenological stage, water availability, or structure may be required in addition to remotely sensed greenness level; the modelling capability to combine this information would also be needed. Ecosystem services that can be estimated through this process include

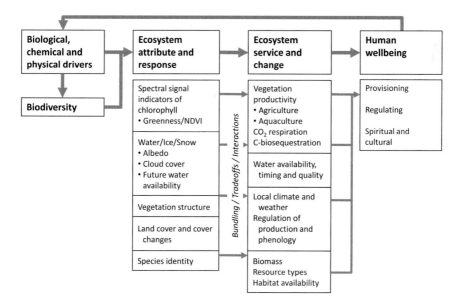

Fig. 8.3 Remote sensing is used indirectly to evaluate changes in ecosystem services

productivity; biomass; water availability, timing and quality; ecosystem regulating processes like those affecting CO_2 and methane release; and bio-sequestration rates (see Fig. 8.3).

8.2.2   Species

Remote sensing can play a particularly important role in helping to understand where species live and in providing measures of diversity such as species richness. Several reviews give more detail and provide references on using RS for biodiversity studies (for example, see Franklin 2010).

8.2.2.1   Mapping Where Species Live

Although it is possible to use RS to directly see and therefore map where some species live, this usually requires expensive, very high resolution data, and is generally only applicable to large organisms or populations of small organisms which are very dominant in the community. And while the status of some populations of large animals (e.g., elephants or whales) can be assessed in this way, traditional methods (e.g., visual surveys from the air) have so far been more cost effective. Mapping the distribution of some tree species may be possible over certain areas but again require very high spatial and/or spectral resolution images. Low-cost (e.g., <US$2000) unmanned aerial vehicles (UAVs) in combination with photogrammetry software are increasingly being adopted as an effective local-scale monitoring tool for applications such as identification of land/benthic cover, wildlife census, and monitoring of illegal activities. Advantages of using UAVs include very high spatial (<5 cm) and temporal resolution that often permit feature identification at the species level. More commonly, in situ observations of the specific locations where a species has been seen are correlated with environmental variables for those locations to develop a model that describes the set of conditions where a species is most likely to be found. This is called species distribution modelling (SDM). Once that model is created it can be combined with environmental data from a broad set of locations to generate a map of the potential distribution of that species—that is, the area where the environment is apparently suitable for that species to live. Many environmental variables, such as topography, land cover, temperature, and precipitation are derived from RS data. Since many are climate-related, SDM has been used to investigate how climate change could affect species distributions by substituting forecasted, instead of historical, climate data into the model. Although useful, correlative SDMs ignore biotic interactions and assume that a species is in equilibrium with environment; such limitations may be especially important when projecting species' distributions into novel environments such as predicting the impacts of climate change, land use change, and invasive species.

8.2.2.2   Plant Functional Types

Because direct detection and mapping of vegetation at the species level can be difficult, an alternative approach is to use "plant functional types"—groups of species having similar functionality. The principle is that species with similar function will have similar physiology and therefore spectral similarities that allow grouping into spectral or phenotypic—and thus functional—types (Ustin and Gamon 2010). More specifically, areas with the same ecosystem functional type have similar energy dynamics and exchange of matter. Creating appropriate combinations of functional types allows upscaling into vegetation communities, and time series of RS datasets can then be used to assess change at the community level, such as that due to natural or human-caused disturbance, succession, or phenology. All global maps of land cover implicitly or explicitly use the concept of plant function types. For instance, a map of "forest cover" is invoking the type "tree". Even global maps usually provide more detail than that—"evergreen" versus "deciduous" and "broadleaf" versus "needleleaf" trees can be reliably discriminated using RS, based on spectral and phenological characteristics. Field calibration and validation is, of course, still required to understand and have confidence in the observed changes.

8.2.2.3   Generating Biodiversity Indices

Measures of community diversity such as species richness and abundance can be derived using RS at landscape to global scales. Remotely sensed measures of productivity, water and nutrient status, vegetation structure, phenology, and biochemical diversity are often correlated with diversity metrics for a variety of taxonomic groups, but especially for plants. For example, biochemical diversity extracted from hyperspectral images and structural diversity derived from multispectral imagery have both been used to estimate tree species richness. Reflectance, surface temperature and NDVI from the MODIS and AVHRR sensors describe patterns of primary productivity that are related to continental-scale patterns of tree species richness.

8.2.3   Genes

Landscape genetics is a way to understand how the landscape affects genetic patterns by looking for spatial discontinuities in genetic variation and correlating them with observed landscape features (e.g., Manel et al. 2003). Remote sensing is a good way to characterise the landscape and identify specific landscape features, and thus provides key inputs to these analyses. For example, a landscape genetics approach might explore how habitat fragmentation patterns (visible using RS) correlate with genetic discontinuities such as differences between individuals within and among habitat patches (measured in situ). Because landscape genetics is spatially explicit, incorporating remotely sensed or other spatial landscape data, the approach is more powerful than traditional population genetic approaches to explore the effects of habitat fragmentation on gene flow.

One common landscape genetics method is called least-cost modelling. This builds on an isolation-by-distance population genetic framework, which examines the correlation between measures of genetic "distance" (e.g., how genetically similar or dissimilar two populations are) and the geographic distance separating populations or individuals (geographic distance can lead to isolation of populations by limiting gene flow between them). In least-cost modelling, landscape features are used when determining the easiest (i.e., "least-cost") routes for gene flow across a landscape (features such as mountains can obstruct gene flow). Least-cost distances thus combine knowledge of species habitat preferences with detailed information on landscape features acquired from RS data, including land use, land cover, or topography. The length of least-cost routes, as well as linear geographic distances, are then correlated with genetic distances. If the correlation is stronger with least-cost distances than with linear distances, then we can infer that something in the landscape has affected the pattern of genetic variation.

In addition to informing studies that explore how habitat fragmentation affects genetic variation, a landscape genetics approach can provide guidance to the selection of conservation areas to maximise adaptive genetic diversity and, thus, future evolutionary potential. Landscape genetics can also be applied in marine realms (seascape genetics) using seawater current, water temperature, or other spatially explicit resource gradient data.

8.3 Marine

Remote sensing of the marine environment is characterised by a number of unique challenges and complexities. Four primary challenges are: (1) the marine environment is profoundly dynamic, with significant change often occurring in sub-daily time steps (e.g., tides, mobile oceanographic features, diel migrations); (2) RS observations generally record only surface conditions, however, biophysical interactions occur throughout the entire water column; (3) the biological entity of interest is often highly mobile and also responds with lags to the physical environment; and (4) the time scales required to properly characterise marine ecosystem processes often do not match RS data collection cycles.

Remote sensing in the marine environment is generally used to identify patterns in biophysical features that can be used to infer ecological processes. Biogeographic patterns in marine data are often the result of multiple interacting processes (e.g., terrestrial runoff, interacting water masses, upwelling nutrient movement, subsurface bathymetric structures, etc.). Often several different RS platforms and measurement methods must be combined to characterise the processes related to biological patterns. As with terrestrial systems, marine biodiversity observing systems are often most successful when RS and in situ measurements are combined. For example, to develop a model that incorporates the feeding habits of a whale species so that the areas where it is likely to be found can be predicted (useful, for example, to re-route ship traffic) requires a lot of location data (aerial and/or ship surveys, or GPS tagging data) as well as RS data collection. The former provides georeferenced data on where the whales have been, while the latter provides information on ocean biophysical condition that the species may be responding to. Gaps in observations are common (e.g., RS images with cloud cover) and can complicate processing and reduce the observations available for model development. However, it is not always necessary to collect new observations because databases containing existing marine species observations are available and can be used as inputs to such species distribution models. For example, one of the largest marine species databases is the Ocean Biogeographic Information System (OBIS) with tens of millions of records and growing, accessible via the web (http://www.iobis.org/).

Ecological patterns and species distributions often respond to biophysical processes at different spatial and temporal scales. Broad, ocean-basin-scale patterns are often best characterised by recording seasonal patterns observed over multiple years or even decades of observations and using relatively low spatial resolution (e.g., ≥ 1 km). These broad scales of analysis may require processing the corresponding RS observations into standardised "climatologies" of oceanographic features such as sea surface temperature (SST), ocean colour (chlorophyll-a), and sea surface height (SSH) or into derived variables such as fronts, eddies or kinetic energy. The construction of such climatologies allows for the selection and averaging of observations across multiple time periods so that data gaps can be mitigated. In contrast, fine-scale observations of ecological patterns or habitat condition often require instantaneous observations or time series of observations (e.g., benthic reef habitats monitoring, harmful algal blooms, feeding or spawning events, etc.) at higher sampling frequencies and spatial resolutions. Such ephemeral ecological features or events require RS observations tied specifically to the appropriate time period. To facilitate discussion of these challenges, RS of the marine environment is divided into three broad categories: habitat extent, habitat condition, and change.

8.3.1 Habitat Extent

Remote sensing offers a unique perspective to map the extent of shallow benthic or intertidal marine habitats and, in some circumstances, identify and inventory individual key marine species. However, the quality and accuracy of the information extracted from the RS data are closely tied to the level of effort that goes into collecting the in situ observations used to validate mapped features. A common method involves direct mapping of the habitat or species using high or medium spatial resolution images. Examples of benthic habitats that have been mapped

(a)

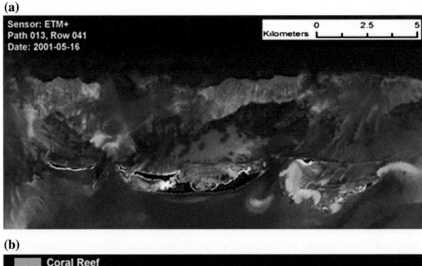

(b)

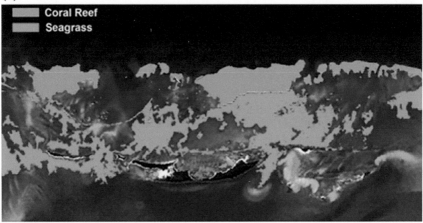

Fig. 8.4 An example of freely-available, medium spatial resolution imagery (Landsat ETM+), here used to map benthic habitats near Abaco Island in the Bahamas. **a** Original image; **b** image overlaid with extracted coral reef and seagrass polygons. The extraction used a tool to assign pixels with similar spectral, spatial, and/or textural characteristics to the same group, which was then assigned to one of the classes by the analyst (*source* Steven Schill)

using RS include coral reefs, seagrass, and kelp forests (Fig. 8.4). However, given the ever-changing environmental state of the ocean, distinguishing underwater habitats can be challenging, requiring careful planning and selection of RS data (i.e., free of sun glint and water column sediment). Recent RS systems such as Landsat 8 have been designed with new spectral bands that improve water column penetration. Compared to benthic habitats, the mapping of intertidal habitats such as mangroves, beaches, mud flats, rocky shores, or salt marshes is less problematic, but requires coordination of data acquisition with the tides to capture their full extent. Both benthic and intertidal habitats can be mapped using high (<5 m) or medium (5–30 m) spatial resolution imagery by applying classification techniques such as those described in the terrestrial section, automated feature extraction algorithms, or on-screen digitisation methods. Deeper ocean habitats cannot be mapped using space-based optical systems, instead requiring submerged active sensors such as multi-beam sonar, buoy-based instrumentation, and gliders.

Although marine species such as sea turtles, sharks and marine mammals can be directly observed using very high resolution imagery (<1 m), such images are expensive and often impractical, particularly for surveying large areas. Similar to terrestrial systems, species distribution modelling using physical variables (e.g., bathymetry, seabed sediment, chlorophyll-a, and SST) can predict the potential range of a particular species, the location of a particular habitat (e.g., areas of high productivity), or when combined with a mechanistic model, a biological response. For example, Fig. 8.5 shows predicted patterns of growth in mussels (*Mytilus edulis*) using chlorophyll-a (and SST, though this is not shown in the figure) as inputs to an energetics model. Similarly, it is sometimes possible to correlate biological diversity with spectral radiance values. This involves extensive georeferenced in situ data coupled with hyperspectral RS data in an attempt to understand biodiversity patterns over large areas. At broader scales, RS data can be used to assess the health and functioning of marine ecosystems at regional and global scales and determine the distribution and spatial variability of several oceanographic

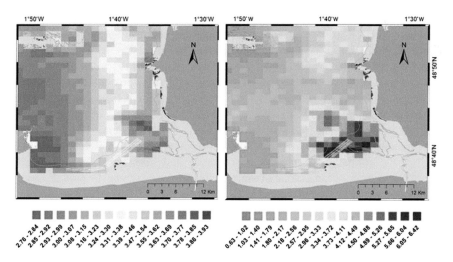

Fig. 8.5 Predictions of mussel growth (cm of shell growth, *left panel*) in Mount Saint-Michel Bay, France, using chlorophyll-a (μg/L, *right panel*, using Sea-WIFS data) and sea surface temperature (not shown, using NOAA-18 data) as inputs to a bioenergetics model predicting growth patterns (*source* Thomas et al. 2011)

phenomena over time. For example, optical sensors such as MODIS, MERIS, and SeaWiFS/OrbView-2 monitor ocean colour, temperature, or primary productivity and can warn of harmful algal blooms or potential coral bleaching events. More recently, RS data from the SMOS and Aquarius/SAC-D sensors are being used to estimate patterns of Sea Surface Salinity, for example measuring the extent of freshwater coastal plumes from rivers and monitoring changes to the global water cycle.

8.3.2 Habitat Condition

Some of the most common RS measurements of marine ecosystems include the extent and density (e.g., sparse versus dense seagrass beds) of habitats, but measuring condition is more difficult to assess. This is partly attributed to the high variability of the marine environment, but also the effort needed to properly gather and couple in situ with RS data to calculate indices that can be used to estimate resource condition. These data are often combined into condition indices. For example, biophysical indices such as benthic complexity or rugosity (roughness) have been developed and compared to species diversity in marine applications, and changes in these indices can be indicative of changes in condition due to a disturbance event. SPOT multispectral images have been used to map the spatial distribution of kelps, which drive the biodiversity of many benthic communities, and decreased kelp bed extent can also indicate a change in condition. More generally, there are a variety of indices of biological diversity that use remotely sensed inputs; since diversity can change with condition this approach can help monitor condition itself. One rather direct measure that can help assess the condition of a coral reef is its reflectance, which is visible in satellite images; a bleaching event, such as caused by excessive temperatures, results in a rapid increase in reflectance as the corals expel their symbiotic zooxanthellae.

Recent calls have been made for the development of indices that reflect the condition of commercially or ecologically important processes, such as the provision of nursery habitat, filtration by suspension-feeding invertebrates, sediment stabilisation in coastal environments by reef-forming species, and conditions that lead to the formation of blooms of nuisance and disease-causing organisms. Advancements in ocean colour product calibration are helping to provide water quality assessments in coastal areas (and large inland fresh or saltwater bodies) and making it easier to understand and monitor ecosystem conditions. Additionally, sophisticated condition indices that include multiple parameters such as temperature, light, pH, wind, seasonality, sediments and nutrients are under development.

8.3.3   Detecting Change and Issues of Scale

In the marine environment, matching the temporal and spatial scale of the environmental data with the ecosystem or species of interest is particularly important because they are very dynamic due to tides and currents. However, the match between the biophysical data and species may be multi- or cross-scale. For example, using species distribution modelling to map the habitat of a migratory predator species during the foraging season may require relatively high spatial resolution at high temporal frequency, while understanding where the species migrates may require lower resolution but larger area coverage. Space-time plots are a good way to help identify the appropriate match between spatial and temporal domains: by creating a diagram with spatial and temporal scales on the x and y axes and plotting ecosystem processes or properties in those coordinates, users can visualise multi- and cross-scale dynamics, which can aid sampling design and analysis approaches.

A variety of historical datasets are available to use in change studies, though not all are RS-based. For example, data from the NOAA AVHRR Pathfinder provides research-quality sea surface temperature (SST) data beginning in 1981 at a resolution of 32 km. Retrospective analyses that draw from in situ monitoring and remotely sensed data provide coverage even farther back in time. To go back even farther, the International Comprehensive Ocean-Atmosphere Data Set (ICOADS) is useful, providing data at a $2° \times 2°$ resolution from the early nineteenth century (obviously, not RS-based). Data sets such as the National Centers for Environmental Prediction (NCEP), NOAA Climate Forecast System Reanalysis (CFSR) and NASA MERRA (Modern Era Retrospective-Analysis for Research and Applications) provide extensive coverage at moderate levels of resolution. And recently, NOAA released an "Optimum Interpolation" (OI) product that provides estimates of SST at $0.25°$ resolution, thus providing an excellent means of detecting rates of change.

Combining RS and in situ observations can be a powerful approach for monitoring and predicting biodiversity and ecosystem function changes because together they provide more parameters over more temporal and spatial resolutions. However, users should be aware that combining both types of observations may lead to false-negatives or mischaracterised change trajectories because changes observable by RS may lag behind other significant changes to an ecosystem. For example, RS can readily observe many structuring species such as large kelps, corals, marsh grasses and intertidal mussel and oyster beds which may either control patterns of diversity in species that use them as habitat, or facilitate other species by acting as ecosystem engineers. However, declines in ecosystem services and functions such as productivity may precede any observable changes in the cover of structuring species. Thus, observing the spatial extent of structuring species may often but not always provide a good measure of biodiversity or ecosystem service change.

8.4 Freshwater

Although they occupy a relatively small portion of the Earth's surface, freshwater systems play a disproportionate role in driving biodiversity, ecological function and ecosystem services, and freshwater biodiversity is perhaps the most heavily impacted by human activities. Remote sensing for freshwater biodiversity observation relies primarily on observation of the processes that drive freshwater biodiversity rather than direct observation of the biodiversity itself. These can be broadly categorised into biophysical/hydrologic factors and landscape/large scale habitat function and structure. Of these processes, the most readily observable by RS are:

- Habitat function & structure

  - Land use and land cover change in the watershed (catchment), including deforestation
  - Area and location of rivers, lakes, impoundments and wetlands and habitats such as submerged or emergent macrophytes and riparian forests
  - Habitat connectivity along the water body and to adjacent water bodies and terrestrial ecosystems

- Biophysical/hydrological

  - Water body extent (a proxy for volume) and retention time
  - Hydro-period (the temporal pattern of high and low water)
  - Water column trophic status, especially eutrophication and sediment load
  - Submerged vegetation
  - Invasive alien species (IAS)

Direct observation of species and habitats is possible in limited cases—with high spatial and/or spectral resolution data it is possible to directly observe riparian, wetland, and submerged macrophytes and IAS. Such images tend to be expensive, however.

8.4.1 Considerations for Remote Sensing of Freshwater Biodiversity

8.4.1.1 Observing Small Systems from Space: Considering Spatial Scale

Because freshwater systems are relatively small and sparse on the Earth's surface, careful consideration of scale and resolution is required. A pixel is said to be "mixed" when there are multiple spectral classes contained in the same pixel (e.g., a 30 m pixel that covers a stream and its banks). Pixel mixing can provide a challenge to users, but not necessarily a barrier. For example, it is not possible to estimate lake water quality from a mixed pixel because it is impossible to say how much of the chlorophyll in the pixel is coming from the water column and how much is coming from the riparian/upland vegetation on the edge of the lake. However, with spectral unmixing techniques or other sub-pixel analyses it is possible to estimate how much of that pixel contains water, which is useful for determining inundation and hydro-period.

8.4.1.2 Observing Dynamic Systems: Considering Observation Extent and Frequency

Remote sensing for freshwater biodiversity observation is possible through the observation of the physical and ecological processes that drive biodiversity. Freshwater systems, like marine systems, are highly dynamic, and inundation period, extent and frequency are some of the primary drivers of biodiversity. This can make RS of biodiversity challenging. For example, a single scene may be useful in creating a land cover map, but this snapshot will not capture the variability that drives biodiversity. However, because satellite RS provides regular systematic observations, users can build time series that capture the dynamism of freshwater systems. Using RS to understand freshwater biodiversity requires that users consider both spatial and temporal scales of biodiversity-related processes.

Figure 8.6 shows the most observable ecological processes that drive freshwater biodiversity and the relevant spatial and temporal scales that should be considered for RS. Multiple processes can be observed using data with similar spatial and temporal scales, such as habitat area and land use and change. However, the most effective observations will make use of multiple datasets with different spatial and temporal characteristics to observe multiple biodiversity drivers (see Box 8.1).

Box 8.1. Freshwater systems detectable from space[1]
The number of freshwater systems resolved by satellite RS is dependent on the spatial resolution of the sensor and the size and geometry of the water body. For example, in Europe, nearly all freshwater systems are detectable using Landsat-type sensors (30 × 30 m), and almost half of Europe's freshwater systems are still detectable using MODIS-type sensors with 250 × 250 m resolution. Whereas in Australia only about 10 % of floodplains and 3 % of lakes and reservoirs can be detected by MODIS type sensors (250 × 250 m) and rivers are not detectable at all. However at the Landsat pixel scale (30 × 30 m), over 70 % of Australian reservoirs, lakes, wetlands and floodplains can be detected.

---

[1]To avoid potential mixed pixels, we considered a freshwater system to be "detectable" if it is 4 times larger than the pixel.

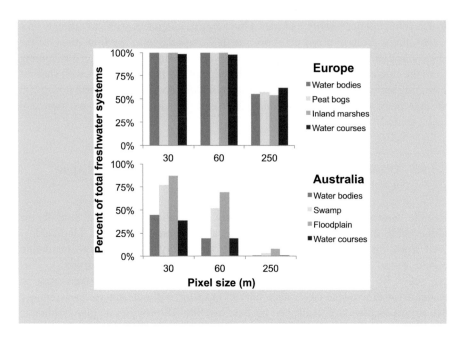

Fig. 8.6 Spatial and temporal scales for freshwater biodiversity processes (*source* Hestir et al. 2015)

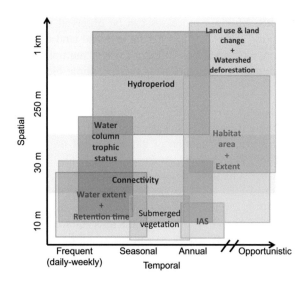

8.4.2 Approaches for Observing Biodiversity Drivers

For freshwater biodiversity process observations, the most important regions of the electromagnetic spectrum are the optical region (visible to shortwave infrared; 350–2500 nm) and the radar region (microwave; 3.75–7.5 cm or 8–4 GHz). Both optical and radar sensors have their advantages and limitations; these are linked to the driver to be measured and to the sensor specifications. Combining both methods typically leads to improved observations.

The level of processing required can vary depending on the biodiversity process and region of the electromagnetic spectrum. Reflectance and radiance measurements from key spectral bands in the optical region can be helpful in determining properties of habitat function and structure. Band ratios and other indexes highlight key reflectance and absorption characteristics, allowing users to better use reflectance and radiance information for mapping habitat area and extent, watershed land use and forest cover and riparian buffers (Adam et al. 2010).

Radiative transfer inversions are algorithms that use models of light and energy interaction with materials (e.g., plant canopies, water column) to estimate the wetland plant biomass and abundance, submerged vegetation and water column clarity and algal concentrations from RS. By combining RS observations with hydrologic models and field-based measurements, full process characterisation through modelling is readily achieved. Fortunately for most biodiversity assessments, regional or global products using many of these approaches already exist, enabling ease of use in incorporating RS products into freshwater biodiversity observing systems.

8.4.2.1 Ecosystems

Remote sensing of biodiversity drivers is most successful at the ecosystem level where many of the processes driving biodiversity are observable.

Habitat Function and Structure

Land use/land cover change and forest cover products are readily available globally or for most regions of the world, and products should be selected based on the suitability for a given region and the application. As mentioned earlier, care should be taken when using such wide-area products to ensure they are sufficiently accurate and appropriate for local conditions; remember that land use and land cover products often have average accuracies hovering around 80 % (local accuracies can be less). Remote sensing can also directly map the area and extent of habitat, typically by mapping wetland, floodplain and riparian vegetation, or by mapping the area of lakes. The latitudinal connectivity of a river—its connection to the floodplain and wetlands—is observed directly from identification of the riparian buffer and from water inundation mapping.

Biophysical/Hydrological Characteristics

Inundation mapping over a time-series of satellite observations provides information about the water extent and retention time and the hydro-period for a freshwater system. Water column trophic status can be determined from estimates of chlorophyll, sediment and coloured dissolved organic matter concentration, or from estimates of water clarity/Secchi depth using optical RS.

Vegetation Community Detection

Vegetation community identification is typically approached through classification procedures similar to those described in the terrestrial section of this chapter. These procedures work because they take advantage of a physical characteristic of green vegetation: strong absorption of red and blue wavelengths by the chlorophyll in the surface layers, and reflectance in the near infrared from the inner cell structure. Measuring reflectance in those wavelengths can be related to vegetation properties such as biomass or stress, which are the first order properties used for mapping the specialised vegetation communities that occupy different wetland zones and are good proxies for habitat diversity.

8.4.2.2   Species and Ecosystem Services

To further detail classification to species level or to identify intrinsic species characteristics or processes, airborne hyperspectral sensors may be needed since current spaceborne systems do not have sufficient spatial or spectral resolution for this, and the small size of many wetland communities can make airborne monitoring practical. In submerged aquatic plant communities species differentiation may be possible because the fine spectral bands measured by a hyperspectral sensor allow for more precise characterisation of individual plant species reflectance. This type of data can then be linked to intrinsic plant physiological processes. For example, it is possible to use hyperspectral reflectance characteristics and stable isotope markers to distinguish native submerged plant species from submerged IAS because they use different photosynthetic pathways. This information can provide insight into IAS adaptation traits for freshwater ecosystems (Santos et al. 2012). Using hyperspectral data, it is also possible to measure the foliar chemistry of inundated plants or other biotic communities such as the cyanobacteria commonly associated with harmful algal blooms (HABs). These allow inferences about the status of freshwater ecosystem services such as safe drinking water, nutrient cycling/eutrophication and carbon cycling. However, additional datasets are usually required for these inferences, such as laboratory samples of foliar chemistry, photosynthetic rates, respiration rates, stable isotope concentrations, biomass, and other measurable properties of plant species.

8.5 Conclusions

Remote sensing has a major role to play in monitoring changes in biodiversity and ecosystems. Space-based RS is typically global and provides periodic, repeat images that make it particularly appropriate for monitoring. Although, historically, cost and ease of access has been a problem these barriers have largely disappeared and continue to diminish. Because of these historical barriers, and because the RS community and the "traditional" conservation community tend to be separate, the use of RS has not yet been fully assimilated into standard biodiversity conservation practices. The RS community needs to continue to reach out to the broader conservation community and to simplify access to images and the derived products that the broader community need. These actions will facilitate greater use and integration and increase the return on the huge investment in RS infrastructure.

Despite its value, RS does have limitations. Its full value is typically only realised when combined with in situ measurements, which tend to be labour intensive and may not be available for a particular area. Given their often sparse nature it is usually necessary to interpolate between points; interpolation allows the generation of a surface depicting the characteristics of interest. Clouds are often an obstacle, particularly in the wet tropics, and although radar sensors can penetrate clouds radar data is not universally accessible (ESA's Sentinel-1 is starting to change this), and radar has its own set of limitations. Hyperspectral data can provide much more detailed information than typical multispectral sensors such as Landsat or Sentinel-2, but such data are, currently, available only on a very limited basis.

Limitations aside, RS currently provides a tremendous amount of information of value for understanding and monitoring biodiversity and how it is changing. Changes in terrestrial ecosystems that RS can monitor include changes in ecosystem extent, forest extent, health (e.g., by monitoring greenness, though estimating health can be a challenge) and in functional type. RS also provides a variety of information that facilitates estimating species distributions, and is an essential input into models that estimate overall biodiversity and how it is changing. It is also used in landscape genetic studies that correlate ecosystem patterns with genetic distributions, and in understanding how ecosystem services change over time. Marine ecosystem monitoring also depends on RS, where it is used to identify patterns in biophysical features that are correlated with ecological processes. While benthic ecosystems cannot be monitored from space, intertidal areas such as mangroves and salt marshes are commonly monitored. As for terrestrial areas, monitoring marine condition has some limitations, though RS is nonetheless a useful input to marine condition indices. Monitoring freshwater systems has its own unique challenges because they tend to be small and are temporally variable; the parameters most easily monitored include land use and land cover change within the watershed, the extent and temporal variability of water bodies, submerged vegetation, water column trophic status, and sediment and dissolved organic carbon content.

While more can be done to increase access to and use by the broader biodiversity conservation community, RS plays an essential role in monitoring terrestrial, marine and freshwater ecosystems. As more and different types of sensors become available and as coordination with that broader community continues to increase, RS will play an ever-increasing role, providing global, periodic data that can improve our understanding of change as well as how society responds.

References

Adam, E., Mutanga, O., & Rugege, D. (2010). Multispectral and hyperspectral remote sensing for identification and mapping of wetland vegetation: A review. *Wetlands Ecology and Management, 18*, 281–296.

Franklin, J. (2010). *Mapping species distributions: Spatial inference and prediction*. Cambridge, UK: Cambridge University Press.

Hestir, E. L., Brando, V. E., Bresciani, M., Giardino, C., Mattta, E., Villa, P., et al. (2015). Measuring freshwater aquatic ecosystems: The need for a hyperspectral global mapping satellite mission. *Remote Sensing of Environment, 167*, 181–195.

Manel, S., Schwarts, M. K., Luikart, G., & Taberlet, P. (2003). Landscape genetics: Combining landscape ecology and population genetics. *Trends in Ecology & Evolution, 18*, 189–197.

Santos, M. J., Hestir, E. L., Khanna, S., & Ustin, S. L. (2012). Image spectroscopy and stable isotopes elucidate functional dissimilarity between native and non-native plant species in the aquatic environment. *New Phytologist, 193*, 683–695.

Thomas, Y., Mazurié, J., Alunno-Bruscia, M., Bacher, C., Bouget, J.-F., & Gohinc, F. (2011). Modelling spatio-temporal variability of *Mytilus edulis* (L.) growth by forcing a dynamic energy budget model with satellite-derived environmental data. *Journal of Sea Research, 66*, 308–317.

Ustin, S. L., & Gamon, J. A. (2010). Remote sensing of plant functional types. *New Phytologist, 186*, 795–816.

Chapter 9
Involving Citizen Scientists in Biodiversity Observation

Mark Chandler, Linda See, Christina D. Buesching,
Jenny A. Cousins, Chris Gillies, Roland W. Kays, Chris Newman,
Henrique M. Pereira and Patricia Tiago

Abstract The involvement of non-professionals in scientific research and environmental monitoring, termed Citizen Science (CS), has now become a mainstream approach for collecting data on earth processes, ecosystems and biodiversity. This chapter examines how CS might contribute to ongoing efforts in biodiversity monitoring, enhancing observation and recording of key species and systems in a standardised manner, thereby supporting data relevant to the Essential Biodiversity Variables (EBVs), as well as reaching key constituencies who would benefit Biodiversity Observation Networks (BONs). The design of successful monitoring or observation networks that rely on citizen observers requires a careful balancing of the two primary user groups, namely data users and data contributors (i.e., citizen

M. Chandler (✉)
Earthwatch Institute, 114 Western Avenue, Boston, MA 02143, USA
e-mail: mchandler@earthwatch.org

L. See
International Institute for Applied Systems Analysis, Schlossplatz 1, 2361 Laxenburg, Austria
e-mail: see@iiasa.ac.at

C.D. Buesching
Wildlife Conservation Research Unit, Department of Zoology, The Recanati Kaplan-Centre, University of Oxford, Tubney House, Abingdon Road, Tubney, Abingdon OX13 5QL, UK
e-mail: christina.buesching@zoo.ox.ac.uk

J.A. Cousins
Earthwatch Institute, Mayfield House, 256 Banbury Road, Oxford OX2 7DE, UK
e-mail: jcousins@earthwatch.org.uk

C. Gillies
The Nature Conservancy Australia, 60 Leicester St., Carlton, Australia
e-mail: chris.gillies@tnc.org

R.W. Kays
North Carolina State University and NC Museum of Natural Sciences, Raleigh, NC, USA
e-mail: rwkays@ncsu.edu

C. Newman
WildCRU, Department of Zoology, The Recanati Kaplan-Centre, University of Oxford, Tubney House, Abingdon Road, Tubney, Abingdon OX13 5QL, UK
e-mail: chris.newman@zoo.ox.ac.uk

© The Author(s) 2017
M. Walters and R.J. Scholes (eds.), *The GEO Handbook on Biodiversity Observation Networks*, DOI 10.1007/978-3-319-27288-7_9

212 M. Chandler et al.

scientists). To this end, this chapter identifies examples of successful CS programs as well as considering practical issues such as the reliability of the data, participant recruitment and motivation, and the use of emerging technologies.

Keywords Citizen science · Essential biodiversity variables · Biodiversity monitoring · Data reliability · Data standards · Emerging technologies

9.1 Citizen Science

The involvement of non-professionals in scientific research and environmental monitoring, termed Citizen Science (CS), has now become a mainstream approach for collecting data on earth processes, ecosystems and biodiversity. Although the term has appeared only more recently as a formal way of referring to these activities, CS actually has a very long history. In the past, amateur scientists have contributed a great deal to science, particularly with networks of weather collectors and ocean monitoring. Famous names such as Alfred Russell Wallace, Thomas Edison and Gregor Mendel are all prime historical examples of citizen scientists.

With recent changes in technology and social media enabling outreach and interaction with a much wider audience than ever before, CS is becoming an increasingly integral part of contemporary scientific research, particularly in terms of data acquisition. With limited budgets to pay for professional scientists, or to support government-sponsored environmental monitoring, engaging citizens to help with ground-based monitoring efforts and the reporting of rare events, makes sense. By achieving hitherto unrealised levels of large-scale monitoring for features which remain invisible to remote sensing, CS is likely the most realistic way of covering much of the planet's biosphere (Pereira and Cooper 2006; Pereira et al. 2010).

H.M. Pereira
German Centre for Integrative Biodiversity Research (iDiv) Halle-Jena-Leipzig, Deutscher Platz 5e, 04103 Leipzig, Germany
e-mail: hpereira@idiv.de

H.M. Pereira
Institute of Biology, Martin Luther University Halle-Wittenberg, Am Kirchtor 1, 06108 Halle (Saale), Germany

H.M. Pereira
Cátedra Infraestruturas de Portugal Biodiversidade, Centro de Investigação em Biodiversidade e Recursos Genéticos (CIBIO/InBIO), Universidade do Porto, Campus Agrário de Vairão, 4485-661 Vairão, Portugal

P. Tiago
Centre for Ecology, Evolution and Environmental Changes, Faculdade de Ciências da Universidade de Lisboa, 1749-016 Lisbon, Portugal
e-mail: pmtiago@fc.ul.pt

This chapter provides examples of how CS can contribute to ongoing efforts in biodiversity monitoring, enhancing observation and recording of key species and systems in a standardised manner, and supporting the collection of Essential Biodiversity Variables (EBVs), as well as reaching key constituencies who would benefit Biodiversity Observation Networks (BONs). Referred to as contributory CS, which is based on a typology developed by Bonney et al. (2009a, b) and Miller-Rushing et al. (2012), involving citizens primarily in data collection is the most common form and probably the simplest starting point for those interested in developing new CS projects. Other forms of CS are also possible, such as through the Earthwatch model (http://earthwatch.org/) where members of the public join research projects; these require more training, direction and supervision of participants to ensure systematic data collection for answering specific scientific research questions.

The design of successful monitoring or observation networks that rely on citizen observers requires a careful balancing of the two primary user groups, namely data users and data contributors (i.e., citizen scientists; Pocock et al. 2015). To this end, this chapter also considers practical issues such as reliability of the data (Buesching et al. 2014), participant recruitment and motivation (Buesching et al. 2015; Silvertown et al. 2013), and the use of emerging technologies. All are important issues that determine whether useable data are collected and how a team of willing and capable participants is maintained.

9.2 Citizen Science and Biodiversity Observation Networks (BONs)

The aim of a BON is to help improve information available on the distribution and change of biodiversity in a given region or associated with a specific theme (e.g., an ecosystem domain or a particular type of monitoring) (GEO BON 2015a, b). BONs obtain baseline data, develop monitoring programs to detect change, publish biodiversity observations, and help identify the factors underlying the observed changes. This supports the modelling communities and the development of ecosystem assessments and future scenarios supporting conservation mitigation strategies (Akçkaya et al. 2016). CS can contribute to this aim in a number of different ways, as outlined below.

9.2.1 Monitoring Biodiversity Over Large Spatial and Temporal Scales

Using citizen scientists in biodiversity monitoring networks significantly expands the spatial and temporal scale of what is possible, because the additional people allows considerably more data to be collected, both in terms of range and quantity. CS can be a practical way to achieve the geographic coverage required to document ecological patterns and address ecological questions at scales relevant to regional population trends, shifts in species range, patterns of migration, impacts of environmental processes like Human Induced Rapid Environmental Change (Sih 2013), spread of infectious disease and invasive species, and national environmental policy assessment. This is especially important for smaller, rarer or more fragmented habitats and species that may be hard to detect in coarse or infrequent surveys, but also for very common and widespread species where the sheer size of the species range may prove challenging to sample (Buesching et al. 2015). Large-scale CS projects are thus valuable when attempting to gather data on large geographical scales, such as engaging participants in national or even global surveys, with participants collecting data in many locations simultaneously. These projects can involve very substantial numbers of contributors, and can persist for a long time, making it possible to map trends. Moreover, CS can lead to the engagement and coordination of an active and long-lasting community around permanent monitoring sites such as those established by existing BONs (e.g., National Ecological Observatory Network (NEON) in the USA through its Citizen Science Academy, Terrestrial Ecosystem Research Network (TERN) sites in Australia who have partnered with Earthwatch) or long-term research and monitoring plots such as Hawkwatch Monitoring North America sites. The results can also be used to inform population management decisions and even international environmental and conservation policy.

9.2.2 Mapping Species Location and Abundance

Most biodiversity-oriented CS programs aim to record the location and abundance of species through time (Table 9.1). These observations are used to monitor population trends and geographic range dynamics (e.g., eBird http://ebird.org/content/ebird, iNaturalist http://www.inaturalist.org, iSpot http://www.ispotnature.org). Indeed, close to 50 % of all species occurrence records in GBIF are published from sources that already publish data collected through CS projects (http://www.gbif.org). Most of these programs contribute largely to collaborative projects, rely on high participation rates to reduce data errors (e.g., by 2015 eBird had over 200 million observations contributed to GBIF; http://www.gbif.org), and in many cases there is little or no formal training required for participation.

Some programs are designed to ensure a balance between providing regular scientific updates on species location and movements while engaging the public in enjoyable, hobby-like activities. Some of these programs have stemmed from rapid biodiversity surveys that involve both researchers and the public, e.g. a BioBlitz (Lundmark 2003). They are often run in association with local museums, naturalist clubs and schools on international days of environmental recognition. BioBlitzes are still immensely popular and continue to contribute to the discovery of new

Table 9.1 Potential contribution of citizen science projects to the candidate Essential Biodiversity Variables

EBV Class	Candidate EBV	Description	Project/Database Name	Website	Country
Genetic composition	Breed and variety diversity	Number of animals of each livestock breed and proportion of farmed area under each local crop variety, at multiple locations	No known CS projects as yet, but CS approaches could be used to study this EBV—see Bohmann et al. (2014)		
Species population	Species distribution	Presence surveys for groups of species easy to monitor, over an extensive network of sites with geographic representativeness. Potential role for incidental data from any spatial locations	iNaturalist	http://www.inaturalist.org/	World
			iSpot	http://www.ispot.org.uk/	World
			Observado	http://www.observado.org/	World
			eBird	http://ebird.org/	World
			eMammal	http://eMammal.org	World
			The Great SunFlower Project	http://www.greatsunflower.org/	US
			Breeding Bird Survey	http://bto.org/bbs/index.htm/	UK
			French Common Bat Monitoring	http://mnhn.fr/vigie-nature/	France
			Pan European Common Bird Monitoring	http://ebcc.info/pecbm.html/	Europe
			FrogWatch USA	http://www.nwf.org/frogwatchUSA/	USA
			Ontario Turtle Tally	http://torontozoo.com/adopttapond/	Canada
			German Butterfly Monitoring	http://science4you.org	Germany
			Spiders Web Watch	http://spiderwebwatch.org/	Canada, US
			Anglers Monitoring Initiative	http://riverflies.org/	UK
			Great Lake Worm Watch	http://nrri.umn.edu/worms/	US
			Plant Watch	http://www.naturewatch.ca/	Canada
			BioDiversity4All	http://www.biodiversity4all.org/	Portugal
			Coral Watch	http://www.coralwatch.org/	World

(continued)

Table 9.1 (continued)

EBV Class	CandidateEBV	Description	Project/Database Name	Website	Country
Species population	Population abundance	Population counts for groups of species easy to monitor and/or important for ecosystem services, over an extensive network of sites with geographic representativeness	eBird	http://ebird.org/	World
			German Butterfly Monitoring	http://science4you.org	Germany
			Garden Moths Count	http://www.mothcount.brc.ac.uk/	UK
			REEF (Coral Reef Fish)	http://www.reef.org/	World
			Custodians for Rare and Endangered Wildflowers	http://www.sanbi.org/biodiversity-science/state-biodiversity/biodiversity-monitoring-assessment/custodians-rare-and-endan	South Africa
Species traits	Phenology	Record timing of periodic biological events for selected taxa/phenomena at defined locations	National Phenology Network	https://www.usanpn.org/	US
			BudBurst	http://windows.ucar.edu/citizen_science/budburst/	US
			Seasons Observatory (plant bloom and migratory birds)	http://www.obs-saison.fr	France
Species traits	Natal Dispersal Distance	Record median/frequency distribution of dispersal distances of a sample of selected taxa	No projects known at present, but advanced CS observers could contribute to professional databases such as TRY		
Species traits	Migratory behaviour	Record presence/absence/destinations/pathways of migrant selected taxa	eBird	http://ebird.org/	World
			Migrant Watch in India	http://www.migrantwatch.in/	India
			Migrant watch in UK	http://butterfly-conservation.org/612/migrant-watch.html	UK

(continued)

Table 9.1 (continued)

EBV Class	CandidateEBV	Description	Project/Database Name	Website	Country
Community composition	Taxonomic diversity	Multi-taxa surveys (including by morphospecies) and metagenomics at selected in situ locations at consistent sampling scales over time	iNaturalist	http://www.inaturalist.org/	World
			iSpot	http://www.ispot.org.uk/	World
			Observado	http://www.observado.org/	World
			BioDiversity4All	http://www.biodiversity4all.org/	Portugal
			Watchers—Intertidal Monitoring	http://www.beachwatchers.wsu.edu/	US
Community composition	Species interactions	Studies of important interactions or interaction networks in selected communities, such as plant-bird seed dispersal systems	Naturalist	http://www.inaturalist.org/	World
			iSpot	http://www.ispot.org.uk/	World
			Observado	http://www.observado.org/	World

species and range extensions. In some countries (e.g., Ireland), nationally organised Bioblitzes have become an important avenue to collecting biodiversity data as well as engaging citizens. Environmentally distributed ecological networks (EDENs) are growing increasingly important in ecology, coordinating research in more disciplines and over larger areas than ever before (Craine et al. 2007).

9.2.3   Timing of Nature's Events

Recently, the potential for broad scale analyses of phenology and migration has increased considerably due to public interest in conservation and particularly the development of several online CS projects (Table 9.1). Ranging from national to international efforts, examples include Nature's Notebook https://www.usanpn.org/natures_notebook, which supports large-scale plant phenology observations to collect ecological data on the timing of leafing, flowering, and fruiting of plants across the USA, attracting thousands of participants, and Project Budburst in the USA http://www.budburst.org/, which also has a strong educational focus. In the UK, Nature's Calendar http://www.naturescalendar.org.uk/ addresses the lack of long term phenological data available, as does BirdTrack http://www.bto.org/volunteer-surveys/birdtrack/about. ClimateWatch http://www.climatewatch.org.au asks volunteers to record the seasonal behaviour and location of over 180 marine and terrestrial animals across Australia. Engaging educators in the program has increased the number of sightings recorded significantly, while raising awareness about the impacts of climate change. Other national systems include Observatory of Seasons http://www.obs-saisons.fr/ in France, NatureWatch https://www.naturewatch.ca/english/ in Canada, and MigrantWatch http://www.migrantwatch.in and SeasonWatch http://www.seasonwatch.in/ in India. Finally, Journey North http://www.journeynorth.org is global in scope, aiming to study wildlife migration and seasonal change via various projects, e.g. the Spring Monarch Butterfly Migration Monitoring project http://www.learner.org/jnorth/monarch/index.html, which allows participants to track monarch butterfly migrations each fall and spring. Collectively, these projects span a vast range of plant and animal species, using web platforms and mobile apps to record data from the field.

9.2.4   Early Detection and Mapping of Pests and Invasive Species

CS projects can contribute to finding and tracking invasive species, which is especially important in detecting early outbreaks of important pests and exotics. At a more local level, apps developed for Outsmart Invasive Species http://masswoods.net/outsmart, and IveGot1 http://www.eddmaps.org/florida/report/index.cfm allow species observations to be submitted directly from the field in order to help detect and map the extent of invasive species in Massachusetts and Florida. CS programs are increasingly working at larger scale such as monitoring marine invasive species along the east coast of North America (Invasive Tracers, http://www. InvasiveTracers.com), with a focus on recently introduced non-native crabs, and tracking the spread of exotic lionfish in the Caribbean (e.g., http://www.reef.org/ programs/exotic/report and http://nas.er.usgs.gov/SightingReport.aspx). In the UK, the Big Seaweed Search http://www.nhm.ac.uk/nature-online/british-natural-history/seaweeds-survey/ is asking citizens to record sightings of 12 species of live seaweed in order to track and monitor the effects of climate change and invasive species such as wireweed (*Polygonum aviculare* L.) on the UK's seaweeds. Larger and collaborative government sponsored initiatives have been developed that bundle together reporting of exotics by CS, verification by experts, automated notification of agencies to act on potential threats, as well as tools to manage exotics (http://www.imapinvasives.org).

9.2.5 Desk Assessment and Field Validation of Imagery

CS can help to process large amounts of digital footage created by the recent explosion of low cost-high resolution video, photographic and satellite imagery. Previously, such footage would have been too cumbersome to analyse in its entirety by any single researcher or institution. While automated software can assist in this process, online crowdsourcing is particularly useful in instances where the human eye performs better than image analysis algorithms. For example, Digital Fishers http://digitalfishers.net/ allows volunteers to analyse deep sea video footage and describe what they see through a web interface that resembles the control panel of a deep sea submersible. As volunteers become more experienced, they are asked to improve their descriptions and are rewarded with new facts about deep sea species. The same video is analysed by multiple volunteers to improve consistency of descriptions. The program provides the public with an opportunity to see underwater habitats and rarely sighted deep sea species. Moreover, it continually provides new 'missions' for volunteers to maintain interest while providing researchers with valuable biodiversity data. Another example is the crowdsourcing of species from photographs taken by a camera trap, e.g. the Zooniverse Wildcam Gorongosa project http://www.wildcamgorongosa.org/ and Snapshot Serengeti (Swanson et al. 2015) as well as the multitude of other Zooniverse projects http://www.zooniverse. org that involve citizens in analysing photographs and images.

Crowdsourcing of digital imagery analysis has been shown to improve existing online data sets such as global land cover. Geo-Wiki http://www.geo-wiki.org/ involves volunteers in clarifying discrepancies between different land cover maps from their observations of Google Earth images. This removes areas of ambiguity for the development of integrated land cover maps and, as a more accurate baseline, to inform integrated assessment models. Other programs such as ForestWatchers http://forestwatchers.net/ ask volunteers to clean satellite images by selecting those with the least cloud cover, or identify areas of deforestation by marking suspect areas on a satellite image using online drawing tools. Moreover, other CS programs such as http://www.tela-botanica.org/page:herbonautes and http://herbariaunited. org/atHome/ are now engaging citizens to assist in interpreting and digitizing their museum collections making historic records accessible to wider audiences.

9.2.6 Linking Citizen Science and Large Scale Biodiversity Monitoring Databases

The global scale of anthropogenic change and the significant variance in its impact across regions has resulted in international environmental agreements, such as the Convention on Biological Diversity (Balmford et al. 2005). GEO BON aims to develop a global observation system that provides regular and timely information on biodiversity change to the CBD and other users. The examples above demonstrate the power of CS in data collection and science communication at both local and regional levels. We think there are three key initiatives that could be developed to scale up CS efforts to a global level:

- Foster data compatibility, standards, quality, storage and sharing of CS data in nationally or internationally recognised databases and support CS programs in choosing which of these are most appropriate for their program. Wiggins et al. (2013) have produced a guide on data management for CS projects that covers the full data management cycle and provides best practice guidance on many CS data issues;
- Identify data that can be collected by CS projects around the world (see Table 9.1) and carry out a gap analysis to determine where existing and future CS programs can best compliment or enhance other global data sets. For example, GEO BON has produced a candidate list of Essential Biodiversity Variables (Pereira et al. 2013) that may be appropriate to be collected by CS; and
- Build capacity globally within organisations to develop, lead and sustain CS programs that achieve sufficient rigor to collect valid data, and meaningfully engage participants over the spatial and time scales needed.

The first initiative could be realised through inviting CS programs that operate at scales larger than the local community (i.e., state, national or international programs) to join the larger scale initiatives (e.g., GEO BON), involving clear linking mechanisms (e.g., a GEO BON representative) providing guidance to this effect. For example, this representative could ensure that the data are standardized internationally, e.g. Darwin Core (GBIF 2012), and assist in identifying the most suitable national or international databases for storage. Such guidance would reduce the costs associated with developing web interfaces and web server costs associated with housing online databases. The outcome would provide global biodiversity observatories and the broader scientific community with access to usable, standardised data and provide a mechanism that can be communicated to the general public. Institutions wishing to support global biodiversity observatories can also assist CS by designing and testing both existing and new methods of data collection, analysis and interpretation, as well as by scaling these to protocols of international standards. Protocols could then be disseminated to other agencies and thereby improve both the research and communication quality of CS programs globally.

The second initiative requires global coordination and mobilisation of efforts. There has been a proliferation of CS programs in recent years (see list on scistarter.com and citsci.org), which means added competition for human resources. CS activities are often small scale and respond to local needs. The strength of CS is to develop and implement new research programs rapidly and can also expose challenges in linking to other programs with common interests. Often these programs are regional variations of the same basic theme (e.g., phenology programs such as Nature's Calendar, ClimateWatch, Project Budburst). Scistarter and others are looking into how best to simplify and serve interested participants who may want to contribute to multiple projects without needing to navigate, sign up and learn how to interact with different interfaces, tools and systems (Azavea and Scistarter 2014). Moreover, there are clearly trade-offs between projects focused on the local level and the needs of larger scale monitoring. A more coordinated approach and global framework to CS, such as the Wiki model (e.g., Geo-Wiki), would better address global issues such as climate change, land use, or introduced pests. Such a global framework would also reduce program operating costs in each participating country significantly, while simultaneously increasing the value of these data and enhance educational benefits that link local actions to global consequences. Such a global framework would also benefit from the identification of gaps where existing and future CS programs could compliment or enhance global data sets. Danielsen et al. (2014) have made progress in this area by examining how different approaches can contribute to the monitoring of the CBD Aichi Targets and 11 other international environmental agreements, including community-based projects but a comprehensive gap analysis is still lacking.

The third area is currently being addressed in part through the development of professional CS associations across the globe that are helping to coordinate and support training and capacity building around the creation and delivery of CS programs. Moreover, a number of CS toolkits (e.g., Roy et al. 2012; Tweddle et al. 2012; http://www.birds.cornell.edu/citscitoolkit/toolkit/steps; https://crowdsourcing-toolkit. sites.usa.gov/) are now available online to assist in the creation of new CS programs. GEO BON is also developing the BON-in-a-Box toolkit, which includes specific tools for CS projects aligned with BON efforts.

The rest of this chapter deals with practical issues around implementing CS programs including data quality, recruitment and motivation of participants, and the role of emerging technologies.

9.3   Enhancing Data Reliability and Reuse

CS projects span a spectrum of citizen engagement, from education and raising awareness on one end, where data collection is not necessarily a key component, to rigorous CS, where the data collected by citizens will be used for scientific research. Below we discuss two key aspects for enhancing data reliability: data quality and data standards.

9.3.1   Data Quality and Control

Accurate species identification including the identification of species through secondary field signs, such as scat surveys, bird song recording, cetacean calls (Buesching et al. 2014), is one of the most common and essential components of many CS projects. Generally, citizen scientists are better at identifying higher taxonomic categories that show a higher difference in physical characteristics and can struggle with genera lacking simple distinguishing characteristics among species. Another tendency is for participants to misidentify rarer species with limited or highly localised distributions. While an increase in data quality has been associated with the length of time and confidence of the person participating in the project (and the more familiar they become with the species monitored; e.g., Buesching et al. 2014), it is often best to leave difficult species to taxonomists. This generalisation does not always hold: for some taxa and in some places, the most reliable identifier may be an experienced and passionate lay person. Some CS systems establish a hierarchy of observers, and use the more experienced and tested observers to assess and moderate data supplied by less experienced observers. On the other hand CS participants are often willing to try and make identifications to a finer level from photographs than taxonomists are. Part of this issue is that keys and identification tools are not necessarily geared to advances in technology (e.g., digital camera and sound recordings) so that CS initiatives may result in a rethink about how tools are constructed by taxonomists, e.g. the use of Bayesian keys for biological identification on mobile devices (Rosewell and Edwards 2009).

In addition to issues of species identification, sources of bias may be present in the data, such as uneven recording intensity over time, uneven spatial coverage, uneven sampling effort per visit, uneven species detectability and variation in the types of data collected (i.e., presence-only versus presence-absence data; Bird et al. 2014; Isaac et al. 2014). Each source of variation has the potential to introduce substantial bias in trend estimates for individual species (Isaac et al. 2014). These concerns have encouraged CS practitioners to maximise data quality through improved sampling protocols and training, data standardisation and database management, and filtering or subsampling data to deal with error and uneven effort (Bird et al. 2014). For large projects or for broadly distributed databases it may be challenging to implement rigid protocols, or to effectively train volunteers or to eliminate all sources of error and bias. In these situations new statistical and high-performance computing tools can help address data-quality issues such as sampling bias, detection, measurement error, identification, and spatial clustering (Bonney et al. 2014). Whilst there are a number of proposed methods in the literature based on filtering the data to remove bias, methods of statistical correction procedure to treat recorder activities are less frequent but have (according to Isaac et al. 2014) a greater variety of mechanisms to control for recorder activity (see Isaac et al. 2014; Bird et al. 2014) for a description of statistical methods). In order to maximise data quality in citizen science, basic principles of data collection, management and analysis need to be carefully planned, and collaborations with statisticians should be considered, potentially leading to the development of new statistical approaches and survey designs for CS (Bird et al. 2014).

Training is essential and can be through online instruction and quizzes, training courses, workshops or field sessions. Face-to-face training is the most effective (e.g., Newman et al. 2003), but it is typically limited to smaller regional projects although larger scale projects can partner with local organisations to hold regional workshops. Videos are a particularly powerful way of training participants (e.g., http://masswoods.net/outsmart-workflow), as they bring a personal feel, and can also be re-watched when volunteers need a refresher. McShea et al. (2015) found that online training with videos was just as effective as in-person training while Newman et al. (2010) found that online training tools improved the quality of citizen observations in measuring percentage plant cover. Aside from introductory training, careful supervision is necessary to minimise observer error and to enhance volunteer performance (Newman et al. 2003; Buesching et al. 2014). This has proven to be particularly important in the initial training period, with follow-up spot-checks and intensive training sessions concentrating on any emergent issues to do with quality (Buesching et al. 2005, 2014).

Online communities of support such as iSpot and iNaturalist can help citizen scientists to reduce errors in their identification by drawing on the experience of others—users upload photographs of a species with a suggested identification and the online community confirms the identification or suggests other possibilities. The development of online communities can take several forms including one where members of the online community can be awarded badges to reflect their individual abilities and for the taxonomic groupings they are best able to identify. The maintenance of these communities of practice through recognition and reward systems is one of the most promising avenues of growth for helping to identify the more challenging species when using crowdsourced CS projects. iSpot provides one of the best developed systems for supporting citizens scientists and uses a multi-dimensional reputation and reward system, which is also used to verify observations (Silvertown et al. 2015).

In the process of submitting data, automated online forms can be used to highlight suspect species identification (i.e., species that are outside their known range) to both the observer as they enter these data and for data users after submission—see eBird http://ebird.org/content/ebird/ and Project FeederWatch http://feederwatch.org/ who use such systems (Bonter and Cooper 2012). Asking volunteers to upload photographs of the species recorded allows experts to carry out spot checks and address common identification issues. Innovative smartphone applications such as Leafsnap http://leafsnap.com/, which uses visual recognition software to help identify tree species from photographs of leaves, can further advance accurate species identification. Camera trap based surveys have the added advantage that all records can be verified by expert review (McShea et al. 2015) or through consensus identification by multiple crowdsourced volunteers (Swanson et al. 2015).

Validation can be further enhanced at the data entry phase, with data being filtered as they are entered in a database, using specific criteria that generate an instantaneous automated evaluation of data submissions, achieved with a checklist of species for a certain area and/or species count limits for a given date and location. Any information added that is inconsistent with predicted values should then be reviewed by an expert, e.g. depending on the type of survey, verification of an observation by photo identification, supported by extra information about the observation from the volunteer including metadata. A subset of these data can also be requested, or a few participants may be accompanied and their measurements observed, thus providing another way to understand how they are following the project protocols.

A number of papers have appeared on the quality of the data collected by citizens. Some suggest that volunteers are able to collect data of a quality similar to professionals (Brandon et al. 2003; Engel and Voshell 2002; Fore et al. 2001) while others showed variable performance; e.g. Gollan et al. (2012) found that volunteers were in less agreement with benchmark measurements compared to scientists but that this varied by individual and attribute while Kelling et al. (2015) examined data from eBird and showed variability in quality between participants. However, those with high quality submissions also tended to be the ones who contributed the most data. Techniques like those outlined above as well the big data approaches of Kelling et al. (2015) are needed to ensure that data quality is controlled for in CS projects.

In addition to the quality of those primary data that are collected (e.g., species identification), the quality of ancillary data should also be considered, e.g. the accuracy of land cover/land use maps and other demographic and ecological data obtained. Mobile apps can be used to help volunteers verify this information or some data may be checked automatically, e.g. by electronic comparison of entries against existing map layers and checklists. Feedback to contributors is essential and can be a valuable component of training or follow-up/refresher training. Statistics on frequent contributors can contribute to detecting inconsistencies in definitions and differing interpretation of instructions.

9.3.2 Data Sharing and Standards

CS projects must adopt data uniform standards if these data are to be shared across multiple projects and networks, nationally or globally. The Darwin Core (DwC) is a commonly used metadata standard for biodiversity applications, which consists of a vocabulary for taxa and their occurrence in nature. The DwC has been adopted by the Global Biodiversity Information Facility (GBIF). GBIF's website (http://tools. gbif.org/) also provides links to a number of tools that can be adopted by CS projects to facilitate the publishing of biodiversity data for further scientific use. iNaturalist, for example, was an early adopter of data standards and they now share their data openly through the GBIF portal.

Another site that promotes the sharing of species and ecosystem data is the NatureServe network (http://www.natureserve.org/) which has operated for almost 30 years. Using a set of standards and protocols referred to as the natural heritage methodology, more than 75 distributed databases have been linked successfully, searchable via a resource discovery tool on the site. NatureServe is also a data provider to GBIF and provides templates that may be of use to CS projects. DataONE is a distributed framework that links together 75 data centres, networks and organisations in order to openly share environmental data. The site includes a data management guide specifically written for the CS community that discusses the eight stages within the data management life-cycle including data discovery and sharing (Wiggins et al. 2013) while more information on standards can be found at https://www.dataone.org/all-best-practices.

9.4   Recruiting, Motivating and Retaining Participants

There are three key issues in developing a committed community of participants that will help CS projects collect reliable data successfully. These are the recruitment of contributors; the importance of considering participant motivation in the project design; and how participation can be retained and supported over the longer term, as well as ensuring that the experience is safe and well-managed. Much has been written on these topics and the reader is referred to a number of good guidance documents and articles (Dickinson et al. 2012; Roy et al. 2012; Pandya 2012; Tweddle et al. 2012; Van den Berg et al. 2012; Silvertown et al. 2013; Buesching et al. 2015).

Searchable databases are available from sites such as CS Central (http://www. birds.cornell.edu/citscitoolkit), SciStarter (http://scistarter.com/) and CS Alliance (http://www.citizensciencealliance.org/) for finding scientists and other project partners. Not surprisingly, these sites are dominated by projects for participants in English. National portals for CS projects also afford important avenues to selecting projects (e.g., Artportalen.org (Sweden), Observation.org (Netherlands), Atlas of Living Australia). Most of these sites also provide many resources and best practice guidelines on CS projects in general. Emerging technologies (see Sect. 9.5) can also play a potentially powerful role in finding partners, developing virtual communities and appealing to those people with a particular interest in technology. Simplifying and enhancing how participants can choose and participate in the right project(s) is an active area of exploration, especially for larger networks such as SciStarter (Azavea and SciStarter 2014) or iNaturalist.

Recruitment is necessary so that citizens become aware of a project's existence. The starting point for recruitment is to determine who the target audience is (e.g., school children vs. bird watchers) and to then tailor the promotion and recruitment process towards this group (Tweddle et al. 2012).

The creation of a safe and meaningful experience requires careful forethought about the nature of the participant's experience, including where, what, when and how the data will be collected, any inherent risks that may arise and how to avoid, mitigate or manage those risks. This is especially important when participants may encounter challenging or hazardous conditions, such as observations which take place on or near waterbodies, from light aircraft, in remote or risky areas or involving dangerous or poisonous species. Addressing these considerations early on with careful planning and a response plan in case problems arise is essential to creating a sustainable CS program. Earthwatch has created templates for planning and managing risk on field-based CS projects as part of a broader approach to developing field-based CS projects (Earthwatch Institute 2013).

For those individuals already engaged with these subjects, promotion and support via e-mail, newsletters, Facebook and Twitter may be sufficient. Other actions might, however, be necessary to recruit new participants, such as through the use of the national, local or regional press or utilizing different types of media (e.g., TV, radio, print, online) and specialist publications. Holding a launch event, or an event at an existing festival or fair, can provide valuable face-to-face contact that will inform potential volunteers about the aims of the project, why their help is important and what they will gain from the project. These types of events also allow citizens to interact directly with the scientists involved and establish close relationships. Word-of-mouth recruitment by existing participants is one of the most powerful means of growing the base of volunteers for a program (Prestopnik and Crowston 2012a, b; Tweddle et al. 2012).

With respect to volunteer motivation, there are many studies (Bramston et al. 2011; Bruyere and Rappe 2007; Buesching et al. 2015; Raddick et al. 2013; Silvertown et al. 2013; Van den Berg et al. 2009) that have examined this aspect of CS projects. Understanding motivation is a critical prerequisite to developing successful CS projects. For example, Van den Berg et al. (2009) surveyed volunteers enrolled in a conservation program, and revealed a number of motivations including: the desire to learn more about the science behind the project; enjoyment of the outdoors; the feeling that they are helping the environment; getting to know other people with similar interests and as a way to make new friends; and having fun. The main motivation found by Raddick et al. (2013) in participating in the Galaxy Zoo CS project was the desire to contribute to science while other motivators included interest in the scientific subject and the possibility of making new scientific discoveries. Although this list of motivations is far from exhaustive, it highlights the need to recognise that individuals are motivated by a number of different drivers and that these may differ across communities and across different demographic groups. Some communities may feel excluded and identifying the barriers to participation is important for finding solutions to widening the participation (Pandya 2012).

Project design will inevitably involve trade-offs between achieving scientific goals, e.g. gathering comprehensive, high quality data according to rigorous scientific protocols, and the ease of data collection. If the data collection is too complex or too time consuming, volunteers often lose their desire to participate and thus understanding and adapting the program to the skills, expectations and interests of the volunteers is critical (Roy et al. 2012).

Motivation is also clearly linked to maintaining participation in the longer term and data quality. Giving rapid feedback and providing regular communication about their contribution and the outcomes from the project are also excellent tools to motivate participants (Rotman et al. 2014). This can be done in different ways, such as through field events, email, phone, newsletters, blogs, discussion forums and various forms of social media.

Volunteers like the idea of knowing that their work is important and that their contributions can help scientists make better and more comprehensive analyses (Rotman et al. 2014). Rewarding citizen scientists is therefore an effective way to encourage and support participation (Tweddle et al. 2012). A reward system can be implemented in several different ways, e.g. highlighting the identity of contributors with observations to acknowledge their contributions explicitly (e.g., in Observado, iSpot and iNaturalist); providing participants with certificates of recognition (Dickinson et al. 2012); thanking participants and acknowledging their role, e.g. through organisation of a closing event, which can also be used to solicit further inputs and present the project results (Tweddle et al. 2012); providing open access to all of the non-sensitive records in the database; holding a competition to encourage participation, e.g. a photography contest (Dickinson et al. 2012); and recognizing the degree of volunteer expertise (e.g., progressing from amateur to expert levels in iSpot). Websites should make an effort to provide easy access to scientific, institutional, managerial and/or legislative products produced from project data, and to summarise these in ways of interest to contributors. It may not be readily apparent to citizens what contributions a few species observations might make collectively, e.g. to alert authorities to the arrival of invasive/pest species that appear on a list published under a national or provincial law. Encouraging these types of outreach and communication activities with citizen scientists may help to increase motivation.

Corporate engagement, fellowships and sponsorship (such as Earthwatch's 'Student Challenge Award Program/Ignite' for teenagers, the Sustainability Leadership Program for senior corporate executives (e.g., HSBC Bank) and the African Fellows program to build capacity among conservation managers) help to fulfil cross-sector participation. Integrating volunteer service directly into educational programs is another effective way to recruit and motivate individuals (Van den Berg et al. 2009). Master Naturalist programs have been established in several states such as California, Virginia, Texas and Florida that partner universities with extension services and wildlife management agencies at the state level while the Conservation Stewards Program has been established in Michigan. These programs provide individuals with a certification and require a certain number of volunteer hours, both as part of the certification and to retain certification in the future. This type of approach caters towards educational motivations for participation in CS projects and encourages longer term engagement (Van den Berg et al. 2009). School children can become highly motivated contributors in the long term to BONs, becoming networks in and of themselves. The GLOBE (Global Learning and Research to Benefit the Environment) network is one very successful example of involving students aged 13–18 in CS (Bowser and Shanley 2013). Enabling features include the development of learning elements that align with relevant core curriculum standards. Partnerships between schools and BONs are likely to become much more important in the future.

9.5 New Tools and Technologies

CS has gained in popularity over the last decade due to the emergence of a number of new tools and technologies. Web 2.0 and the Internet of Things have radically changed the way that individuals interact, collaborate and share data online. Good overviews of the technology available for CS along with the strengths and weaknesses are provided in Roy et al. (2012) and Newman et al. (2012). Here we briefly outline the potential of a range of new tools and technologies that can be used in CS projects.

9.5.1 Websites and Portals

Websites are now an established media for disseminating information, where many CS projects have online forms for data collection. Some projects also provide visualisation and analysis tools and facilities to download the data (see, for example, eBird http://ebird.org).

In some countries, national level web portals exist, which provide the ability to customise local projects to suit the needs and interests of key stakeholders (i.e., project leads, participants) at the same time as feeding into larger databases using standardised data collection and curation protocols. Moreover, these web portals provide extensive training and support to prospective and ongoing programs. Examples include Artportalen in Sweden, the Norwegian Biodiversity Information Centre, Observation.org (Netherlands), National Biodiversity Network (UK), Atlas for Living Australia, India Biodiversity Portal among others. These portals create a bridge between the needs of large BONs and addressing local needs by reducing many of the barriers that would facilitate data flow. Namely, these portals provide many of the tools, systems access to expertise, feedback and other resources that otherwise make connecting local projects to global programs challenging.

9.5.2 Mobile Devices

Smartphones and tablets have fundamentally altered CS. Through software applications or 'apps' developed specifically for these devices, training materials can be disseminated and data collection on the ground is now much easier. Since most of these mobile devices have an integrated GPS (Global Positioning System), these data can be spatially referenced automatically, with a specified degree of accuracy. Constant internet connectivity is not required as these data, collected while in the field, can be stored locally and then uploaded to a server once a wireless connection is available. With the high quality cameras that are now a common feature of many mobile devices, photographic evidence can readily accompany observations, which makes the verification of species possible. In the context of biodiversity monitoring, there are many different species identification apps available, e.g. the iNaturalist (http://www.inaturalist.org/) and iSpot (http://www.ispot.org.uk/) apps, which cover a broad geographical area, as well as more localised apps to address a specific issue, e.g. the US Department of Agriculture provides a list of apps for reporting invasive species locally (http://www.invasivespeciesinfo.gov/toolkit/monitoringsmart.shtml). Other apps include phenological information for key species (http://www. climatewatch.org.au; http://www.budburst.org).

9.5.3 Sensors

Mobile devices can also act as sensors for measuring environmental variables, e.g. the built-in microphone in these devices can be used to measure noise levels (e.g., the NoiseTube project; http://noisetube.net/) while new sensors have emerged that can measure environmental variables where the sensor communicates directly with the mobile devices using Bluetooth and other wireless technologies, e.g. SenseBox, which is a DIY sensor box for measuring environmental variables such as weather and air quality (http://www.sensebox.de/). Citizens can also wear or transport many of these new devices and take measurements as they move around in space during their daily routine. In the EU, a number of environmental citizen observatories have been developed to measure air quality, air pollution, water quality and flooding (http://www.citizen-obs.eu/). In the USA, Public Lab is a non-profit initiative to allow communities to develop and mobilise low cost, open source sensors for environmental monitoring (https://publiclab.org/). Their first project involved mapping the BP oil spill on the Gulf Coast using balloons, kites and digital cameras and they now have several ongoing community-led projects. As the Internet of Things continues to become more prevalent, sensors will become a common part of everyday citizen life.

9.5.4 Camera Traps

Camera traps are motion-sensitive sensors that record a photograph or video when an animal passes in front of it. The photographs can be verified by experts for accurate species identification. Used by scientists since the 1920s, recent developments in digital photography and the cost reduction resulting from mass commercial production have finally made them an appropriate tool for citizen use. Camera traps are used to record which species live where, to estimate their abundance, to establish rarity in the endangered species context, to capture interesting behaviours or rare events and to potentially put off poachers. Choosing a camera model can be complicated because they are constantly improving with better technology becoming available. The website http://trailcampro.com provides an annual test of commercial units in their 'trail camera shootout'. Swann et al. (2011) provide a good overview of different types of cameras, the most frequent types of problems encountered and a framework for assessing needs, while other guides are available for Australian and Malaysian contexts (Ancrenaz et al. 2012; Meek et al. 2012). The eMammal project (http://emammal.si.edu/) has developed robust cyber-infrastructure and software to have volunteers process and upload pictures directly to a digital archive at the Smithsonian. In their first year, volunteers processed over 1.5 million pictures from 1200 camera locations (McShea et al. 2015). The Snapshot Serengeti project used scientists to set cameras in Africa, but recruited citizens to help them identify the animals in their 1.2 million pictures (Swanson et al. 2015). Live image transmission from cameras via phone networks is relatively expensive, but offers a powerful way to engage the public through the unpredictable flow of animal pictures to their screen. This has been used, for example, by the Instant Wild project (http://www.edgeofexistence.org/instantwild/), which asks volunteers to use a smartphone app or website to identify animals that have been photographed from camera traps in remote places such as Kenya, Sri Lanka and Indonesia. @Camtrap live is a similar Twitter feed that streams live images and commentary from two cameras in the USA.

9.5.5 Social Media and Social Networking

There has been considerable growth in social media and social networking sites. In 2015, Facebook was estimated to have 1.55 billion active monthly users worldwide, with 1.31 billion accessing the application through their mobile devices (Statista 2015), while Twitter was estimated to have more than 320 million monthly active users (Twitter 2015). Instagram, which is another popular social media site, had more than 182.5 million users who uploaded around 58 million photos per day based on statistics for September 2015 (Statistic Brain Research Institute 2015). There has been a recent trend away from smaller, local social platforms to these large global sites, which has implications for CS projects wanting to establish a presence via social media. Social networking sites represent a very powerful way for building and maintaining CS communities and for providing virtual support mechanisms to a wide geographical audience. Many CS projects already provide integration via Facebook (e.g., iSpot), while Twitter is used to report sightings of invasive species in Ontario, (e.g., to @invspecies). Discussion forums and blog sites have been around for longer but also represent effective methods of virtual communication while Skype is now being used by teachers live from the field to reach out to children in their schools.

9.5.6 Gaming

Another approach used in CS for generating participation is 'gamification', or the addition of game elements to existing applications (Deterding et al. 2011). This approach can help to improve volunteer motivation as a tangible form of recognition by linking their contributions to levels of achievement or badges of expertise. For example, the iSpot project allows individuals to progress to 'expert' status as they identify more species, as well as a quiz to test oneself (http://www.ispotnature. org/quiz/try). The Biotracker app, which is used to contribute phenology data to Project Budburst, and uses badges and a leader board, was shown to attract an additional user group referred to as Millennials, which is the younger, technologically experienced generation (Bowser et al. 2013). Other examples of gamification in CS include Tiger Nation, which tracks the movements of tigers (Mason et al. 2012), and Happy Soft, which uses gamification in species identification (Prestopnik and Crowston 2012a, b).

9.5.7 Cyber-Infrastructure and Networked Databases

Cyber-infrastructure refers to the IT systems that support various data and system functions and ensures interoperable data exchange via networked databases. Functions include support for data storage and management, geospatial analysis tools, visualisation capability, social networking tools, quality control and training. Newman et al. (2011) provide a framework that advises CS project managers in developing and/or selecting data management systems based on the scope, scale, activities and the system approach taken within a given project. They have also developed the CitSci.org cyber-infrastructure system as a flexible open source solution (Wang et al. 2015). Other available cyber-infrastructure systems are compared by system features in Newman et al. (2011), which may help guide the choice of a system to meet project needs. More recently, some CS projects have begun to provide the otherwise expensive cyber-infrastructure to help facilitate scaling up. For example, iNaturalist lets you create a group within their program, which allows use of their cyber-infrastructure to record the location and time of any sub-group of biodiversity desired, and they offer their code as open access. Zooniverse has developed a platform for setting up CS projects, which can then be showcased on the Zooniverse platform for tapping into the Zooniverse network of users. eMammal is providing the same service for camera traps. Finally, SciStarter is preparing to upgrade their system to serve as a better basic sign-up infrastructure for simpler projects.

9.6   Challenges and Opportunities for the Future

CS provides many opportunities for increased data collection and greater involvement of citizens in scientific research across many areas that are of relevance to BONs. Indeed every day, new CS programs are launched in every corner of the globe offering people new opportunities to monitor or track species or environmental events. While this proliferation of projects offers great opportunity, there are also a number of challenges that will need to be resolved.

There are trade-offs between localised, customised projects focusing on a restricted taxonomic group or location where the advantages are more local buy-in, ownership and control, versus more interconnected or networked larger scale efforts, where there are economies of scale with data that are often more accessible and shared. How are participants to choose between similar sounding programs? How can localised programs feed into larger scale initiatives, and vice versa? Resolving questions around data standards, interoperability of systems, and attribution will be important in creating a more coherent 'marketplace' of CS opportunities. Two promising avenues are opening up. One explores how to simplify the choice of projects and reduce the barriers to learning new tools and systems for citizen scientists by improving the front end of engagement by participants (Azavea and Scistarter 2014). The other is the development of web portals that simplify much of the data management, processing and sharing across many projects. These web portals may be national in scope such as Artportalen (Sweden), the National Biodiversity Network (UK), Atlas of Living Australia and the India Biodiversity portal; taxonomic in scope (e.g., eBird), observation tool based (e.g., iNaturalist, iSpot); or EBV based (e.g., National Phenology Network). While many of these programs are mainly focused on species occurrence data, they bring together tools, processes and systems that link the local with the large scale databases.

There are also trade-offs between the collection of rigorous or reliable data gathered in a systematised fashion, on the one hand, and the ease of use or accessibility of CS programs, on the other (Pocock et al. 2015). Easing data collection protocols and reducing the number of variables collected can reduce barriers and increase or broaden involvement. Environmental education and other engagement goals are important but they can simultaneously act to increase the volume of data collected. Yet, verifiable and reliable data are often seen as essential for management decision making and scientific research outcomes. Moreover, ensuring data quality is important in attracting more scientists to use and engage with CS programs. More explicit statements about a CS program's goals, whether they seek more rigorous science or a broader environmental education effort is an important step in avoiding confusion, in expectations and outcomes, among participants and scientists alike (Pocock et al. 2015). Secondly, the development and adoption of more robust statistical approaches can help programs reduce sampling error, allowing a better balance between quantity and quality of data collected (Isaac et al. 2014; van Strien et al. 2013).

A key challenge in the next few decades is to extend the reach of CS into places where it has not had a prominent role in the past. Current CS networks are predominantly active in Europe, North America and some former colonies, such as Australia, New Zealand and South Africa. Africa, Latin America and Asia are under-represented. Growing wealth and education in these areas, along with near-universal penetration of internet services and cell phones, creates an opportunity to extend CS into these biodiversity-rich regions. The motivations and social mechanisms to do so may differ from those found in 'western' societies, but there is nevertheless a rich vein of traditional knowledge and interest in biodiversity which can be tapped.

CS is already playing an important role in ground-based monitoring, complementing and corroborating the global satellite-based observations and more focused government or institution led efforts. This chapter outlined some of the tools and opportunities for building on existing and developing new CS initiatives to help BON efforts increase our understanding of the status and trends of biodiversity. Perhaps most importantly, a growth of CS programs that engage a broader constituency of people collecting biodiversity information will build the essential social equity and foster the necessary dialogue that stimulates the political will to make the decisions necessary for a sustainable and biodiverse planet.

References

Akçakaya, H. R., Pereira, H. M., Canziani, G., Mbow, C., Mori, A., Palomo, M. G., et al. (2016). Improving the rigour and usefulness of scenarios and models through ongoing evaluation and refinement. In S. Ferrier, K. N. Ninan, P. Leadley, R. Alkemade, L. Acosta-Michlik, H. R. Akcakaya, L. Brotons, W. Cheung, V. Christensen, K. H. Harhash, J. Kabubo-Mariara, C. Lundquist, M. Obersteiner, H. M. Pereira, G. Peterson, R. Pichs, C. Rondinini, N.

Ravindranath, B. Wintle (Eds.), IPBES, 2016: Methodological assessment of scenario analysis and modelling of biodiversity and ecosystem services. IPBES Secretariat, Bonn, Germany.

Ancrenaz, M., Hearn, A. J., Ross, J., Sollmann, R., & Wilting, A. (2012). *Handbook for wildlife monitoring using camera-traps.* Kota Kinabalu, Malaysia: BBEC II Secretariat. http://www.bbec.sabah.gov.my/japanese/downloads/2012/april/camera_trap_manual_for_printing_final.pdf

Azavea and SciStarter. (2014). *Citizen science data factory. A distributed data collection platform for citizen science. Part 1: Data collection platform evaluation.* http://www.azavea.com/index.php/download_file/view/1368/

Balmford, A., Bennun, L., ten Brink, B., Cooper, D., Côté, I. M., Crane, P., et al. (2005). The convention on biological diversity's 2010 target. *Science, 307,* 212–213.

Bird, T. J., Bates, A. E., Lefcheck, J. S., Hill, A., Thomson, R. J., Edgar, G. J., et al. (2014). Statistical solutions for error and bias in global citizen science datasets. *Biological Conservation, 173,* 144–154. http://dx.doi.org/10.1016/j.biocon.2013.07.037

Bohmann, K., Evans, A., Gilbert, M. T. P., Carvalho, G. R., Creer, S., Knapp, M., et al. (2014). Environmental DNA for wildlife biology and biodiversity monitoring. *Trends in Ecology & Evolution, 29,* 358–367.

Bonney, R., Ballard, H., Jordan, R., McCallie, E., Phillips, T., Shirk, J., & Wilderman, C. C. (2009a). *Public Participation in Scientific Research: Defining the Field and Assessing its Potential for Informal Science Education (A CAISE Inquiry Group Report).* Center for Advancement of Informal Science Education (CAISE), Washington DC, USA. http://caise.insci.org/uploads/docs/PPSR%20report%20FINAL.pdf

Bonney, R., Cooper, C. B., Dickinson, J., Kelling, S., Phillips, T., Rosenberg, K. V., et al. (2009b). Citizen science: A developing tool for expanding science knowledge and scientific literacy. *BioScience, 59,* 977–984.

Bonney, R., Shirk, J. L., Phillips, T. B., Wiggins, A., Ballard, H. L., Miller-Rushing, A. J., & Parrish, J. K. (2014). Next steps for citizen science. *Science, 343,* 1436–1437.

Bonter, D. N., & Cooper, C. B. (2012). Data validation in citizen science: a case study from Project FeederWatch. *Frontiers in Ecology and the Environment, 10,* 305–307.

Bowser, A., Hansen, D., He, Y., Boston, C., Reid, M., Gunnell, L., & Preece, J. (2013). Using gamification to inspire new citizen science volunteers. In *Proceedings of Gamification 2013* (pp. 18–25). Presented at Gamification 2013. Stratford, ON, Canada: ACM Press. doi:10.1145/2583008.2583011

Bowser, A., & Shanley, L. (2013). *New visions in citizen science (Case Study Series).* Washington, D.C., USA: The Woodrow Wilson Center.

Bramston, P., Pretty, G., & Zammit, C. (2011). Assessing environmental stewardship motivation. *Environment and Behavior, 43,* 776–788.

Brandon, A., Spyreas, G., Molano-Flores, B., Carroll, C., & Ellis, J. (2003). Can volunteers provide reliable data for forest vegetation surveys? *Natural Areas Journal, 23,* 254–262.

Bruyere, B., & Rappe, S. (2007). Identifying the motivations of environmental volunteers. *Journal of Environmental Planning and Management, 50,* 503–516.

Buesching, C. D., Newman, C., & Macdonald, D. W. (2005). Volunteers in ecological research: amateur ecological monitors: The benefits and challenges of using volunteers. *Bulletin of the British Ecological Society, 36,* 20–22.

Buesching, C. D., Newman, C., & Macdonald, D. W. (2014). How dear are deer volunteers: the efficiency of monitoring deer using teams of volunteers to conduct pellet group counts. *Oryx, 48,* 593–601.

Buesching, C. D., Slade, E. M., Newman, C., Ruitta, T., Riordan, P., & Macdonald, D. W. (2015). Many hands make light work—But do they? A critical evaluation of citizen science. In D. W. Macdonald, R. Feber (Eds.), *Wildlife Conservation on Farmland Volume 2: Conflict in the Countryside.* Oxford and New York: Oxford University Press, 293–317.

Craine, J. M., Battersby, J., Elmore, A. J., & Jones, A. W. (2007). Building EDENs: The rise of environmentally distributed ecological networks. *BioScience, 57,* 45–54.

Danielsen, F., Pirhofer-Walzl, K., Adrian, T. P., Kapijimpanga, D. R., Burgess, N. D., Jensen, P. M., et al. (2014). Linking public participation in scientific research to the indicators and needs of international environmental agreements: Monitoring environmental agreements. *Conservation Letters, 7*, 12–24.

Deterding, S., Sicart, M., Nacke, L., O'Hara, K., & Dixon, D. (2011). Gamification. using game-design elements in non-gaming contexts. In *Proceedings of Gamification 2013* (p. 2425). Presented at Gamification 2013. Stratford, ON, Canada: ACM Press.

Dickinson, J. L., Shirk, J., Bonter, D., Bonney, R., Crain, R. L., Martin, J., et al. (2012). The current state of citizen science as a tool for ecological research and public engagement. *Frontiers in Ecology and the Environment, 10*, 291–297.

Earthwatch Institute. (2013). *Eartwatch field manual* (2nd ed.). http://earthwatch.org/Portals/0/Downloads/Research/scientist-materials/earthwatch-field-manual.pdf

Engel, S. R., & Voshell, J. R. (2002). Volunteer biological monitoring: Can it accurately assess the ecological condition of streams? *American Entomologist, 48*, 164–177.

Fore, L. S., Paulsen, K., & O'Laughlin, K. (2001). Assessing the performance of volunteers in monitoring streams. *Freshwater Biology, 46*, 109–123.

GBIF. (2012). Darwin Core Quick Reference Guide (version 1.3). Copenhagen, Denmark: Global Biodiversity Information Facility (GBIF). http://www.gbif.org/resource/80633

GEO BON. (2015a). GEO BON Biannual Progress Report 2014-2015. GEO BON Secretariat, Leipzig, Germany. http://geobon.org

GEO BON (2015b). National, regional and thematic Biodiversity Observation Networks (BONs): Background and criteria for endorsement. GEO BON Secretariat, Leipzig, Germany. Available at http://geobon.org

Gollan, J., de Bruyn, L. L., Reid, N., & Wilkie, L. (2012). Can volunteers collect data that are comparable to professional scientists? A study of variables used in monitoring the outcomes of ecosystem rehabilitation. *Environmental Management, 50*(5), 969–978.

Isaac, N. J. B., van Strien, A. J., August, T. A., de Zeeuw, M. P., & Roy, D. B. (2014). Statistics for citizen science: Extracting signals of change from noisy ecological data. *Methods in Ecology and Evolution, 5*, 1052–1060.

Kelling, S., Fink, D., La Sorte, F. A., Johnston, A., Bruns, N. E., & Hochachka, W. M. (2015). Taking a "Big Data" approach to data quality in a citizen science project. *Ambio, 44*(S4), 601–611. doi:10.1007/s13280-015-0710-4.

Lundmark, C. (2003). BioBlitz: Getting into backyard biodiversity. *BioScience, 53*(4), 329.

Mason, A. D., Michalakidis, G., & Krause, P. J. (2012). *Tiger Nation: Empowering citizen scientists* (pp. 1–5). IEEE.

McShea, W. J., Forrester, T., Costello, R., He, Z., & Kays, R. (2015). Volunteer-run cameras as distributed sensors for macrosystem mammal research. *Landscape Ecology, 31*(1), 55–66.

Meek, P., Ballard, G., & Fleming, P. (2012). An introduction to camera trapping for wildlife surveys in Australia. Invasive Animals Cooperative Research Centre, Canberra. 95 p.

Miller-Rushing, A., Primack, R., & Bonney, R. (2012). The history of public participation in ecological research. *Frontiers in Ecology and the Environment, 10*(6), 285–290.

Newman, C., Buesching, C. D., & Macdonald, D. W. (2003). Validating mammal monitoring methods and assessing the performance of volunteers in wildlife conservation—"Sed quis custodiet ipsos custodies?". *Biological Conservation, 113*(2), 189–197.

Newman, G., Crall, A., Laituri, M., Graham, J., Stohlgren, T., Moore, J. C., et al. (2010). Teaching citizen science skills online: Implications for invasive species training programs. *Applied Environmental Education and Communication, 9*(4), 276–286.

Newman, G., Graham, J., Crall, A., & Laituri, M. (2011). The art and science of multi-scale citizen science support. *Ecological Informatics, 6*, 217–227.

Newman, G., Wiggins, A., Crall, A., Graham, E., Newman, S., & Crowston, K. (2012). The future of citizen science: emerging technologies and shifting paradigms. *Frontiers in Ecology and the Environment, 10*, 298–304.

Pandya, R. (2012). A framework for engaging diverse communities in citizen science in the US. *Frontiers in Ecology and the Environment, 10*, 314–317.

Pereira, H., & Cooper, D. (2006). Towards the global monitoring of biodiversity change. *Trends in Ecology & Evolution, 21*(3), 123–129.

Pereira, H. M., Belnap, J., Brummitt, N., Collen, B., Ding, H., Gonzalez-Espinosa, M., et al. (2010). Global biodiversity monitoring. *Frontiers in Ecology and the Environment, 8*(9), 459–460.

Pereira, H. M., Ferrier, S., Walters, M., Geller, G. N., Jongman, R. H. G., Scholes, R. J., et al. (2013). Essential biodiversity variables. *Science, 339*, 277–278.

Pocock, M. J. O., Newson, S. E., Henderson, I. G., Peyton, J., Sutherland, W. J., Noble, D. G., et al. (2015). Developing and enhancing biodiversity monitoring programmes: A collaborative assessment of priorities. *Journal of Applied Ecology, 52*(3), 686–695.

Prestopnik, N., & Crowston, K. (2012a). *Purposeful gaming & socio-computational systems: A citizen science design case* (p. 75). ACM Press.

Prestopnik, N. R., & Crowston, K. (2012b). Citizen science system assemblages: understanding the technologies that support crowdsourced science. In *Proceedings of the 2012 iConference* (pp. 168–176). ACM Press.

Raddick, M. J., Bracey, G., Gay, P. L., Lintott, C. J., Cardamone, C., Murray, P., et al. (2013). Galaxy Zoo: Motivations of citizen scientists. *Astronomy Education Review, 12*(1), 010106.

Rosewell, J., & Edwards, M. (2009). *Bayesian keys: Biological identification on mobile devices.* Presented at the ICL2009, Villach, Austria.

Rotman, D., Hammock, J., Preece, J., Hansen, D., Boston, C., Bowser, A., & He, Y. (2014). Motivations affecting initial and long-term participation in citizen science projects in three countries. In *iConference 2014 Proceedings* (pp. 110–124). Presented at the iConference 2014, iSchools.

Roy, H. E., Pocock, M. J. O., Preston, C. D., Roy, D. B., Savage, J., Tweddle, J. C., & Robinson, L. D. (2012). *Understanding Citizen Science & Environmental Monitoring. Final Report on behalf of UK-EOF.* NERC Centre for Ecology & Hydrology and Natural History Museum, UK. http://www.ceh.ac.uk/products/publications/documents/citizensciencereview.pdf

Sih, A. (2013). Understanding variation in behavioural responses to human-induced rapid environmental change: A conceptual overview. *Animal Behaviour, 85*(5), 1077–1088.

Silvertown, J., Buesching, C. D., Jacobson, S., Rebello, T., & Birtles, A. (2013). Citizen science and nature conservation. *Key Topics in Conservation Biology, 2*, 127–142.

Silvertown, J., Harvey, M., Greenwood, R., Dodd, M., Rosewell, J., Rebelo, T., et al. (2015). Crowdsourcing the identification of organisms: A case-study of iSpot. *ZooKeys, 480*, 125–146.

Statista. (2015). The Statistics Portal: Number of monthly active Facebook users worldwide as of 3rd quarter 2015 (in millions). http://www.statista.com

Statistic Brain Research Institute. (2015). Instagram Company Statistics. http://www.statisticbrain.com/instagram-company-statistics/

Swann, D. E., Kawanishi, K., & Palmer, J. (2011). Evaluating types and features of camera traps in ecological studies: A guide for researchers. In A. O'Connell, J. D. Nichols, & K. U. Karanth (Eds.), *Camera traps in animal ecology: Methods and analyses* (pp. 27–44). Dordrecht, Heidelberg: Springer.

Swanson, A., Kosmala, M., Lintott, C., Simpson, R., Smith, A., & Packer, C. (2015). Snapshot Serengeti, high-frequency annotated camera trap images of 40 mammalian species in an African savanna. *Scientific Data, 2*, 150026.

Tweddle, J. C., Robinson, L. D., Pocock, M. J. O., & Roy, H. E. (2012). *Guide to citizen Science: developing, implementing and evaluating citizen science to study biodiversity and the environment in the UK.* Natural History Museum and NERC Centre for Ecology & Hydrology for UK-EOF. http://www.ceh.ac.uk/products/publications/documents/CitizenScienceGuide.pdf

Twitter. (2015). Twitter. http://www.twitter.com

Van den Berg, H. A., Dann, S. L., & Dirk, J. M. (2009). Motivations of adults for non-formal conservation education and volunteerism: Implications for Programming. *Applied Environmental Education and Communication, 8*, 6–17.

van Strien, A. J., van Swaay, C. A. M., & Termaat, T. (2013). Opportunistic citizen science data of animal species produce reliable estimates of distribution trends if analysed with occupancy models. *Journal of Applied Ecology, 50*(6), 1450–1458.

Wang, Y., Kaplan, N., Newman, G., & Scarpino, R. (2015). CitSci.org: A new model for managing, documenting, and sharing citizen science data. *PLoS Biology, 13*(10), e1002280.

Wiggins, A., Bonney, R., Graham, E., Henderson, S., Kelling, S., Littauer, R., et al. (2013). *Data management guide for public participation in scientific research.* DataONE, Albuquerque, NM. https://www.dataone.org/sites/all/documents/DataONE-PPSR-DataManagementGuide.pdf

Chapter 10
Biodiversity Modelling as Part of an Observation System

Simon Ferrier, Walter Jetz and Jörn Scharlemann

Abstract Modelling provides an effective means of integrating the complementary strengths of biodiversity data derived from in situ observation versus remote sensing. The use of modelling in biodiversity change observation, or monitoring, is just one of a number of roles that modelling can play in biodiversity assessment. These roles place different levels of emphasis on explanatory versus predictive modelling, and on modelling across space alone, versus across both space and time, either past-to-present or present-to-future. One of the most challenging, yet vitally important, applications of modelling to biodiversity monitoring involves mapping change in the distribution and retention of terrestrial biodiversity. Unlike many structural and functional attributes of ecosystems, most biological entities at the species and genetic levels of biodiversity cannot be readily detected through remote sensing. Estimating change in these levels of biodiversity across large spatial extents is therefore benefiting from advances in both species-level and community-level approaches to model-based integration of in situ biological observations and remotely sensed environmental data.

Keywords Biodiversity · Modelling · Monitoring · Remote sensing · Explanatory · Predictive · Species-level · Community-level

S. Ferrier (✉)
CSIRO Land and Water, Australian Capital Territory 2601, PO Box 1700,
Canberra, Australia
e-mail: Simon.Ferrier@csiro.au

S. Ferrier · J. Scharlemann
United Nations Environment Programme World Conservation Monitoring Centre,
219 Huntingdon Road, Cambridge CB3 0DL, UK
e-mail: J.Scharlemann@sussex.ac.uk

W. Jetz
Department of Ecology and Evolutionary Biology, Yale University, New Haven,
CT, USA
e-mail: walter.jetz@yale.edu

J. Scharlemann
School of Life Sciences, University of Sussex, Brighton BN1 9QG, UK

© The Author(s) 2017
M. Walters and R.J. Scholes (eds.), *The GEO Handbook on Biodiversity Observation Networks*, DOI 10.1007/978-3-319-27288-7_10

10.1 Introduction

Data on changes in the state of biodiversity on our planet come mostly from two broad sources: (1) in situ observation of organisms, or attributes of these organisms, obtained directly through application of various on-ground or in-water biological survey techniques, or through collection of museum specimens; and (2) remote sensing of biophysical characteristics of the planet's surface detected by various satellite-borne or airborne sensors. These two sources of data have complementary strengths and weaknesses (Ferrier 2011). In situ observation provides direct information on a rich array of relevant biological entities and attributes, but the spatial coverage of such surveys is often very sparse—i.e., sampled locations are typically separated by expanses of unsurveyed land or ocean. Remote sensing, on the other hand, provides complete spatial coverage, but has limited capability to reliably detect or measure many of the biological entities or attributes of interest in biodiversity monitoring. These complementarities have, over recent decades, stimulated extensive efforts to develop change-observation methodologies that better integrate the respective strengths of in situ observation and remote sensing (Turner 2014).

While approaches to achieving this integration are many and varied, we here make an initial distinction between two broad strategies, based largely on the extent to which the biodiversity variable of interest is detectable, and therefore measurable, through remote sensing. Some types of variables are much easier to measure using remote sensing than others. For example, variables relating to ecosystem-level structural properties or functional processes—e.g., percent tree cover, canopy height, biomass, gross primary productivity (Smith et al. 2014)—tend to be more amenable to remote measurement than variables relating to species-level or genetic-level composition (Skidmore et al. 2015). Where variables, or suitable proxies, can be estimated through reasonably direct analysis or modelling of raw data from remote sensing, integration with in situ data focuses mainly on calibration and validation—i.e., using ground-based observations of the same variable as that measured remotely to calibrate (or train) the interpretation of remote data, and to test the accuracy of mapping (Baccini et al. 2007). However, in situ/remote sensing integration becomes considerably more challenging if the biological entity or attribute of interest cannot be detected readily through remote sensing—as is the case for most elements of biodiversity at the species and genetic levels.

Imagine, for example, setting out to map change in the distribution of a small forest-dwelling bird species. Unlike a variable such as percent tree cover, the presence of this species cannot be estimated directly from remote sensing. This situation demands an approach to integration that focuses less on linking in situ and remotely-sensed estimates of the same variable, and more on modelling the relationship between a variable of interest, measurable only through in situ observation, and one or more remotely mapped variables thought to be potential drivers of this variable. In this case modelling might, for example, be used to predict (or infer) change in the bird's distribution as a function of the observed relationship between in situ data on this species and remote mapping of climate and land-cover change. In reality the measurability of variables through remote sensing forms a continuous spectrum, requiring a gradation of approaches to in situ/remote sensing integration. At one end of this spectrum in situ data are used purely to calibrate and validate estimates of a variable derived directly from remote sensing. At the other end of the spectrum, estimation of a variable of interest is made possible only by integrating in situ and remotely-sensed data through modelling, because the variable is not directly measurable through remote sensing. In this latter situation in situ data are used to calibrate and validate a model predicting the variable of interest, rather than for calibrating and validating observations of the variable itself.

Several other chapters in this book discuss applications of modelling in different fields of biodiversity monitoring—e.g. for tracking change in freshwater biodiversity in Chap. 7, in terrestrial species in Chap. 4, in genetic diversity in Chap. 5, and in ecosystem services in Chap. 3; and for adding value to remote sensing of change in Chap. 8. This chapter complements these other treatments by exploring in greater depth: (1) how the use of modelling in biodiversity monitoring relates to, and should therefore link with, the broader set of roles that modelling plays in biodiversity assessment (Sect. 10.2); and (2) the importance of matching employed modelling techniques to the particular needs of different applications in biodiversity monitoring, using as a case study the challenge of mapping change in biodiversity composition (Sect. 10.3).

10.2 Broad Roles of Modelling in Biodiversity Assessment

The use of modelling in biodiversity change observation, or monitoring, is just one of a number of roles that modelling can play in biodiversity assessment. To make better sense of this diversity of roles it is useful to first define more precisely what is actually meant by 'modelling' in a biodiversity context. In simple terms, a model is a set of mathematical equations (e.g., $y = a + bx$), or logical rules (e.g., if $x > c$ then $y = 1$), that link a biodiversity variable of interest (the 'y' in these examples; referred to variously as the 'dependent', 'response' or 'outcome' variable) to one or more other variables (e.g., environmental drivers) thought to be of importance in determining, or influencing, this response (referred to variously as 'independent', 'predictor', or 'explanatory' variables). When publications or reports on biodiversity talk about 'modelling' they can be referring to either one, or sometimes both, of two quite different activities. The first of these is what we will call here 'explanatory modelling' (Shmueli 2010). This activity is essentially a form of data analysis, and involves using available data (observations) both for the biodiversity response variable of interest, and for the relevant predictor variables, to generate or fit a model that assesses, and describes, the relationship between these two sets of variables. In other words, known information on predictor and response variables is used to derive a model that did not exist prior to this activity. The second activity, which we here call 'predictive modelling' (Shmueli 2010), instead presumes that a model describing the relationship of interest is already known, as are observed or estimated values of the relevant predictor variables, and therefore combines these to predict previously unknown values of the biodiversity response variable. The model used to make such predictions can be either an 'inductive model' derived through data analysis, in which case the activities of data analysis and prediction are integrally linked, or a 'deductive model' built directly from existing expert knowledge of the relationship between response and predictor variables (Corsi et al. 2000; Overmars et al. 2007; Tuanmu and Jetz 2014).

10.2.1 Modelling Across Space Alone

Both explanatory and predictive modelling can be conducted either across geographical space, or across time, or across both space and time. The various roles played by modelling in biodiversity assessment involve different combinations of these possibilities (Figs. 10.1 and 10.2). The most basic roles are those in which modelling is conducted across space alone, at a single point in time (usually the present). Explanatory modelling of correlations, or associations, between a biodiversity response variable observed at a sample of geographical locations, and a set

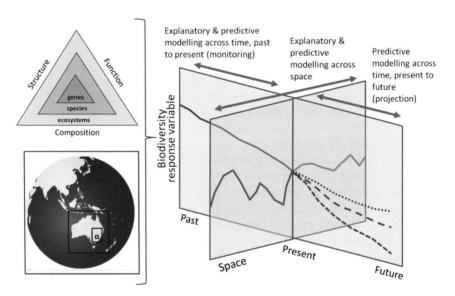

Fig. 10.1 Major roles of modelling in biodiversity assessment: explanatory versus predictive modelling across space versus time (past-to-present and present-to-future), for biodiversity response variables relating to different levels and dimensions of biological organisation, and different spatial scales

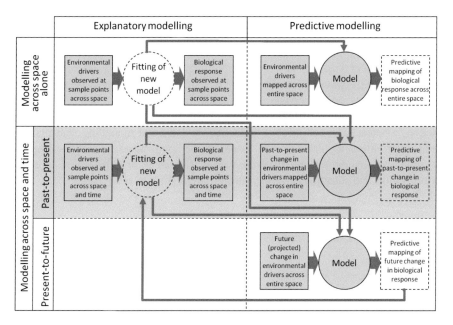

Fig. 10.2 Major modes of biodiversity modelling, distinguishing between explanatory and predictive modelling, and between modelling across space alone and modelling across space and time (either past-to-present or present-to-future). The shaded portion highlights those modes of most relevance to biodiversity change observation and monitoring

of predictor variables measured, or estimated, at these same locations, can help to shed light on the relative importance of different drivers in determining spatial patterns in biodiversity, and on the form (shape) of these relationships. The fitting of correlative species distribution models (SDMs) relating observations of presence, presence-absence, or abundance of a given species to multiple environmental variables (e.g., climate, terrain, soil, land-use variables) is probably the best known, and most widely applied, manifestation of such data analysis (Elith and Leathwick 2009). Other examples include statistical analyses of community-level, or ecosystem-level, attributes (e.g., species richness, functional diversity) measured at field sites distributed across different classes of land use or management (de Baan et al. 2013; Newbold et al. 2015).

Explanatory modelling of drivers affecting the spatial distribution of biodiversity may be all that is required for some applications—e.g., to inform development of government policy to reduce the detrimental impact of a particular form of land use or management. However if the environmental variables used in model fitting are also mapped across an entire region of interest (e.g., as grids in a GIS) then a model derived through data analysis can, in turn, provide the foundation for prediction across geographical space (Miller et al. 2004). In the case of an SDM, this involves combining the fitted model with environmental values for each grid-cell in the region to predict occurrence within that cell, thereby producing a complete map of the predicted distribution of the species of interest. Predictive modelling of biodiversity response variables across geographical space can also be undertaken using models developed through means other than correlative data analysis. For example, the distribution of a species of interest might be predicted using a simple deductive model relating the presence or abundance of that species to mapped vegetation (or land-cover) types and/or classes of land use or management, based on expert knowledge (Stoms et al. 1992; Pearce et al. 2001; Jetz et al. 2012). Spatial prediction, whether achieved through inductive or deductive modelling, can make a vital contribution to planning and management applications requiring complete geographical mapping of biodiversity values (Guisan et al. 2013)—e.g., for the prioritisation and selection of new protected areas (Ferrier et al. 2002).

10.2.2 Modelling Across Space and Time, Present to Future

Other applications of modelling to policy development, planning and management require explanatory and/or predictive modelling to be performed not only across space, but also across time. The use of modelling to predict potential changes in biodiversity into the future, often referred to as 'forecasts' or 'projections' (Coreau et al. 2009), as a function of ongoing impacts of environmental drivers (e.g., climate and land-use change), has gained particular prominence in recent years (Pereira et al. 2010; Cook et al. 2014). Such modelling poses special challenges, as there is usually considerable uncertainty associated with the future trajectories of relevant environmental drivers, which themselves will be affected by socio-economic events and decisions that are yet to occur, and are therefore highly unpredictable. These uncertainties are often addressed through the use of scenarios—i.e., multiple plausible trajectories for environmental drivers, that account for the reality that not just one, but many, futures are possible (van Vuuren et al. 2012). Model-based biodiversity projections under plausible scenarios of change in key drivers can contribute significantly to policy agenda setting, by helping to characterise and communicate the potential magnitude of ongoing change in biodiversity, and therefore the need for action. By extending scenarios to further consider the effects of alternative policy or management interventions, such projections can also play an important role in decision support—i.e., helping policy-makers, planners and managers to choose between possible actions for addressing the problem at hand, by modelling the difference that each of these alternatives is expected to make to projected outcomes for biodiversity (Cook et al. 2014).

As for predictive modelling across geographical space, projections of biodiversity change into the future can be based on either inductive or deductive modelling (Pereira et al. 2010). When inductive models are employed for future projection, these are most often derived from correlative data analysis (i.e., explanatory modelling) of relationships between biodiversity and environmental drivers observed across space alone, rather than across time. Using such models to project changes across time involves space-for-time substitution. This assumes that the correlation observed across space between a given biological response variable, and one or more environmental variables (e.g., between the presence of a species, and climate and land use), will also hold across time, and can therefore be used to predict future changes in this response variable as a function of changing environmental conditions. While there is often little choice but to rely on space-for-time substitution for projecting future change in biodiversity, questions are increasingly being raised and examined around the robustness of this approach (Bonthoux et al. 2013; Araujo and Peterson 2012; Blois et al. 2013).

10.2.3   Modelling Across Space and Time, Past to Present

Modelling change in biodiversity across time is not limited to future projection, but is also crucially important for observing and analysing change in biodiversity that has already occurred (past to present). Modelling plays two broad roles in biodiversity change observation and monitoring, aligned directly with the distinction between explanatory modelling and predictive modelling introduced above. Where changes both in a biodiversity response of interest, and in relevant environmental drivers, are observed over both space and time, explanatory modelling of driver-response correlations can be taken to a level of rigour beyond that of modelling based on observations from across space alone (Kery et al. 2013). In addition to direct provision of stronger policy-relevant evidence for the impact of drivers on biodiversity, explanatory modelling based on temporal observations is also vital to achieving more effective integration of biodiversity monitoring (past to present) and projection (present to future). Inductive models derived through analysis of observed changes in biodiversity and environmental drivers over time are likely to provide a stronger foundation for projecting future change than projections based purely on space-for-time substitution (Santika et al. 2014). This is because models fitted to temporal data have potential to better distinguish actual drivers of change from environmental variables simply exhibiting spatial autocorrelation with these drivers, and to better account for the effects of dynamic processes that may be difficult to detect and describe based on spatial data alone (e.g., the phenotypic plasticity of species in the face of environmental change). Even more importantly, using explanatory modelling to analyse future observations generated by ongoing biological and environmental monitoring initiatives offers a powerful means of testing projections made over the same time period, thereby informing adaptive refinement of models underpinning policy and decision-making into the future (Ferrier 2012; Rapacciuolo et al. 2014).

   The second major role that modelling plays in relation to biodiversity change observation and monitoring is predictive, rather than explanatory, in nature. Rather than projecting potential changes in biodiversity into the future (as described in Sect. 10.2.2), model-based prediction is used here to help fill spatial and temporal gaps in the coverage of direct observations of biodiversity change past-to-present. As noted earlier, many biological entities or attributes of interest from a biodiversity monitoring perspective can be detected only through in situ observation. Locations at which changes in these variables are measured directly therefore tend to be distributed very sparsely, and often unevenly, across the planet's surface. In contrast, changes in environmental drivers—e.g. climate, land use—are often more amenable to detection through remote sensing, and therefore potentially mappable across large geographical extents. For applications in policy, planning or management that require complete geographical coverage of information on biodiversity change, predictive modelling can play a valuable role in translating mapped changes in key drivers, generated through remote sensing, into expected changes in biodiversity (Lung et al. 2012; Soberon and Peterson 2009). The models underpinning such translation can, again, be either deductive or inductive, with the latter derived from explanatory modelling of biological and environmental data distributed either across space alone (and therefore constituting another form of space-for-time substitution), or across both space and time.

The remainder of this chapter explores, in greater depth, this last role of modelling in biodiversity monitoring—i.e., the use of predictive modelling to help map past-to-present changes in the distribution of biodiversity across large spatial extents.

10.3   A Key Modelling Challenge: Mapping Change in the Distribution and Retention of Terrestrial Biodiversity

Unlike many structural and functional attributes at the ecosystem level, most biological entities at the species and genetic levels of biodiversity cannot be readily detected through remote sensing. Notable exceptions include the emerging use of very high spatial resolution imagery to identify individual organisms of certain large-bodied, and conspicuous, animal species (e.g., penguins; Fretwell et al. 2012), and the use of hyperspectral sensors to detect variation in plant species composition in the top layer of vegetation communities (Leutner et al. 2012). These developments offer considerable potential for direct derivation of spatially-complete mapping of temporal change from remote sensing, for at least a subset of biological entities. However this still leaves a very large proportion of our planet's biological diversity that is effectively invisible to satellite-borne remote sensing, both at the species level and, even more so, at the genetic level. In situ monitoring of change in these components of diversity at selected locations may provide all the information that is needed for some applications—e.g., for monitoring the performance of local-scale management actions (Lindenmayer et al. 2012). Estimating change across large spatial extents—e.g., across a whole ecoregion, country or continent, or across the entire planet—poses a much greater challenge for in situ monitoring, particularly if these changes need to be mapped at relatively fine spatial resolution across the entire extent of interest (Ferrier 2011; Jetz et al. 2012; Pereira and Cooper 2006). We here explore how various modelling approaches can be used to help address this challenge, by integrating the respective strengths of data generated through in situ and remote sensing observation techniques.

Our focus is mostly on the terrestrial realm, although many of the modelling approaches discussed below are also applicable in freshwater and marine systems. We first consider 'species-level approaches' that model and map changes in the distribution of individual species, and then move on to examine 'community-level approaches' that focus instead on modelling and mapping changes in the distribution and retention of biological diversity within whole communities, without providing explicit information on the individual species comprising this diversity.

10.3.1 Species-Level Approaches

Interest in techniques for modelling, and thereby mapping, distributions of individual species as a function of remotely mapped environmental variables has grown rapidly over the past 30 years. Particularly strong attention has been directed towards correlative species distribution modelling (SDM) which uses statistical model-fitting, or machine learning, to derive explanatory models linking in situ observations of species occurrence to environmental predictors (Elith and Leathwick 2009). This largely inductive approach has been complemented, to a lesser extent, by deductive modelling of distributions based on expert knowledge of the environmental or habitat requirements of species (Jetz et al. 2012), or by more mechanistic modelling based on independently acquired evidence of ecophysiological limits or understanding of other relevant ecological factors and processes shaping species distributions (Kearney et al. 2010).

While the scientific literature on species distribution modelling is now very extensive (Guisan et al. 2013; Ahmed et al. 2015), a large proportion of these studies have focused on using such modelling to predictively map distributions across space alone, or to project potential changes in distribution into the future under alternative global-change scenarios. The use of this modelling paradigm in biodiversity monitoring—i.e., to help map past-to-present changes in species distributions—is surprisingly rare relative to these other applications. A number of options are nevertheless available for making effective use of species distribution modelling in monitoring (Fig. 10.3). To simplify the explanation of these options we will here focus on just two of the main drivers of ongoing changes in biological distributions—i.e., habitat loss or degradation (linked to changes in land cover and use) and climate change—both of which are amenable to spatially-complete change detection and mapping through remote sensing. If remotely-sensed variables relating to habitat loss or degradation are included as predictors in explanatory models fitted inductively to species occurrences observed across space alone, then such models can be used to predictively map distributional changes as a function of observed change in these variables, through simple space-for-time substitution (Lung et al. 2012). Alternatively, deductive modelling based on expert knowledge

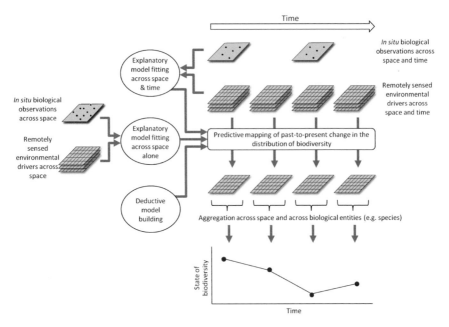

Fig. 10.3 General framework for using modelling to integrate in situ and remotely sensed observations for mapping change in the distribution and retention of terrestrial biodiversity

of the association between a given species and classes of land cover, or broad habitat type, can be used to predict changes in the distribution of that species as a function of remote mapping of these classes over time (Jetz et al. 2007, 2012). Considerable scope also exists to combine inductive and deductive modelling approaches in this context—e.g., by using inductive species distribution modelling, and available occurrence records, to map the 'natural' or 'original' distribution of a species as a function of mapped abiotic environmental variables (climate, terrain, soils etc.), and then using remote sensing and simple deduction to map changes over time in the portions of this distribution lost through habitat transformation (Barrows et al. 2008; Rios-Munoz and Navarro-Siguenza 2009).

Using species distribution modelling to predictively map changes in the distribution of species in response to remotely observed changes in climate is rather more challenging than for changes in land use or cover. There is potential to again employ space-for-time substitution for this purpose, by using explanatory models describing relationships between species occurrence and climate across space to predict changes in distribution across time as a function of observed changes in climate, mapped either directly from remote sensing, or through model-based integration of remotely-sensed and in situ climate observations. Many unanswered questions remain, however, regarding the transferability of climatic associations of species between space and time (Araujo and Peterson 2012). The impact of a given change in climate over time—e.g., a 0.5 °C increase in mean annual temperature—

on the occurrence of a given species at a particular location may be substantially less (or in some cases more) than that observed over space, due to complicating factors such as time lags in response, capacity for phenotypic plasticity, genetic adaptation, and biological interactions.

These difficulties point to the desirability of, wherever possible, fitting explanatory models relating species' occurrence to climate (and, for that matter, to land use or cover or abiotic variables such as soil type) using observations gathered across both time and space, rather than across space alone. Rapidly growing interest is now being directed towards extending standard techniques for species distribution modelling to more effectively consider the temporal dimension of observations (Kharouba et al. 2009; Porzig et al. 2014)—e.g., through the use of dynamic occupancy modelling (Kery et al. 2013; Tingley and Beissinger 2009). In an ideal world the fitting of explanatory models to biological and environmental observations obtained over time at a sample of locations, and the use of these fitted models to predictively map changes in biological distributions across an entire region of interest, would occur in parallel (as depicted in Fig. 10.3).

The process described so far is focused on predicting past-to-present change in the occurrence of a given species at a given location (e.g., grid cell), and thereby mapping change across all locations within a region of interest. For some applications this raw spatio-temporal information may need to be subjected to further aggregation or synthesis to address questions regarding, for example, changes in the overall state of a species, or of a whole group of species, within a given region. The most straightforward approach to deriving such aggregate measures is through simple summation, or averaging, of the predicted occurrence of a species across all locations (grid cells) in the region and, in turn, across all species in the group of interest. However it is worth noting in passing that other options exist for incorporating additional factors into this process of aggregation and synthesis—e.g., the use of metapopulation modelling to consider the effects of spatial configuration of predicted occurrence on the overall persistence of a species (Drielsma and Ferrier 2009), or the incorporation of information on phylogenetic relationships or functional traits into aggregate measures of the state of biodiversity across multiple species (Fenker et al. 2014).

10.3.2 Community-Level Approaches

In the species-level approaches discussed above, modelling is used to predictively map changes in the distribution of individual species. We now turn our attention to so-called 'community-level approaches' to modelling, and thereby mapping, changes in the distribution and retention of biodiversity within whole communities, without providing explicit information on the individual (named) species comprising this diversity. These approaches have particular utility in situations where the number of species in a biological group of interest is so high, and/or the average amount of information available for each of these species is so low, that species-level approaches start to lose tractability—e.g., for arthropods or plants in tropical forests. To appreciate the role that such approaches can play in mapping biodiversity change, let us start with a relatively basic challenge—i.e., estimating the loss (or, conversely, retention) of biological diversity at a single location (grid cell) as a function of remotely-sensed habitat loss or degradation. If remote sensing is used to classify the natural habitat within each grid cell in a region as being either intact or removed (Hansen et al. 2013), then simple deduction may be all that is required to predict the impact of this state on local biodiversity within that cell—i.e., it can be assumed that most of the species that were dependent on this habitat will no longer occur at this particular location. Alternatively, remote sensing can be used to classify locations into multiple classes of land use or habitat condition/intactness (Martinez and Mollicone 2012). These classes are expected to have varying levels of impact on local biodiversity. Prediction of these impacts should ideally be based on explanatory modelling of biological data gathered from the different classes, either across space alone (Souza et al. 2015) or, preferably, across both space and time (Casner et al. 2014). A particularly noteworthy example of this application of explanatory modelling is the PREDICTS initiative, which has undertaken an extensive meta-analysis of land-use impacts on local biodiversity (change in species richness) based on data for 27,000 species at over 11,000 sites globally (Newbold et al. 2015).

Linking explanatory models such as this to remotely-sensed land-use change opens up considerable potential to predictively map past-to-present change in local biodiversity across all grid cells in a region, or even across the entire planet. Change in local biodiversity is, however, not the only aspect of change that needs to be considered by community-level approaches to modelling biodiversity change. The total diversity—e.g., of species—occurring on our planet is a function not just of the number of species occurring at individual locations (alpha diversity), but also of differences in the composition of species between these locations (beta diversity) (McGill et al. 2015). To properly interpret the impacts of habitat loss (and, in turn, climate change) on retention of overall biodiversity it is therefore highly desirable to factor beta diversity into any model-based interpretation of remotely-sensed environmental change. Two broad strategies are available for achieving this, one using discrete classes to represent spatial pattern in beta diversity, and the other accounting for beta diversity through modelling of continuous patterns of spatial turnover in species composition (Ferrier 2011).

10.3.2.1 Discrete Community-Level Approaches

Many different types of discrete classification of communities can be employed in this context (Ferrier et al. 2009). The only real constraint is that the relevant classes are mapped across the entire region of interest, and that these classes provide a reasonable representation of major spatial patterns expected in the distribution of biodiversity in the absence of habitat loss or degradation. The last part of this constraint is particularly important. If the effects of habitat degradation are reflected in the classification itself (e.g., an area of rainforest cleared for domestic grazing is treated as a grassland rather than a forest) then it ceases to provide a logical basis for incorporating consideration of beta diversity into the interpretation of remotely-sensed environmental change. Mapped ecoregions may serve this purpose well at coarser spatial scales (Giam et al. 2011), as may mapping of the 'natural' extent (prior to anthropogenic alteration) of vegetation communities at finer scales (Keith et al. 2009).

With recent advances in the availability and resolution of abiotic environmental layers globally (for climate, terrain, soils etc.) another option growing in popularity is to derive environmental classes by integrating these layers—either by generating all unique combinations of expert-defined categories for each environmental variable (Ferrier and Watson 1997; Sayre et al. 2014), or through some form of automated numerical classification (Mackey et al. 2008). If sufficient biological data are available—i.e., in situ records for multiple species, well distributed across the region of interest—then various community-level modelling techniques can also be used to automatically derive and map environmental classes that best fit observed biological patterns (Ferrier and Guisan 2006).

Assuming that a mapped classification has been generated using one of the above approaches, this can be combined with remote mapping of habitat loss or degradation to estimate change in the retention of biodiversity. Where remote sensing yields a binary habitat versus no-habitat measure for each grid cell, then the changing state of a given class (e.g., an ecoregion) can be most simply expressed as the proportion of cells in that class with intact habitat. If remote sensing instead yields multiple levels of habitat condition/intactness—e.g., land-use classes translated into proportional losses of local species richness using results from the PREDICTS meta-analysis (described above)—then weighted averaging of these levels across all cells in a class can be used to derive an effective proportion of habitat remaining in that class (Scholes and Biggs 2005; Pereira and Daily 2006). In some cases this effective proportion is further adjusted to account for the effects of the spatial configuration of habitat—e.g., a cell with a given condition value located within a small isolated habitat fragment is assigned less weight than a cell of the same value located within a large well-connected area of habitat (Drielsma et al. 2014; Ferrier and Drielsma 2010).

Estimation of the proportion, or effective proportion, of habitat remaining in a class can be further used to predict the proportion of species, originally occurring within that class, that are expected to persist if this proportion of habitat is retained over the longer term. Such prediction is most commonly undertaken using some form of species-area relationship (SAR) (Ferrier 2002; Pereira and Daily 2006). SAR-based approaches typically assume that all classes are equally rich in species, and treat each mapped class (e.g., an ecoregion) as if it is a closed system—i.e., it is assumed that the species occurring within this class do not also occur in any of the other classes. The overall proportion of species predicted to be retained within an entire region of interest is therefore calculated as a simple average of the predicted proportions of species retained when the SAR is applied separately to each of the classes within the region (Faith et al. 2008; Proenca and Pereira 2013). Where estimates are available of the relative species richness of classes, and of the level of overlap in species composition between classes (e.g., the proportion of species occurring in ecoregion m that also occur in ecoregion n) then techniques exist for incorporating this information directly into SAR-based prediction of the overall proportion of species retained in a region as a function of remotely-sensed proportions, or effective proportions, of habitat retained in each class (Turak et al. 2011; Leathwick et al. 2010; Faith et al. 2008).

10.3.2.2 Continuous Community-Level Approaches

In the discrete community-level approaches described above, each location (e.g., grid cell) in the region of interest is viewed as belonging to a discrete class of locations that are assumed to be equally similar to one another, and equally different from locations in other classes, in the species they support. Real-world patterns of spatial change, or turnover, in species composition are, however, often more complex than can be effectively represented by a discrete classification with hard boundaries between mapped classes. Continuous community-level approaches attempt to address this reality by treating the composition of species occurring at each individual location as being unique, and the proportional overlap, or conversely distinctiveness, in composition between this location and any other given location within the region of interest as varying in a continuous manner (Ferrier et al. 2009).

One approach to applying this continuous community-level perspective to predictive mapping of change in biodiversity, as a function of remotely-sensed changes in habitat and/or climate, is through the use of generalised dissimilarity modelling (GDM) (Ferrier et al. 2007). GDM employs in situ occurrence records for all species in a given biological group (e.g., all plants, reptiles, or land snails) to fit a non-linear statistical model relating the dissimilarity in species composition observed between two locations to environmental differences based on remotely-mapped predictors (climate, terrain, soil etc.). Models fitted with GDM effectively weight and scale these environmental variables, thereby transforming multidimensional environmental space in such a way that distances within this transformed space match observed compositional dissimilarities as closely as possible. Using fitted GDM models to interpret remotely-sensed change in the distribution and condition of habitat can be achieved in various ways, but one of the most straightforward solutions is an extension of the SAR-based approach described above for the discrete community-level situation. In this extended approach the proportion, or effective proportion, of habitat remaining is estimated separately for each individual grid-cell within a region. This is calculated as a weighted average of habitat condition in all cells environmentally similar to the cell of interest, with each cell weighted by the level of similarity predicted by the fitted GDM. SAR-based estimates of the proportion of species retained relative to each cell can then be aggregated into an overall estimate of the proportion of species retained within the region as a whole (or within any required subset of this) factoring in GDM-predicted compositional dissimilarities between these cells (Ferrier et al. 2004; Allnutt et al. 2008).

Because continuous community-level approaches, such as GDM, incorporate abiotic environmental variables directly into the modelling of beta-diversity patterns, this opens up potential to further predict changes in the distribution and retention of biodiversity as a function of remotely-observed changes in climate. This can be achieved by invoking space-for-time substitution in a similar manner to that described earlier for species distribution modelling (Fitzpatrick et al. 2011; Prober et al. 2012). However it should be noted that employing space-for-time substitution in community-level approaches is also affected by many of the same complicating factors identified for species-level applications—e.g., time lags in response, capacity for phenotypic plasticity, genetic adaptation, and biological interactions (Blois et al. 2013). This again points to the desirability of fitting explanatory models relating patterns of biological distribution (in this case, turnover in species composition) to climate and habitat using observations gathered across both time and space, rather than across space alone. As for species distribution modelling, interest is now growing in extending existing community-level modelling approaches to more effectively consider the temporal dimension of biological observations.

10.4 Conclusion

As outlined in this chapter, modelling can play a crucial role in biodiversity monitoring by enabling more effective integration of in situ biological data with remotely-observed changes in key environmental drivers. This integration can involve both explanatory modelling—i.e., assessing and describing the effect of drivers on biodiversity through analysis of relationships between observed changes in biological and environmental data; and predictive modelling—i.e., using modelled relationships to predictively map change in biodiversity across whole regions as a function of remotely-sensed environmental change.

The most significant challenge now facing applications of modelling to biodiversity monitoring is to reduce reliance on models fitted to in situ biological observations gathered across space alone by making more extensive and effective use of observations from across both space and time. Recent escalation of interest in, and uptake of, citizen science initiatives (see Chap. 9) for collecting large quantities of spatially- and temporally-explicit biological observations offers considerable potential in this regard. In many cases incorporating data generated by such initiatives into biodiversity modelling will require extension of existing modelling techniques, or development of whole new techniques (Bird et al. 2014; Isaac et al. 2014; van Strien et al. 2013). These advances are likely to significantly strengthen links between explanatory and predictive modelling within the context of biodiversity monitoring. They are also likely to help strengthen links with applications of modelling to the projection of future biodiversity outcomes, by providing a more rigorous foundation both for fitting models employed in such projections, and for ongoing testing of the accuracy of these projections.

References

Ahmed, S. E., McInerny, G., O'Hara, K., Harper, R., Salido, L., Emmott, S., et al. (2015). Scientists and software—surveying the species distribution modelling community. *Diversity and Distributions, 21*, 258–267.

Allnutt, T., Ferrier, S., Manion, G., Powell, G., Ricketts, T., Fisher, B., et al. (2008). A method for quantifying biodiversity loss and its application to a 50-year record of deforestation across Madagascar. *Conservation Letters, 1*, 173–181.

Araujo, M. B., & Peterson, A. T. (2012). Uses and misuses of bioclimatic envelope modeling. *Ecology, 93*, 1527–1539.

Baccini, A., Friedl, M. A., Woodcock, C. E., & Zhu, Z. (2007). Scaling field data to calibrate and validate moderate spatial resolution remote sensing models. *Photogrammetric Engineering and Remote Sensing, 73*, 945–954.

Barrows, C. W., Preston, K. L., Rotenberry, J. T., & Allen, M. F. (2008). Using occurrence records to model historic distributions and estimate habitat losses for two psammophilic lizards. *Biological Conservation, 141*, 1885–1893.

Bird, T. J., Bates, A. E., Lefcheck, J. S., Hill, N. A., Thomson, R. J., Edgar, G. J., et al. (2014). Statistical solutions for error and bias in global citizen science datasets. *Biological Conservation, 173*, 144–154.

Blois, J. L., Williams, J. W., Fitzpatrick, M. C., Jackson, S. T., & Ferrier, S. (2013). Space can substitute for time in predicting climate-change effects on biodiversity. *Proceedings of the National Academy of Sciences of the United States of America, 110*, 9374–9379.

Bonthoux, S., Barnagaud, J. Y., Goulard, M., & Balent, G. (2013). Contrasting spatial and temporal responses of bird communities to landscape changes. *Oecologia, 172*, 563–574.

Casner, K. L., Forister, M. L., O'Brien, J. M., Thorne, J., Waetjen, D., & Shapiro, A. M. (2014). Contribution of urban expansion and a changing climate to decline of a butterfly fauna. *Conservation Biology, 28*, 773–782.

Cook, C. N., Inayatullah, S., Burgman, M. A., Sutherland, W. J., & Wintle, B. A. (2014). Strategic foresight: How planning for the unpredictable can improve environmental decision-making. *Trends in Ecology & Evolution, 29*, 531–541.

Coreau, A., Pinay, G., Thompson, J. D., Cheptou, P. O., & Mermet, L. (2009). The rise of research on futures in ecology: Rebalancing scenarios and predictions. *Ecology Letters, 12*, 1277–1286.

Corsi, F., de Leeuw, J., & Skidmore, A. (2000) Modelling species distribution with GIS. In L. Boitani & T. Fuller (Eds.), *Research techniques in animal ecology* (pp. 389–413). Columbia University Press.

de Baan, L., Alkemade, R., & Koellner, T. (2013). Land use impacts on biodiversity in LCA: A global approach. *International Journal of Life Cycle Assessment, 18*, 1216–1230.

Drielsma, M., & Ferrier, S. (2009). Rapid evaluation of metapopulation persistence in highly variegated landscapes. *Biological Conservation, 142*, 529–540.

Drielsma, M., Ferrier, S., Howling, G., Manion, G., Taylor, S., & Love, J. (2014). The biodiversity forecasting toolkit: Answering the 'how much', 'what', and 'where' of planning for biodiversity persistence. *Ecological Modelling, 274*, 80–91.

Elith, J., & Leathwick, J. R. (2009). Species distribution models: Ecological explanation and prediction across space and time. *Annual Review of Ecology Evolution and Systematics, 40*, 677–697.

Faith, D., Ferrier, S., & Williams, K. (2008). Getting biodiversity intactness indices right: Ensuring that 'biodiversity' reflects 'diversity'. *Global Change Biology, 14*, 207–217.

Fenker, J., Tedeschi, L. G., Pyron, R. A., & Nogueira, C. D. (2014). Phylogenetic diversity, habitat loss and conservation in South American pitvipers (Crotalinae: *Bothrops* and *Bothrocophias*). *Diversity and Distributions, 20*, 1108–1119.

Ferrier, S. (2002). Mapping spatial pattern in biodiversity for regional conservation planning: Where to from here? *Systematic Biology, 51*, 331–363.

Ferrier, S. (2011). Extracting more value from biodiversity change observations through integrated modeling. *BioScience, 61*, 96–97.

Ferrier, S. (2012). Big-picture assessment of biodiversity change: Scaling up monitoring without selling out on scientific rigour. In D. Lindenmayer & P. Gibbons (Eds.), *Biodiversity monitoring in Australia* (pp. 63–70). Canberra: CSIRO Publishing.

Ferrier, S., & Drielsma, M. (2010). Synthesis of pattern and process in biodiversity conservation assessment: A flexible whole-landscape modelling framework. *Diversity and Distributions, 16*, 386–402.

Ferrier, S., Faith, D., Arponen, A., & Drielsma, M. (2009) Community-level approaches to spatial conservation prioritization. In A. Moilanen, H. Possingham & K. Wilson (Eds.), *Spatial conservation prioritization: Quantitative methods and computational tools*. Oxford University Press.

Ferrier, S., & Guisan, A. (2006). Spatial modelling of biodiversity at the community level. *Journal of Applied Ecology, 43*, 393–404.

Ferrier, S., Manion, G., Elith, J., & Richardson, K. (2007). Using generalized dissimilarity modelling to analyse and predict patterns of beta diversity in regional biodiversity assessment. *Diversity and Distributions, 13*, 252–264.

Ferrier, S., Powell, G., Richardson, K., Manion, G., Overton, J., Allnutt, T., et al. (2004). Mapping more of terrestrial biodiversity for global conservation assessment. *BioScience, 54*, 1101–1109.

Ferrier, S., & Watson, G. (1997) *An evaluation of the effectiveness of environmental surrogates and modelling techniques in predicting the distribution of biological diversity*. Canberra: Environment Australia. http://www.environment.gov.au/archive/biodiversity/publications/technical/surrogates/index.html

Ferrier, S., Watson, G., Pearce, J., & Drielsma, M. (2002). Extended statistical approaches to modelling spatial pattern in biodiversity in northeast New South Wales. I. Species-level modelling. *Biodiversity and Conservation, 11*, 2275–2307.

Fitzpatrick, M. C., Sanders, N. J., Ferrier, S., Longino, J. T., Weiser, M. D., & Dunn, R. (2011). Forecasting the future of biodiversity: A test of single- and multi-species models for ants in North America. *Ecography, 34*, 836–847.

Fretwell, P. T., LaRue, M. A., Morin, P., Kooyman, G. L., Wienecke, B., Ratcliffe, N., et al. (2012). An emperor penguin population estimate: The first global, synoptic survey of a species from space. *PLoS ONE, 7*, 11.

Giam, X. L., Sodhi, N. S., Brook, B. W., Tan, H. T. W., & Bradshaw, C. J. A. (2011). Relative need for conservation assessments of vascular plant species among ecoregions. *Journal of Biogeography, 38*, 55–68.

Guisan, A., Tingley, R., Baumgartner, J. B., Naujokaitis-Lewis, I., Sutcliffe, P. R., Tulloch, A. I. T., et al. (2013). Predicting species distributions for conservation decisions. *Ecology Letters, 16*, 1424–1435.

Hansen, M. C., Potapov, P. V., Moore, R., Hancher, M., Turubanova, S. A., Tyukavina, A., et al. (2013). High-resolution global maps of 21st-Century forest cover change. *Science, 342*, 850–853.

Isaac, N. J. B., van Strien, A. J., August, T. A., de Zeeuw, M. P., & Roy, D. B. (2014). Statistics for citizen science: Extracting signals of change from noisy ecological data. *Methods in Ecology and Evolution, 5*, 1052–1060.

Jetz, W., McPherson, J. M., & Guralnick, R. P. (2012). Integrating biodiversity distribution knowledge: Toward a global map of life. *Trends in Ecology & Evolution, 27*, 151–159.

Jetz, W., Wilcove, D. S., & Dobson, A. P. (2007). Projected impacts of climate and land-use change on the global diversity of birds. *PLoS Biology, 5*, 1211–1219.

Kearney, M. R., Wintle, B. A., & Porter, W. P. (2010). Correlative and mechanistic models of species distribution provide congruent forecasts under climate change. *Conservation Letters, 3*, 203–213.

Keith, D. A., Orscheg, C., Simpson, C. C., Clarke, P. J., Hughes, L., Kennelly, S. J., et al. (2009). A new approach and case study for estimating extent and rates of habitat loss for ecological communities. *Biological Conservation, 142*, 1469–1479.

Kery, M., Guillera-Arroita, G., & Lahoz-Monfort, J. J. (2013). Analysing and mapping species range dynamics using occupancy models. *Journal of Biogeography, 40*, 1463–1474.

Kharouba, H. M., Algar, A. C., & Kerr, J. T. (2009). Historically calibrated predictions of butterfly species' range shift using global change as a pseudo-experiment. *Ecology, 90*, 2213–2222.

Leathwick, J., Moilanen, A., Ferrier, S., & Julian, K. (2010). Complementarity-based conservation prioritization using a community classification, and its application to riverine ecosystems. *Biological Conservation, 143*, 984–991.

Leutner, B. F., Reineking, B., Muller, J., Bachmann, M., Beierkuhnlein, C., Dech, S., et al. (2012). Modelling forest alpha-diversity and floristic composition—on the added value of LiDAR plus hyperspectral remote sensing. *Remote Sensing, 4*, 2818–2845.

Lindenmayer, D. B., Gibbons, P., Bourke, M., Burgman, M., Dickman, C. R., Ferrier, S., et al. (2012). Improving biodiversity monitoring. *Austral Ecology, 37*, 285–294.

Lung, T., Peters, M. K., Farwig, N., Bohning-Gaese, K., & Schaab, G. (2012). Combining long-term land cover time series and field observations for spatially explicit predictions on changes in tropical forest biodiversity. *International Journal of Remote Sensing, 33*, 13–40.

Mackey, B. G., Berry, S. L., & Brown, T. (2008). Reconciling approaches to biogeographical regionalization: A systematic and generic framework examined with a case study of the Australian continent. *Journal of Biogeography, 35*, 213–229.

Martinez, S., & Mollicone, D. (2012). From land cover to land use: A methodology to assess land use from remote sensing data. *Remote Sensing, 4*, 1024–1045.

McGill, B. J., Dornelas, M., Gotelli, N. J., & Magurran, A. E. (2015). Fifteen forms of biodiversity trend in the Anthropocene. *Trends in Ecology & Evolution, 30*, 104–113.

Miller, J. R., Turner, M. G., Smithwick, E. A. H., Dent, C. L., & Stanley, E. H. (2004). Spatial extrapolation: The science of predicting ecological patterns and processes. *BioScience, 54*, 310–320.

Newbold, T., Hudson, L. N., Hill, S. L. L., Contu, S., Lysenko, I., & Senior, R. A. et al. (2015) Global effects of land use on local terrestrial biodiversity. *Nature, 520*, 45–50.

Overmars, K. P., de Groot, W. T., & Huigen, M. G. A. (2007). Comparing inductive and deductive modeling of land use decisions: Principles, a model and an illustration from the Philippines. *Human Ecology, 35*, 439–452.

Pearce, J., Cherry, K., Drielsma, M., Ferrier, S., & Whish, G. (2001). Incorporating expert opinion and fine-scale vegetation mapping into statistical models of faunal distribution. *Journal of Applied Ecology, 38*, 412–424.

Pereira, H. M., & Cooper, H. D. (2006). Towards the global monitoring of biodiversity change. *Trends in Ecology & Evolution, 21*, 123–129.

Pereira, H. M., & Daily, G. C. (2006). Modeling biodiversity dynamics in countryside landscapes. *Ecology, 87*, 1877–1885.

Pereira, H. M., Leadley, P. W., Proenca, V., Alkemade, R., Scharlemann, J. P. W., Fernandez-Manjarres, J. F., et al. (2010). Scenarios for global biodiversity in the 21st century. *Science, 330*, 1496–1501.

Porzig, E. L., Seavy, N. E., Gardali, T., Geupel, G. R., Holyoak, M., & Eadie, J. M. (2014). Habitat suitability through time: Using time series and habitat models to understand changes in bird density. *Ecosphere, 5*, 16.

Prober, S. M., Hilbert, D. W., Ferrier, S., Dunlop, M., & Gobbett, D. (2012). Combining community-level spatial modelling and expert knowledge to inform climate adaptation in temperate grassy eucalypt woodlands and related grasslands. *Biodiversity and Conservation, 21*, 1627–1650.

Proenca, V., & Pereira, H. M. (2013). Species-area models to assess biodiversity change in multi-habitat landscapes: The importance of species habitat affinity. *Basic and Applied Ecology, 14*, 102–114.

Rapacciuolo, G., Roy, D. B., Gillings, S., & Purvis, A. (2014). Temporal validation plots: Quantifying how well correlative species distribution models predict species' range changes over time. *Methods in Ecology and Evolution, 5*, 407–420.

Rios-Munoz, C. A., & Navarro-Siguenza, A. G. (2009). Effects of land use change on the hypothetical habitat availability for Mexican parrots. *Ornitologia Neotropical, 20*, 491–509.

Santika, T., McAlpine, C. A., Lunney, D., Wilson, K. A., & Rhodes, J. R. (2014). Modelling species distributional shifts across broad spatial extents by linking dynamic occupancy models with public-based surveys. *Diversity and Distributions, 20*, 786–796.

Sayre, R., Dangermond, J., Frye, C., Vaughan, R., Aniello, P., & Breyer, S. et al. (2014) *A new map of global ecological land units—An ecophysiographic stratification approach.* Washington, D.C.: Association of American Geographers. http://www.aag.org/galleries/default-file/AAG_Global_Ecosyst_bklt72.pdf

Scholes, R. J., & Biggs, R. (2005). A biodiversity intactness index. *Nature, 434*, 45–49.

Shmueli, G. (2010). To explain or to predict? *Statistical Science, 25*, 289–310.

Skidmore, A. K., Pettorelli, N., Coops, N. C., Geller, G. N., Hansen, M., Lucas, R., et al. (2015). Agree on biodiversity metrics to track from space. *Nature, 523*, 403–405.

Smith, A. M. S., Falkowski, M. J., Greenberg, J. A., & Tinkham, W. T. (2014). Remote sensing of vegetation structure, function, and condition: Special issue. *Remote Sensing of Environment, 154*, 319–321.

Soberon, J., & Peterson, A. T. (2009). Monitoring biodiversity loss with primary species-occurrence data: Toward national-level indicators for the 2010 target of the convention on biological diversity. *AMBIO, 38*, 29–34.

Souza, D. M., Teixeira, R. F. M., & Ostermann, O. P. (2015). Assessing biodiversity loss due to land use with life cycle assessment: Are we there yet? *Global Change Biology, 21*, 32–47.

Stoms, D. M., Davis, F. W., & Cogan, C. B. (1992). Sensitivity of wildlife habitat models to uncertainties in GIS data. *Photogrammetric Engineering and Remote Sensing, 58*, 843–850.

Tingley, M. W., & Beissinger, S. R. (2009). Detecting range shifts from historical species occurrences: new perspectives on old data. *Trends in Ecology & Evolution, 24*, 625–633.

Tuanmu, M. N., & Jetz, W. (2014). A global 1-km consensus land-cover product for biodiversity and ecosystem modelling. *Global Ecology and Biogeography, 23*, 1031–1045.

Turak, E., Ferrier, S., Barrett, T., Mesley, E., Drielsma, M., Manion, G., et al. (2011). Planning for the persistence of river biodiversity: Exploring alternative futures using process-based models. *Freshwater Biology, 56*, 39–56.

Turner, W. (2014). Sensing biodiversity. *Science, 346*, 301–302.

van Strien, A. J., van Swaay, C. A. M., & Termaat, T. (2013). Opportunistic citizen science data of animal species produce reliable estimates of distribution trends if analysed with occupancy models. *Journal of Applied Ecology, 50*, 1450–1458.

van Vuuren, D. P., Kok, M. T. J., Girod, B., Lucas, P. L., & de Vries, B. (2012). Scenarios in global environmental assessments: Key characteristics and lessons for future use. *Global Environmental Change-Human and Policy Dimensions, 22*, 884–895.

Chapter 11
Global Infrastructures for Biodiversity Data and Services

Wim Hugo, Donald Hobern, Urmas Kõljalg, Éamonn Ó Tuama
and Hannu Saarenmaa

Abstract GEO BON regards development of a global infrastructure in support of Essential Biodiversity Variables (EBVs) as one of its main objectives. To realise the goal, an understanding of the context within which such an infrastructure needs to operate is important (for instance, it is part of a larger drive towards research data infrastructures in support of open science?) and the information technology applicable to such infrastructures needs to be considered. The EBVs are likely to require very specific implementation guidelines once the community has defined them in detail. In the interim it is possible to anticipate the likely architecture for a GEO BON infrastructure, and to provide guidance to individual researchers, institutions, and regional or global initiatives in respect of best practice. The best practice guidelines cover general aspects applicable to all research infrastructures, the use of persistent identifiers, interoperability guidelines in respect of vocabularies, data services and meta-data management, and advice on the use of global infrastructure services and/or federated, standards-based implementations.

Keywords Interoperability · Research · Infrastructure · Architecture · Best practice · Guideline · Persistent identifier · Biodiversity · Informatics

W. Hugo (✉)
South African Environmental Observation Network, P.O. Box 2600,
Pretoria 0001, South Africa
e-mail: wim@saeon.ac.za

D. Hobern · É.Ó. Tuama
Global Biodiversity Information Facility, Universitetsparken 15,
2100 Copenhagen, Denmark
e-mail: dhobern@gbif.org

U. Kõljalg
Institute of Ecology and Earth Sciences, University of Tartu, Ülikooli 18,
50090 Tartu, Estonia
e-mail: urmas.koljalg@ut.ee

H. Saarenmaa
Digitarium/University of Eastern Finland, P.O. Box 111, 80101 Joensuu, Finland
e-mail: hannu.saarenmaa@helsinki.fi

© The Author(s) 2017
M. Walters and R.J. Scholes (eds.), *The GEO Handbook on Biodiversity Observation Networks*, DOI 10.1007/978-3-319-27288-7_11

11.1   An Emerging Culture of Data Sharing, Publication and Citation

It has been widely accepted that the future usability and availability of research outputs, and specifically data, will be enhanced by proper description of these outputs using standardised metadata schemes, supplemented by deposit of the data in trusted repositories. Despite this, such outputs continue to be poorly described in practice. In addition, it is also commonly reported that the data supporting scholarly publication quickly becomes inaccessible or lost (Vines et al. 2014; Goddard et al. 2011). This disparity between what is seen as desirable behaviour, and reality is about to change, due to three significant drivers:

- Data publication and citation is gaining momentum (Chavan and Penev 2011). For a comprehensive review, see the report by a CoDATA[1] Task Group (Socha 2013).
- Funders are increasingly demanding the preservation of and continued open access to tax-funded research outputs.[2,3,4]
- Controversy in respect of reproducibility of scientific claims[5] have led to insistence by journals[6] that the data underpinning articles should be made available.

We believe these drivers will rapidly increase the availability of well-described, well-preserved, and sometimes standardised data services in the future.

11.1.1   Research Infrastructures

The drive towards data publication and citation requires support, hence the growth and proliferation of Research Data Infrastructures. These are supplemented strongly by voluntary, community-driven initiatives, and by member-funded bodies that support standardisation and interoperability.

Infrastructure operates on several levels: it provides governance and collaboration infrastructure (for example, the Belmont Forum[7] and Future Earth[8]),

---

[1]http://www.codata.org/.

[2]Berlin Declaration: http://www.berlin9.org/about/declaration/.

[3]OECD: http://www.oecd.org/sti/sci-tech/oecdprinciplesandguidelinesforaccesstoresearchdatafrom publicfunding.htm.

[4]USA: http://www.whitehouse.gov/sites/default/files/microsites/ostp/ostp_public_access_memo_ 2013.pdf.

[5]http://www.economist.com/news/briefing/21588057-scientists-think-science-self-correcting-alarming-degree-it-not-trouble.

[6]PLOS: http://blogs.plos.org/everyone/2014/02/24/plos-new-data-policy-public-access-data-2/.

[7]Belmont Forum: http://igfagcr.org/index.php/about-us.

[8]Future Earth: http://www.icsu.org/future-earth/media-centre/relevant_publications/future-earth-initial-design-report.

architecture and standards infrastructure (e.g., Research Data Alliance[9]—RDA, TDWG,[10] OGC,[11] GEO[12]), and physical, centralised or federated infrastructure (GBIF,[13] EUDAT,[14] and GEOSS[15]). Some global and regional initiatives span all of these (for example, the ICSU World Data System,[16] and GEO itself), and some are focused more narrowly on regional or domain-specific infrastructures (for example, DataOne,[17] EU BON,[18] Lifewatch,[19] and others).

It is worth noting that one of the motivations for the Research Data Alliance is to provide a cross-disciplinary, global exchange to minimise duplication of effort and divergence. Hence the landscape is at once characterised by divergent initiatives resulting from the nature of competitive grant funding and efforts to converge the impacts of funding these efforts. This is necessary, since divergence results in multiplicity of approaches, standards, protocols, and vocabularies—not supportive of interoperability.

11.1.2 Persistent Identifiers and Linked Open Data

Establishment of access to research outputs, either directly or via standardised services, requires a critical element: the ability to reliably find such objects in the web. This implies a *persistent identifier*, and several mechanisms are available to achieve this.

The biodiversity informatics community requires an identifier architecture that is capable of resolving two overlapping requirements—that of permanently identifying resources (data, services, and other web-based resources), and that of permanently identifying concepts (taxons, biomes, etc.).

There are several services available for either hosting or providing a minting framework for persistent identifiers (PIDs). Services that are general in nature, and allow hosting of PIDs on behalf of anyone, include the foundational Handle System.[20] This service can be used directly, but is also packaged and mediated, for

---

[9]RDA: https://rd-alliance.org/about.html.

[10]TDWG: http://www.tdwg.org/about-tdwg/.

[11]OGC: http://www.opengeospatial.org/.

[12]GEO: https://www.earthobservations.org/index.shtml.

[13]Global Biodiversity Information Facility: http://www.gbif.org/.

[14]EUDAT: http://www.eudat.eu/.

[15]GEOSS: https://www.earthobservations.org/geoss.shtml.

[16]http://www.icsu-wds.org.

[17]DataONE: http://www.dataone.org/.

[18]EU BON: http://eubon.eu/.

[19]LifeWatch: http://www.lifewatch.eu/.

[20]Handle System: http://www.handle.net/factsheet.html.

example by the members of the International Digital Object Identifier (DOI) Consortium[21]—allowing value-added services. DOI-based services that are important to our community include DataCite (linking published data sets and meta-data through DOIs to journal articles for purposes of citation tracking) and, CrossRef (more focused on linking DOI-based references across different journals), and GBIF (allocating DOIs for all published datasets and for search results). Several other biodiversity-focused initiatives exist, and these are discussed in the section on 'Specific Implementation Guidelines' (Barcode of Life,[22] Life Sciences Identifier, and similar, with identifiers.org[23] providing an aggregation of such services).

The availability of persistent identifiers assists the construction of Linked Open Data[24] (LOD) networks—making a significant contribution to the Semantic Web.[25]

11.1.3   Free and Open Data: Licensing and Policy

Delivering interoperable, open access to data and services involves (1) the implementation of applicable policies and (2) appropriate supporting licenses.

There are likely to be as many policies as there are data custodians and providers, but this is not really an issue as long as there is general compliance with the principles of free and open access—as documented by various global programmes such as the ICSU World Data System,[26] GEO,[27] and others.

Licenses, however, do need to be standardised, since machine-readability is a prerequisite for automated processing of data and services in the web. The most widely adopted candidates for this are the Creative Commons[28] family of licenses. These have been tested in multiple jurisdictions. Note that issues still under discussion include:

- 'Legal Interoperability' (how different licenses combine in automated processes, and what the resulting license is) (Uhlir 2013),
- Conditions or exceptions to be added to licenses to address legitimate concerns in respect of privacy, ethics, publication embargoes, endangered species, and similar.

---

[21]Digital Object Identifier: http://www.doi.org/doi_handbook/1_Introduction.html.

[22]http://www.barcodeoflife.org/.

[23]http://identifiers.org/.

[24]Linked Open Data: http://linkeddata.org/.

[25]https://www.w3.org/standards/semanticweb/.

[26]ICSU-WDS Data Policy: http://icsu-wds.org/services/data-policy.

[27]GEO Data Sharing Principles: https://www.earthobservations.org/geoss_dsp.shtml.

[28]Creative Commons and Data: http://wiki.creativecommons.org/Data.

GEO BON, being part of GEO, will adopt the GEOSS Data Sharing Principles (currently under review and likely to be modified slightly). In short, these are:

- There will be full and open exchange of data, metadata and products shared within GEOSS, recognising relevant international instruments and national policies and legislation;
- All shared data, metadata and products will be made available with minimum time delay and at minimum cost;
- All shared data, metadata and products being free of charge or no more than cost of reproduction will be encouraged for research and education.

11.1.4 Data Citation and Publication

Many of the institutional, technical, and legal hurdles that impeded the growth of data citation and publication have been addressed, and there is a broad consensus amongst journal publishers, data centres, and scientists in general on implementation (Socha 2013). CoDATA[29] and RDA[30] have played (and continue to play) a significant enabling role in this process.

Scientists should note that future research would be subject to:

- Planning for deposit and description (through metadata) of research output in a Trusted Digital Repository[31]—increasingly required by funders;
- Allocating persistent identifiers to such outputs, as appropriate.

Global coordinated research programmes, such as Future Earth, also attempt to align their funded outputs with the requirements of free and open access, and to promote a culture supportive of data publication and citation.

11.1.5 Big Data, Citizen Science, Crowdsourcing, and Proliferating Sensors

The field of biodiversity observation and monitoring is subject to rapid change both in regard to the variety of sources and to the volume size of the data that needs to be described, visualised, understood, preserved, and processed. This is due to a number of interrelated factors:

---

[29]CoDATA Task Group: http://www.codata.org/taskgroups/TGdatacitation/index.html.

[30]RDA Working Group: https://rd-alliance.org/working-groups/data-citation-wg.html.

[31]Trusted Digital Repository Checklists: http://www.crl.edu/archiving-preservation/digital-archives/metrics-assessing-and-certifying-0.

- *Growing Diversity and Productivity of Observation Channels*: Increasing availability of sensor channels lead to larger volumes of usable data. Traditional channels (remote sensing, gene sequencing, field observation) are increasingly supplemented by crowd-sourced observations, and the rapidly growing number of connected *smart devices* in the internet (Hugo et al. 2011).
- Methods using automated markup for metadata and data mining of existing or future publications contribute to increasing volumes (Agosti and Egloff 2009).
- *Storing Observations*: It is becoming increasingly affordable to store and process large volumes of data.
- *Less Expensive Platforms*: It is becoming very affordable to deploy observation platforms such as aerial drones[32] and underwater guided cameras, leading to large, multidimensional data sets at low cost of acquisition. Similarly, cost reductions are set to deliver significant and growing volumes of environmental genomic data addressing aspects of biodiversity which until now have been inadequately recorded.

These factors all combine to put pressure on the traditional architecture, standards, and infrastructure arrangements that have evolved to deal with a less demanding situation. The implications of this growth need to be accommodated in requirements for a scalable architecture.

11.2   The Network of the Future

GEO BON is by definition a network, and it is important to recognise that the concept of a network applies on multiple levels: on an institutional and personal level; as a collaboration network; and with the support of an infrastructure network. This infrastructure includes networks defined physically through protocols, schematically and syntactically through registries and catalogues, and semantically in emergent knowledge networks, ontologies, and vocabularies.

Any future networks, and resulting research data infrastructure, will likely be a combination of all of these and require governance, best practice conventions, standards, and reference implementations to work.

11.2.1   A Vision for Future Data and Services

The vision for a future network extends work done earlier by GEO BON (Scholes et al. 2012), and includes ideas about the generic use cases that it should support. This is summarised largely in the GEO BON Manifesto[33] (Hugo et al. 2013), which

---

[32]UNEP: http://www.unep.org/pdf/UNEP-GEAS_MAY_2013.pdf.

[33]Agreed by GEO BON Workgroup 8 at the Asilomar All Hands meeting, December 2012.

highlights a set of functions that are expected to be available. These, in turn, influence architecture and standards that are required to support such a network. The GEO BON Data Working Group (Working Group 8) has focused on these, and on developing a working implementation demonstrating the generic use case.

The Manifesto, as set out in updated form below, addresses description, discovery, assessment, access, analysis, and application or reporting, by stating that it is the interest of any specific community to do the following:

- Ensure that scientific data and services are described properly, preserved properly, and discoverable;
- Once discovered, the utility, quality, and scope of data can be understood, even if the data sets are large;
- Once understood; the data can be accessed freely and openly;
- Once accessed, the data can be included within distributed processes, and collated—preferably automatically (Hernandez et al. 2009a, b), and on large scales (the 'Model Web') (Nativi et al. 2013);
- Once processed, the associated mediations and annotations, usefulness, and knowledge gathered can be re-used.

All of this needs to be implemented against the backdrop of:

- Due recognition to the creators of the data, models, and services;
- The push to extend formal metadata with Linked Open Data and persistent identifiers;
- The increased availability of crowd-sourced and citizen contributions;
- A proliferation of devices and sensors; and
- The construction of knowledge networks.

11.2.2   The Role of Standards and Specifications

Standards and specifications are intended, from a formal systems engineering perspective, to *reduce the risk of failure*. The basic aim of this approach is 'Predictable Assembly from Certifiable Components' (Wallnau 2003). The risk of failure is lowered because assembly is made from components certified to meet the specifications and standards. In the type of scalable, open architecture envisaged for GEO BON, the ability of *third parties* to assemble larger systems from components using well-defined interfaces is critical as a contributor to the goal of interoperability and scalability.

Data standards in biodiversity are primarily defined by the Biodiversity Informatics Standards organisation. It is better known by its earlier name 'Taxonomic Databases Working Group'[34] (TDWG). TDWG works with other

---

[34]http://www.tdwg.org/.

standards bodies, such as Open Geospatial Consortium (OGC), and has been recognised by them.

11.2.3   A Scalable, Interoperable Architecture

A realistic, shorter-term expression of the goals implied by the manifesto can be summarised as follows (Saarenmaa et al. 2014):

- Allow for data flow from observations through various aggregation and processing/modelling services, supporting evaluation of EBVs and derived indicators;
- Automated and streamlined, as appropriate;
- Using a plug-and-play (service-oriented) approach, supported by robust service provider organisations;
- Coordinated through a GEO BON registry system and linked to the GEOSS Common Infrastructure;
- Transparent to users through multiple channels, portals and applications.

11.2.3.1   General Requirements for a Biodiversity Information Architecture

Scalability, access, security, user concurrency and data reliability must be considered. For scalability, it is expected that tens of thousands of data sources will ultimately be integrated through GEO BON. They will be hosted in a smaller number of data repositories. Additionally:

- The infrastructure must incorporate a federated architecture which will allow many data centres, initiatives, and infrastructures to co-exist and participate;
- While a minimum set of standards is desirable, pragmatism and reliance on brokering and mediation will be the norm for a considerable time to come;
- Human resource, financial, scalability, and institutional constraints will necessitate building the infrastructure using many small contributions in addition to a few large, global ones.

The main components in the information architecture can be divided into three main functions, corresponding to the tasks of (i) data publishing, (ii) data discovery, and (iii) data access. As a fourth function, various applications and uses can be envisaged, and for all functions mediation may be required between services and clients in cases where standardisation of services and vocabularies are not perfect.

There are two options for interoperability architecture, both essentially 'service-oriented', with varying degrees of rigour required for implementation. Firstly, the model proposed by EU BON and others, is based on an Enterprise

Service Bus (ESB), and allows automation of asynchronous workflow and distributed processing as envisaged by the Model Web. Secondly, one can serve a significant proportion of needs with less complex synchronous orchestration, using mostly RESTful Services. These architectures are not mutually exclusive and are likely to co-exist within a systems-of-systems environment.

11.2.3.2 Option 1: SOA and ESB

The Service Oriented Architecture (SOA) is a model, which has achieved 'best practice' status within the Open Geospatial Consortium (OGC). Building on SOA has been recommended also for GEO BON (Ó Tuama et al. 2010) and EU BON (Saarenmaa et al. 2014). In an SOA, different functionalities are packaged as component services that can be orchestrated for specific tasks. An Enterprise Service Bus (ESB), which is a virtual private connector over the Internet, would connect external data sources using various SOA standards (WSDL,[35] SOAP,[36] REST[37] and BPEL,[38] among others). The use of an ESB facilitates the interactions among data sources, working in a message-centred interaction and providing the ability to orchestrate web services through the use of workflow handling technology (e.g., Kepler,[39] Taverna[40]).

11.2.3.3 Option 2: Synchronous, RESTful Services

Some applications do not require orchestration of services to take account of long-running, asynchronous processes, and may not require authentication if data services are in the public domain. In these cases, RESTful HTTP calls, stored in OGC Web Context Documents (XML files defining a collection of RESTful services and their roles) should be adequate to collate information in support of a user requirement. The role that each service plays to achieve the collective outcome will have to be captured for future use, and can potentially be stored in OGC Web Context Documents (XML files defining a collection of RESTful services and their roles), but other methods may also be used.

---

[35]Web Services Description language (WSDL); http://www.w3.org/TR/wsdl20/.

[36]Simple Object Access Profile; http://www.w3.org/TR/soap12-part1/.

[37]Representational State Transfer; http://www.ibm.com/developerworks/webservices/library/ws-restful/.

[38]Business Process Execution Language; http://docs.oasis-open.org/wsbpel/2.0/OS/wsbpel-v2.0-OS.html.

[39]https://kepler-project.org/.

[40]http://www.taverna.org.uk/.

11.3   Considerations in Respect of Best Practice

11.3.1   Sources of Data and Its Classification

11.3.1.1   Essential Biodiversity Variables

The Essential Biodiversity Variables (EBVs) (Pereira et al. 2013), under development by GEO BON, provide a critical use case for determining requirements for information systems. An EBV is defined as 'a measurement required for study, reporting, and management of biodiversity change'. EBVs provide focus in two important ways:

- promote harmonised monitoring by stipulating how variables should be sampled and measured;
- facilitate integration of data by acting as an abstraction layer between the primary biodiversity observations and the indicators.

For example (Fig. 11.1), we could build up an aggregated population trend indicator (for multiple species and locations) from an EBV which estimates population abundances for a group of species at a particular place and which, in turn, is derived from the primary, raw data which can involve different sampling events and methodologies.

GEO BON has identified six EBV classes. These are listed in Table 11.1 with some candidate EBV examples. By analysing the variables/measurements associated with each EBV, appropriate data standards can be proposed or recommended, or new and enhanced standards proposed. Of particular relevance are the EBV definitions and how an EBV is measured. For example, the three EBVs listed for the Species Populations class, can be broken down as illustrated in Table 11.2. In fact, the Species Population class EBVs are possibly the most tractable given the current status of biodiversity informatics, and could act as the initial test case.

In addition to suitable data exchange standards, there is a need to identify appropriate communication protocols for messaging and data flow between systems, and, as part of the architecture design, how to automate the data flows for the EBVs.

The EBV on abundances and distributions would need to be measured using 'counts or presence surveys for groups of species easy to monitor or important for

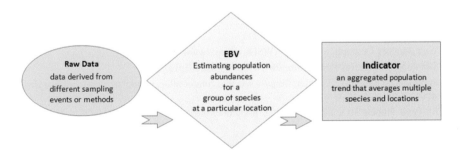

Fig. 11.1 An EBV acts as an intermediate layer between raw data and indicators

Table 11.1 EBV classes with examples

EBV Class	Genetic composition	Species populations	Species traits	Community composition	Ecosystem structure	Ecosystem function
EBV example	Allelic diversity	Abundances and distributions	Phenology	Taxonomic diversity	Habitat structure	Nutrient retention

Source Adapted from Pereira et al. (2013)

Table 11.2 The three EBVs of class species populations with their definitions and variables/measurements

Class	EBV	Definition	How to measure in marine, terrestrial, freshwater (spatial, temporal, taxonomic)
Species populations	Species occurrence	Presence/absence of a given taxon or functional group at a given location	Quantify number/biomass/cover at a sample of selected taxa (or functional groups) at extensive suite of sites (selected from stratified random sample or building on existing networks)
	Population abundance	Quantity of individuals or biomass of a given taxon or functional group at a given location	
	Population structure by age/size class	Quantity of individuals or biomass of a given demographic class of a given taxon or functional group at a given location	

ecosystem services, over an extensive network of sites, complemented with incidental data'. Such an EBV would be updated at intervals from 1 to 10 years. EBVs have not yet been implemented, but need to be piloted.

Implementation of these specific EBVs calls for integration of data from sites such as those of LTER, and other regular surveys, and from historical and recent data published through GBIF. Integration implies processing services that would compute abundance trends and changes in distribution for these two types of data: surveys and incidental. These are shown in Fig. 11.2 as 'ecological' and 'occurrence' domains. Software tools and web services are available to do these computations, for instance from the TRIM,[41] BioVeL,[42] and EUBrazilOpenBio[43] projects. Recent developments within GBIF include support for additional core data elements from survey data,[44] indicating the possibility of incorporating all of these data sources within a single access infrastructure.

---

[41] www.cbs.nl/en-GB/menu/themas/natuur-milieu/methoden/trim.

[42] www.biovel.eu.

[43] www.eubrazilopenbio.eu/.

[44] www.gbif.org/sites/default/files/gbif_IPT-sample-data-primer_en.pdf.

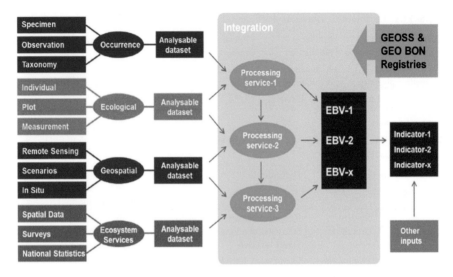

Fig. 11.2 The GEO BON vision of automated, streamlined data flow, end-to-end, from observations to Essential Biodiversity Variables (EBVs), using a plug-and-play service-oriented approach, coordinated through the GEO BON registry system and linked to the GEOSS Common Infrastructure, and transparent to users through portals. *Source* Hugo et al. (2013); modified by Hoffman et al. (2014)

The computation of an EBV of this class involves data cleansing and normalisation and interpolation of values to offer a modelled data surface. Such EBVs could be visualised in a portal, which would allow selecting the data sources and species in question, showing the intermediate steps, and presenting the trend and change of distribution for individual species or whole groups of organisms.

11.3.1.2 Protocols for Observation

The two largest domains of biodiversity observation are specimen occurrences and biological (natural resource) surveys. The former is frequently based on sporadic, opportunistic collection or observation activity, while the latter consists of repeated sampling at known sites, locations and follows a known protocol from which quantitative estimates of abundance, and at times additional information, can be derived. Hence, the latter method is most appropriate for observing change, but the former can also be used, if the observations sets are large enough and sampling biases can be eliminated by computation (Ariño 2010). Data potentially available through both of these domains are very large. GBIF, which already represents the occurrence domain, currently has mobilised more than 15,000 data sets and is

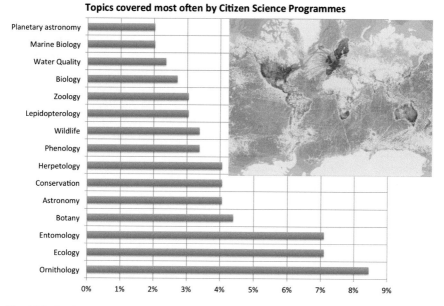

Fig. 11.3 Topics covered most often by Citizen Science Programmes (https://en.wikipedia.org/wiki/List_of_citizen_science_projects#Active_citizen_science_projects). Inset—distribution of GBIF observation data, a large proportion of which originates from volunteer contributions (http://www.gbif.org/occurrence)

expanding to index and integrate data from survey datasets. ILTER, which represents the ecosystem monitoring domain, has 25,000 data sets. Both have the potential of growing at least ten-fold. In particular, for ecosystem monitoring, much data exists in government agencies for the environment, forestry, fisheries, and agriculture, which in many cases have not yet started any data sharing activities.

Biodiversity observation is unique in that for species occurrence, most observations are made by volunteers. The EUMON project[45] estimates that 80 % of biodiversity monitoring data comes from volunteers. In Finland, for example, there are 60 different biodiversity monitoring programmes in which 250 person years are spent annually, and 70 % of this is voluntary work. This pattern is similar to some extent many other countries—a summary prepared based on a listing of such volunteer programmes is shown in Fig. 11.3. In the top 15 topics, only astronomy is unrelated to biodiversity.

Volunteer contributions pose a special challenge in respect of introduction of observer bias and strict adherence to observation protocols, and may be used in special circumstances to derive additional EBVs (Kery et al. 2010; Hui and McGeogh 2014).

---

[45]http://EuMon.ckff.si/index1.php#2.

11.3.1.3 Generic Data Families

The GEO BON working group on data integration and interoperability has developed a classification of generic data families and their interoperability requirements (Fig. 11.4). Data families are grouped according to variations in their spatial, temporal and semantic coverages with each unique combination of these, supported by a vocabulary/ontology, is considered a generic data family. As an example: occurrence, genome, and ecosystem data families all include a reference to a particular place and time, but differ in that occurrence data also references a taxon, genome data references a sequence and ecosystem data references biological phenomena.

The different types of coverage (spatial, temporal and semantic) and their attributes are:

- Spatial Coverage: **XYZ**
- Temporal Coverage: **T** (continuous or near-continuous); **t** (discrete)
- Topic or Semantic/Ontological Coverage

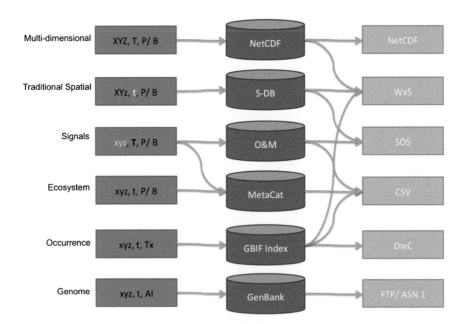

Fig. 11.4 Example generic data families and interoperability requirements. The abbreviations are: S-DB: spatial database; WxS: OGC (Open Geospatial Consortium web services); O&M: OGC Observations and Measurements model; SOS: OGC Sensor Observation Service; CSV: comma separated value; DwC: Darwin Core. The *leftmost* boxes represent typical data families and their dimensions, the *centre* shows typical data storage technology, and the *rightmost* boxes typical services whereby such data is exchanged. Some data storage technologies support multiple service standards. *Source* Hugo et al. (2013)

– **P**: Phenomenon

  mostly physical, chemical, or other contextual data

– **B**: Biological
– **Tx**: Species and Taxonomy (with some extensions)
– **Al**: Allele/Genome/Phylogenetic.

The dimension of a sampling event or specimen applies to all data families.

11.3.2 Published Advice and Guidance

The recommendations from published material discussed here have been incorporated into the 'Specific Implementation Guidance' later in the chapter, as appropriate.

Recent advances in the availability of standards include the development of 'Biological Collections Ontology' (BCO) and the 'Population and Community Ontology' (PCO) (Walls et al. 2014)—bridging a gap in the availability of vocabularies derived from formal ontology to describe the collection of biodiversity data, and to formulate more complex relationships between primary data elements such as evolutionary processes, organismal interactions, and ecological experiments.

11.3.2.1 Research Data Alliance (RDA)

The Research Data Alliance (RDA) produces community consensus on important aspects of research data infrastructure in general, and includes representation from biodiversity and ecosystem data infrastructures.[46] This interest group envisages work in respect of name (vocabulary) services standardisation, with a focus on taxonomy, and the support of improved interoperability. In more general terms, RDA has recently endorsed its first sets of formal outputs, and some of these have a bearing on biodiversity informatics:

- The Data Citation Working Group[47] has produced a clear set of guidelines in respect of implementation of persistent identifiers for data sets.
- The Data Type Registries Working Group[48] aims to standardise the description of complex data types—which in principle includes the 'data families' that can be identified for GEO BON EBVs. This enables processes, visualisations, and other tools to reliably be linked to data services.

---

[46]https://rd-alliance.org/groups/biodiversity-data-integration-ig.html.

[47]https://rd-alliance.org/groups/data-citation-wg.html.

[48]https://rd-alliance.org/groups/data-type-registries-wg.html.

- The Metadata Standards Catalog Working Group[49] has produced a set of principles, and aim in future to develop a canonical set of metadata elements that can serve as a broker between different metadata schemas in use by communities.
- The Practical Policies Working Group[50] has published its first recommendations in respect of 11 important practical policies for repository management, based on a survey of the research repository community.

11.3.2.2 Global Biodiversity Informatics Conference (GBIC)

The Global Biodiversity Informatics Conference (Copenhagen, 2012)[51] assessed the state of Biodiversity Informatics across four focus areas (Understanding, Evidence, Data, and Culture), and provided a community consensus on the desirable futures for the elements in each of these focus areas (Hobern et al. 2012).

11.3.2.3 GEO Data Management Principles

The GEO Data Management Principles[52] were adopted in short form by the organisation in April 2015, and in full form by the GEO Plenary in November 2015. The 10 principles deal with aspects of discoverability, accessibility, usability, preservation, and curation.

11.3.2.4 EU BON

EU BON published a review and guidelines for its proposed architecture (Saarenmaa et al. 2014) that contains a portfolio of recommendations. These recommendations (39 in all) are strongly supportive of existing projects and initiatives (Lifewatch, BioVEL, EBONE, INSPIRE, LTER, GBIF, to name a few) and provide guidance in respect of service-bus type implementation in a service-oriented architecture.

---

[49]https://rd-alliance.org/groups/metadata-standards-catalog-working-group.html.
[50]https://rd-alliance.org/groups/practical-policy-wg.html.
[51]http://www.biodiversityinformatics.org/.
[52]https://www.earthobservations.org/documents/dswg/201504_data_management_principles_long_final.pdf.

11.3.2.5   CReATIVE-B and GLOBIS-B

The CReATIVE-B project[53] (2011–2014) dealt with the 'Coordination of Research e-Infrastructures Activities Toward an International Virtual Environment for Biodiversity'. CReATIVE-B enabled collaboration between the European LifeWatch/ESFRI Research Infrastructure and other large-scale research infrastructures on biodiversity and ecosystems in other parts of the world. The project published an integrated Roadmap in 2014 and this serves as high-level guidance in respect of biodiversity infrastructure and data management activities.

GLOBIS-B has as its main aim the definition of research needs and infrastructure services required to calculate EBVs, and will do so by fostering collaboration between scientists, global infrastructure operators, and legal interoperability experts. GLOBIS-B has produced its first outputs, and a recent publication (Kissling et al. 2015) details thoughts on interoperability in support of EBVs. GLOBIS-B correctly identifies a scientific challenge (definition of EBVs) and a technical one (legal and information technology considerations) that need to be addressed.

11.3.2.6   EarthCube and DataONE

These are primarily US-based initiatives, though DataONE has participating data providers from outside the US, and EarthCube has formal collaboration with EU partners. DataONE publishes and maintains best practice in respect of data management,[54] which was reviewed for inclusion into our guidance, and EarthCube has recently published a roadmap[55] and a supporting architecture[56] that also contributed input by way of principles.

11.4   Specific Implementation Guidelines

References quoted in the following sections are available in the supplementary materials on the Springer Website. Supplementary materials are also hosted and maintained on the GEO BON website at http://dataintegration.geobon.org/guidance.

---

[53]http://www.slideshare.net/dmanset/20140909creativeb-roadmap-interactive.

[54]https://www.dataone.org/sites/all/documents/DataONE_BP_Primer_020212.pdf.

[55]http://earthcube.org/sites/default/files/doc-repository/ECRoadmapv6%203%201.pdf.

[56]https://docs.google.com/document/d/10OhZntRpizn-KaYECXtGY_tcVbanG2kR0OFJ7JZpnWw/edit#.

11.4.1 Recommended Data Management Approaches

This section proposes guidelines for biodiversity data management from three perspectives: that of (1) individual researchers, (2) institutions, projects, or initiatives (such as regional BONs), and (3) from the broader community and GEO BON's perspective. It focuses on the information technology aspects of the challenge to provide an infrastructure in support of EBV calculation. The guidelines support both architectures described above.

For all of these end user categories, we recommend that

- General guidelines in respect of data management be followed (Section A below, and elaborated in supplementary materials), with indications of deficiencies that may exist;
- Specific guidelines to foster semantic interoperability are followed (Section B below). These are also supplemented by online materials and deficiencies are highlighted;
- As a first choice, data be shared in global repositories that serve a specific data family and is well established (Section C below);
- Other data be published and catalogued using widely adopted interoperable service standards and content schema—while recognizing that the community, and especially GEO BON, should play a role in extending such content schema where deficiencies exist (Section D).

Content schema and vocabularies in support of specific EBVs will be required once the community has adopted definitions—GEO BON has a critical role in developing these, and the GLOBIS-B project will make a direct contribution to this effort.

11.4.2 Section A: General Considerations

These considerations apply to all research data infrastructures (Table 11.3).

11.4.3 Section B: Semantic Interoperability

Guidelines in respect of the use of name services (vocabularies, ontologies, and persistent identifiers), and development of a knowledge network as it applies primarily to biodiversity informatics (Table 11.4).

Table 11.3 General guidelines applicable to all research infrastructures

Aspect	Guidance			Reference
	For individual researchers	For institutions and projects	For the community/GEO BON	
Open access	Select open licenses, preferably the most open suitable Creative Commons license by preference possible (CC0, otherwise CC-BY or, if necessary CC-BY-NC—avoid the ND, no-derivates and SA, share-alike options), for all published data sets unless one of a specific set of exceptions apply	Develop data policies in support of open access and open science, and standardise on Creative Commons licenses for all but specific exceptions	Support Creative Commons licenses, and work towards machine-readable, multi-jurisdiction licenses for the valid exceptions not supported by Creative Commons	[1, 4, 5, 8, 10, 39, 40, 42, 43, 46–48]
Federated identity	Make use of globally available resources in this regard, such as EduRoam	If available, use EduRoam as a basis for service and system authentication, and ensure that researchers have access to it	Work with RDA to facilitate a globally available identity resolution framework that can be used by system and service developers	[1, 41, 42]
Data citation	Ensure that data sets are published with a persistent identifier, and make use of persistent identifiers when citing others	Ensure that mechanisms are available for persistent identifiers to be minted for data set publication, and that sufficient infrastructure spending is available for implementation of RDA guidelines in respect of data citation	Promote a culture of data citation and license respect/maintenance. Contribute use cases to RDA working groups on data citation to ensure that the needs of the biodiversity community are included in guidance	[2, 8, 10, 40, 42]
Data types (data families)	Use the guidance below to select a data family appropriate to the data being published. Bear in mind that the publication format is not the same as the format in which the data is best applied in your own context	Make an effort to ensure that data type registries are supported once these become available. Ensure that tools and processing routines are designed and implemented in such a way that the data type registry can be supported		[3, 8, 10]

(continued)

Table 11.3 (continued)

Aspect	Guidance			Reference
	For individual researchers	For institutions and projects	For the community/GEO BON	
Metadata interoperability	Use the guidance below to select an appropriate metadata standard for the data to be published, and ensure that maximal use is made of name services (vocabularies, ontologies, registries of permanent identifiers)	Work towards adoption of metadata standards within the institution. Ensure that catalogues of institutional data offer harvesting end points supported by widely accepted protocols (see below for guidance)	Work towards adoption of metadata standards within the community and develop best practice/guidelines in respect of name service usage, mandatory elements, quality, protocol and lineage description, and other elements supporting re-use of the data	[4, 8, 42, 43]
Name service interoperability	Support appropriate name services (vocabularies, ontologies, and PID registries) wherever these are indicated. Refer to the guidance below in this regard	Develop institutional best practice/guidance in respect of name service usage	Contribute to an support RDA efforts to improve name service interoperability, and actively promote the use of such services by the community	[4, 9, 42, 43]
Data interoperability	Use the guidance below to select a data family appropriate to the data being published. Specifically work towards making sure that the data is not only available in a standardised schema (format), but that it is also available via a standardised protocol (web service)	Ensure that infrastructure exists for implementation of appropriate web services to enable access to standardised data sets	Seek consensus and endorsement within standards bodies of especially content standards for all data families identified below as requiring attention	[4, 8, 42, 43]
Trusted repositories and reliable future access	Make sure that your data is published and archived in a Trusted Digital Repository, with long-term curation policies and contingency planning in place	Take steps to accredit your institutional repository/data service with one of the recommended global initiatives, and register the repository with re3data. Implement recommendations of the RDA Practical Policies Working Group	Support the principle of deposit in Trusted Digital Repositories	[4–8, 42]

Table 11.4 Guidelines in respect of semantic interoperability

Aspect	Guidance			Reference
	For individual researchers	For institutions and projects	For the community/GEO BON	
Use of persistent identifiers	Use persistent identifiers for identification of data sets (see Data Citation above), and for referencing of important dimensions as described below	Ensure that the mechanisms for obtaining PIDs for data sets are available and affordable for researchers in the institution	Assist, within RDA and other initiatives, with the development of a suite of integrated services for PID resolution	[1, 2, 4, 10, 42]
Knowledge networks	Ensure that dimensions of data (see below) make use of recommended name services, and use such name services for provision of keywords in metadata	Adopt institution-endorsed name services and best practice in respect of implementation	Develop standards and infrastructure that allows individual data element annotation, and encourage the use of formal vocabularies and ontologies for such annotation	[4]
Persons and individuals	Obtain an ORCID for use as a persistent identifier in metadata and data	Encourage the use of ORCID within the institution and community		[1]
Sample	Ensure that individual samples (physical samples or biological specimens/tissue, video, audio, images, signals) are assigned a persistent identifier so that analysis and resulting data from multiple sources can be collated. Consider the use of BCO (Biological Collections Ontology)	Develop institutional best practice in respect of sample identifiers, and make use of global identifier services appropriate to the sample type	Support international efforts, such as now emerging in RDA, EU BON, and in GBIF, to explicitly identify and link samples to observations. Assist with the adoption of BCO as a community standard	[4, 10–13, 45]
Protocols and lineage of data	Make use of published and citable protocols and methodology where possible. Publish own protocols independently and assign a persistent identifier. Use these as references in metadata and describe data lineage properly	Encourage the use of published protocols and the publication of institutional or community of practice protocols	Within GEO BON, work towards the development of published and peer-reviewed protocols for monitoring of all EBVs at all relevant scales. Consider hosting a registry of protocols for EBVs	[4, 10, 42]

(continued)

Table 11.4 (continued)

Aspect	Guidance — For individual researchers	For institutions and projects	For the community/GEO BON	Reference
Location, spatial coverage, and stratum	Use a standardised vocabulary for referencing locations in data. If institutional or community guidance is not available, use GeoNames.org as a definitive reference for locations on earth. Provide point or bounding box coordinates—preferably in WGS 84 Lat-Long projection—for study areas defined in metadata	Develop institutional guidelines aligned with national or regional directives, while taking cognisance of international standards that may emerge in this respect	Collaborate towards specific community standards (for example using extensions to Darwin Core) to explicitly indicate and reference plots and their coverages/strata in data and metadata. Develop a robust guideline for location, spatial coverage, sampling plot, and stratum references	[1, 10, 54]
Time	Use UTC (Coordinated Universal Time) to denote events within the present, recent past or future (±100 years). Adhere to guidelines for denoting time on historical, paleo/geological and far future scales	Promote institutional guidelines in respect of time, and implement protocols for synchronisation of automated data sensor date and time stamps	Work towards a definitive community consensus for referencing time in the immediate (±100 years) observation space, historical, paleo/geologic, and far future time scales	[1]
Molecular sequence and genetic data	Implement the guidelines and standards promoted by the Genomics Standards Consortium (GSC), including MIGS, MIMS, and MIMARKS—depending on the data type	Promote the guidelines published by GSC within the institution	Continue the current collaboration between GEO BON and GSC with a view to widespread adoption of the standards and its continuous improvement	[4, 10]
Taxonomy	Use the services registered with the Global Names Architecture (GNA) in the first instance to verify taxonomy. Use widely reviewed sources such as Catalogue of Life. Make use of automated services, such as Plazi for taxonomic data mining	Ensure that taxonomy guidelines for data and metadata are aligned with regional directives and guidelines	Develop best practice in respect of taxonomy referencing, considering use cases that involve changes in taxonomic reference	[1, 4, 10, 14, 37, 38, 50, 51, 52, 57]

(continued)

Table 11.4 (continued)

Aspect	Guidance			Reference
	For individual researchers	For institutions and projects	For the community/GEO BON	
Traits and functional diversity	Use one of a number of ontologies/vocabularies aiming to standardise descriptions of traits (Structured Descriptive Data (SDD), the Plinian Core, the Phenotypic Quality Ontology, and the Animal Natural History ontology)	Agree on institutional use of a specific vocabulary or ontology	Mobilise the community to develop interoperability or brokering between the main trait vocabularies and ontologies. Encourage publication of trait datasets via Encyclopaedia of Life TraitBank	[4, 10, 45]
Habitat, biome, biogeographic and biotope classification	Use the descriptions of biomes and biogeographic regions as promoted or directed by national or regional authorities, or ENVO	It is likely that institutional guidelines will be subject to national or regional directives in this regard	Work towards a brokering or interoperability arrangement to align regionally and nationally adopted biome and bioregion descriptions, and define relationships between them	[10, 33]
Life stage	No definitive vocabulary or ontology for life stages is available. Use the approach proposed by MorphoBank to create a checklist of characteristics and states that cannot be duplicated within your own body of work	Develop institutional best practice to guide data published by researchers	Develop an authoritative vocabulary within a standards body such as TDWG as a community consensus	[1]
Species relationship and biological interaction	Use high-level classification—such classification is less contentious—as well as lower level classifications pertinent to the data at hand. Authoritative services in this regard are not yet available. Consider use of PCO (Population and Community Ontology). Make use of new techniques in metagenomics	Develop institutional best practice to guide data published by researchers	Develop an authoritative vocabulary within a standards body such as TDWG as a community consensus. Use the PCO as a basis of such development	[1, 45, 55]

(continued)

Table 11.4 (continued)

Aspect	Guidance		Reference
	For individual researchers	For institutions and projects	
		For the community/GEO BON	
Ecosystem functions and services	No specific guidance available at present. The best general ontologies to use include the NASA SWEET Ontology and ENVO, if applicable	Work towards development of standardised vocabularies and ontologies for description of ecosystem functions and services	[1, 32, 33]

Table 11.5 Guidelines applicable to data families for which global infrastructures exist

Data family	Metadata and catalogue services	Sustainable international infrastructures	Schematic and syntactic interoperability—service protocols and content standards	Reference
Electronic samples and specimens	A globally available publication platform for audio, video, and image media used as the basis of species identification and traits/character annotation	MorphoBank	Metadata: Site-specific Data Content: SDD or NEXUS/TNT/NeXML Data Deposit: Any valid media file Services: Portal-based search and discovery	[10]
Presence/absence, occurrence data, species survey data	GBIF Indexing is the most appropriate metadata and discovery mechanism, although INSPIRE in Europe also makes provision for such data	GBIF OBIS	Metadata: Darwin Core/ABCD Data Content: Darwin Core/ABCD Data Deposit: IPT/BioCASE Services: Multiple API options provided by GBIF	[10, 19–21, 54, 56]
Allele/genomic	Services as provided by INSDC Consortium members	GenBank DNA Databank of Japan European Molecular Biology Laboratory	MetaData: MIxS compliant Data Content: Multiple upload tools are available, GCDML	[10, 16]

(continued)

Table 11.5 (continued)

Data family	Metadata and catalogue services	Sustainable international infrastructures	Schematic and syntactic interoperability—service protocols and content standards	Reference
Functional genomics/transcriptomics	Services as provided by GEO and ArrayExpress	Gene Expression Omnibus (GEO) ArrayExpress	MetaData: MIAME-compliant Data Content: MIAME-compliant Data Deposit: FTP Upload Services: JSON/FTP	[10, 17, 18]
Phylogenetic data	Metadata is provided by way of a peer-reviewed article—in other words all data submissions are supported by a published article. An OAI-PMH interface is available for metadata harvesting	TreeBASE	Metadata: Published Article Data Content: NEXUS Data Deposit: via web portal Services: OAI-PMH for metadata, portal and RESTful API for data access	[10, 23]
Micro-CT	Service-specific metadata is gathered on submission, no harvestable or machine-readable endpoints	MorphoSource	Metadata: gathered manually on submission Data Content: files produced by scanners Data Deposit: via portal Services: portal search and browse facility	[10, 24]

Table 11.6 Guidelines applicable to data families for which distributed systems and federated access will apply

Data family	Metadata and catalogue services	Aggregating global or regional infrastructures	Schematic and syntactic interoperability—service protocols and content standards	Reference
Traditional spatial data (raster and vector)	OGC Catalogue Services for the Web (CS/W) or OAI-PMH Aggregation to GEOSS Broker	EU BON GEOSS GCMD Biodiversity Catalogue Consider IPT feed to GBIF in respect of species occurrence	Metadata: ISO 19115 preferred, FGDC supported Data Content: domain-dependent Data Deposit: not required—distributed Services: Publish data via OGC WxS services	[19, 25, 26, 34, 35]
Signals and time series observation data	OGC Catalogue Services for the Web (CS/W) or OAI-PMH Aggregation to GEOSS Broker	EU BON GEOSS Biodiversity Catalogue	Metadata: ISO 19115 preferred Data Content: domain-dependent but based on Sensor Markup Language Data Deposit: not required—distributed Services: Publish data via OGC Sensor Observation Services	[25, 26, 34, 35]
Model outputs and multidimensional data	THREDDS and OPeNDAP Aggregation to GEOSS Broker	EU BON GEOSS Biodiversity Catalogue	Metadata: THREDDS crosswalk to ISO 19115 preferred Data Content: domain-dependent Data Deposit: Not required—federated Services: NetCDF/OPeNDAP queries or mapping to WMS	[25–28, 34, 35]
All other tabular data	OAI-PMH serving Dublin Core or EML Metadata	EU BON GEOSS DataOne, KNB, LTER Consider IPT feed to GBIF in respect of species occurrence	Metadata: EML Data Content: domain-dependent DataDeposit: any compatible format Services: download via API	[19, 29, 31, 34, 53]

(continued)

Table 11.6 (continued)

Data family	Metadata and catalogue services	Aggregating global or regional infrastructures	Schematic and syntactic interoperability— service protocols and content standards	Reference
Any other digital object	Media files, grey literature, code, and similar: provide a DataCite metadata record to DataCite and obtain a DOI	DataCite	Metadata: DataCite Data Content: any digital object Data Deposit: not required— distributed Services: DataCite API	[30]

11.4.4 Section C: Specialised Global Infrastructure

For some data types and families, it is best practice to publish data and make it available via established global infrastructures (Table 11.5).

11.4.5 Section D: Aggregators and Open Federated Infrastructures

The data families and types listed below are best published in a federated manner, using standardised service protocols and content standards, with reliance on aggregation of standard metadata implementations to improve accessibility. GEO BON might consider hosting its own metadata aggregator as a component of the GEOSS Common Infrastructure (Table 11.6).

11.5 Conclusions

Biodiversity informatics is inherently a global initiative. With a multitude of organisations from different countries publishing biodiversity data, the foremost challenge is to make the diverse and distributed participating systems interoperable in order to support discovery and access to data. A common exchange technology, e.g. the widely used XML or JSON over HTTP, may allow the syntactic exchange of data blocks, but participating systems also need to understand the schema and semantics of the data being delivered in order to process it meaningfully. Unless the data share a common reference model, the exchange implies brokering, mediation, or other semantic processing.

The challenge, then, from the perspective of GEO BON, is largely one of agreeing appropriate content (schematic and semantic) standards for the main data families appropriate to each EBV. This will not address all requirements, but should go a long way towards creating successful interoperability precedents and simplify the broadening of the scope of application.

11.5.1 What Is Already Achievable?

Researchers, institutions, and regional or global infrastructures or initiatives that follow the guidelines published in the chapter will already make an immense contribution to the components of an interoperable, federated system of systems as envisaged by GEO.

11.5.2 What Needs to Be Improved?

The guidance has indicated for each aspect what role GEO BON can play in coordinating the solutions to non-ideal situations and development of community-endorsed standards, and in general this remains a significant requirement.

If one considers the more specific goal of EBV interoperability: the majority of EBVs still need to be defined by the GEO BON community, and guidance in respect of interoperability standards and software to support these is dependent on these definitions. In practical terms, the tasks at hand are:

- Review the guidance presented here as more EBVs are formalised;
- Identify the main deficiencies in respect of the available interoperability standards that can be used for GEO BON supported EBVs across data families;
- Define extended content standards for the major data exchange service protocols (IPT, OGC WxS, NetCDF, Sensor Observation Services), using patterns and resources that already exist;
- Build mediation tools for mapping of non-standardised data sets, such as those found routinely in MetaCAT and PlantNet repositories, to services that are schematically and semantically interoperable; and
- Build schematic translation tools to serve any content standard over any service syntax.

It remains unclear how large data sets will be made available and included into an interoperable, orchestrated workflow in an open, free environment—the costs and time involved in sub-setting and processing the data may prove to be prohibitive, and it should be appreciated that the concept of having a suite of EBVs available within a distributed, interoperable global system of systems is constrained in many countries by availability of data sets and resources to gather and maintain such data sets.

Despite these constraints, GEO BON hopes to make steady progress in respect of extending the scope of content standards and services that implement them— leading to a set of EBVs available to a variety of end users from a variety of distributed contributors.

References

Agosti, D., & Egloff, W. (2009) Taxonomic information exchange and copyright: The Plazi approach. *BMC Research Notes, 2*, 53. doi:10.1186/1756-0500-2-53

Ariño, A. H. (2010). Approaches to estimating the universe of natural history collections data. *Biodiversity Informatics, 7*, 81–92.

Catapano, T. (2010). TaxPub: An extension of the NLM/NCBI journal publishing DTD for taxonomic descriptions. In *Proceedings of the Journal Article Tag Suite Conference (JATS-Con) 2010 [Internet]*. Bethesda, MD: National Center for Biotechnology Information (US). http://www.ncbi.nlm.nih.gov/books/NBK47081/.

Chavan, V., & Penev, L. (2011). The data paper: A mechanism to incentivize data publishing in biodiversity science. *BMC Bioinformatics, 12*(Suppl 15), S2. doi:10.1186/1471-2105-12-S15-S2.

Goddard, A., Wilson, N., Cryer, P., & Yamashita, G. (2011). Data hosting infrastructure for primary biodiversity data. *BMC Bioinformatics, 12*(Suppl 15), S5. Published online December 15, 2011. doi:10.1186/1471-2105-12-S15-S5.

Hagedorn, G., Mietchen, D., Agosti, D., Penev, L., Berendsohn, W., & Hobern, D. (2011). Creative Commons licenses and the non-commercial condition: Implications for the re-use of biodiversity information. *ZooKeys, 150*, 127–149.

Hardisty, A., Roberts, D., & The Biodiversity Informatics Community. (2013). A decadal view of biodiversity informatics: Challenges and priorities. *BMC Ecology, 13*, 16. http://www.biomedcentral.com/1472-6785/13/16.

Hernandez, V., Poigné, A., Giddy, J., & Hardisty, A. (2009a). Data and modelling tool structures reference model. *Lifewatch Deliverable 5.1.3.*

Hernandez, V., Poigné, A., Giddy, J., Hardisty, A., Voss, A., & Voss, H. (2009b). Towards a reference model for the Lifewatch ICT infrastructure. *Lecture Notes in Informatics 154.*

Hobern, D., et al. (2012). Global Biodiversity Informatics Outlook. *GBIF*. http://www.biodiversityinformatics.org/download-gbio-report/.

Hoffman, A., Penner, J., Vohland, K., Cramer, W., Doubleday, R., Henle, K., et al. (2014). The need for an integrated biodiversity policy support process—Building the European contribution to a global Biodiversity Observation Network (EU BON). 20 p.

Hugo, W., Saarenmaa, H., & Schmidt, J. (2013). Development of extended content standards for biodiversity data. European Geosciences Union (EGU) General Assembly, Vienna, April 8–12, 2013. In *Geophysical Research Abstracts, EGU 2013* (Vol. 15, p. 6968).

Hugo, W., Jensen, S., Onsrud, H., & Ziegler, R. (2011). White Paper 3: Crowdsourcing and Environmental Science. *Eye On Earth Summit, Abu Dhabi*, September 2011. http://www.eyeonearthsummit.org/sites/default/files/WG3_WP3_formatted_Dec5_Final%20check_.pdf.

Hui, C., & McGeoch, M. A. (2014). Zeta diversity as a concept and metric that unifies incidence-based biodiversity patterns. *American Naturalist, 184*(5), 684–694. doi:10.1086/678125

Inspire Thematic Working Group Species Distribution. (2013). Data specification on species distribution—*Draft Technical Guidelines v.3.0rc3*.

Kery, M., Royle, A., Schmid, H., Schaub, M., Volet, B., Häfliger, G., et al. (2010). Site-occupancy distribution modeling to correct population-trend estimates derived from opportunistic observations. *Conservation Biology, 24*, 1388–1397.

Kissling, W., Hardisty, A., Alonso García, E., Santamaria, M., De Leo, F., Pesole, G., et al. (2015). Towards global interoperability for supporting biodiversity research on essential biodiversity variables (EBVs). *Biodiversity, 16*(2–3), 99–107. doi:10.1080/14888386.2015.1068709.

Michener, W. K., Brunt, J. W., Helly, J. J., Kirchner, T. B., & Stafford, S. G. (1997). Nongeospatial metadata for the ecological sciences. *Ecological Applications, 7*(1), 330–342.

Nativi, S., Craglia, M., & Pearlman, J. (2012). The brokering approach for multidisciplinary interoperability: A position paper. *International Journal of Spatial Data Infrastructures Research, 7*, 1–15.

Nativi, S., Mazzetti, P., & Geller, G. (2013). Environmental model access and interoperability: The GEO Model Web initiative. *Environmental Modelling & Software, 39*, 214–228, January 2013. http://dx.doi.org/10.1016/j.envsoft.2012.03.007.

Ó Tuama, É., Saarenmaa, H., Nativi, S., Bertrand, N., van den Berghe, E., Scott, L., et al. (2010). Principles of the GEO BON information architecture. *Group on Earth Observations (Geneva),* 42 p. http://www.earthobservations.org/documents/cop/bi_geobon/geobon_information_architecture_principles.pdf.

Pereira, H. M., Ferrier, S., Walters, M., Geller, G. N., Jongman, R. H. G., Scholes, R. J., et al. (2013). Essential biodiversity variables. *Science, 339*, 277–278. doi:10.1126/science.1229931.

Saarenmaa, H., et al. (2014). Architectural design, review and guidelines for standards. *Deliverable 2.1 (D2.1)—EU BON Project, FP 7 Grant 308454*. http://eubon.eu/documents/1/.

Scholes, R. J., Walters, M., Turak, E., Saarenmaa, H., Heip, C. H. R., Ó Tuama, É., et al. (2012). Building a global observing system for biodiversity. *Current Opinion in Environmental Sustainability, 4*, 139–146. http://dx.doi.org/10.1016/j.cosust.2011.12.005.

Socha, Y. (Ed.). (2013). Out of cite—Out of mind—The current state of practice, policy, and technology for the citation of data, CODATA-ICSTI task group on data citation standards and practices. *Data Science Journal, 12*, September 13, 2013. https://www.jstage.jst.go.jp/article/dsj/12/0/12_OSOM13-043/_pdf.

Uhlir, P. (2013). The Legal Interoperability of Data, National States Geographic Information Council, Public Resources. http://www.nsgic.org/public_resources/02_Uhlir_Legal-Interoperability-of-Data_NSGIC-Conf_Feb13.pdf.

Vines, T., et al. (2014). The availability of research data declines rapidly with article age. *Current Biology, 24*, 94–97, January 6, 2014. http://dx.doi.org/10.1016/j.cub.2013.11.014.

Wallnau, K. (2003). Introducing Predictable Assembly from Certifiable Components (PACC), News at SEI, Library, Carnegie-Mellon Institute. http://www.sei.cmu.edu/library/abstracts/news-at-sei/architect2q03.cfm.

Walls, R. L., Deck, J., Guralnick, R., Baskauf, S., Beaman, R., et al. (2014). Semantics in support of biodiversity knowledge discovery: An introduction to the biological collections ontology and related ontologies. *PLoS ONE, 9*(3), e89606. doi:10.1371/journal.pone.0089606.

Wieczorek, J., Bloom, D., Guralnick, R., Blum, S., Döring, M., De Giovanni, R., et al. (2012). Darwin Core: An evolving community-developed biodiversity data standard. *PLoS ONE, 7*(1), e29715. doi:10.1371/journal.pone.0029715.

Wooley, J. C., Godzik, A., & Friedberg, I. (2010). A primer on metagenomics. *PLoS Computational Biology, 6*, 1000667. http://dx.doi.org/10.1371%2Fjournal.pcbi.1000667.

Web Links and References Used in the Guidance Tables 11.3, 11.4, 11.5 and 11.6

[1] Hardisty et al. (2013): See Reference section.
[2] Research Data Alliance: Data Citation Working Group: https://rd-alliance.org/groups/data-citation-wg.html.
[3] Research Data Alliance: Data Type Registries Working Group: https://rd-alliance.org/groups/data-type-registries-wg.html.
[4] Hobern et al. (2012): See Reference section.
[5] Refer to supplementary material for a review of licenses and exceptions to open licenses: http://dataintegration.geobon.org/.
[6] Research Data Alliance: Repository Audit and Certification DSA–WDS Partnership Working Group: https://rd-alliance.org/groups/repository-audit-and-certification-dsa%E2%80%93wds-partnership-wg.html.
[7] Research Data Alliance: Practical Policy Working Group: https://rd-alliance.org/groups/practical-policy-wg.html.
[8] GEO Data Management Principles: https://www.earthobservations.org/documents/dswg/201504_data_management_principles_long_final.pdf.
[9] Research Data Alliance: Vocabulary Services Interest Group: https://rd-alliance.org/groups/vocabulary-services-interest-group.html.
[10] Saarenmaa et al. (2014): See Reference section.
[11] Darwin Core and Archive Extensions: http://www.gbif.org/resource/80636.
[12] International Geo Sample Number (IGSN): http://schema.igsn.org/description/.
[13] Research Data Alliance: Management and Curation of Physical Samples: https://rd-alliance.org/bof-management-and-curation-physical-samples-ig.html.
[14] Research Data Alliance: Biodiversity Data Integration Interest Group: https://rd-alliance.org/groups/biodiversity-data-integration-ig.html.
[15] Genomic Standards Consortium: http://gensc.org/mixs/.
[16] Genbank and INSDC Members: http://www.ncbi.nlm.nih.gov/genbank/.
[17] Gene Expression Omnibus (GEO): http://www.ncbi.nlm.nih.gov/geo/.
[18] ArrayExpress: https://www.ebi.ac.uk/arrayexpress/.
[19] Global Biodiversity Information Facility (GBIF): http://www.gbif.org/.
[20] Map of Life: https://www.mol.org/about.
[21] Ocean Biogeographic Information System: http://www.iobis.org/.
[22] MorphoBank: http://morphobank.org/index.php/Documentation/Index#d5e1285.
[23] TreeBASE: https://treebase.org/treebase-web/home.html.
[24] MorphoSource: http://morphosource.org/.
[25] Open Geospatial Consortium Standards: http://www.opengeospatial.org/standards.
[26] Global Change Master Directory: http://gcmd.nasa.gov/.
[27] THREDDS—ISO 19115 conversion: https://geo-ide.noaa.gov/wiki/index.php?title=NetCDF_Attribute_Convention_for_Dataset_Discovery#Open_Geospatial_Consortium_Catalog_Service_for_the_Web_.28CSW.29.
[28] ncWMS—University of Reading: http://www.resc.rdg.ac.uk/trac/ncWMS/.
[29] DataOne: https://www.dataone.org/.
[30] DataCite: http://www.datacite.org.
[31] International Long Term Ecological Research Network (ILTER): http://www.ilternet.edu/.
[32] NASA SWEET Ontology: https://sweet.jpl.nasa.gov/.
[33] EnvO: http://www.environmentontology.org/.
[34] Biodiversity Catalogue: https://www.biodiversitycatalogue.org/services.
[35] GEOSS Broker: http://www.eurogeoss.eu/broker/default.aspx.
[36] EU BON: http://www.eubon.eu/.
[37] Catalogue of Life: http://www.catalogueoflife.org/.

[38] Encyclopedia of Life: http://www.eol.org/.

[39] Plazi: http://plazi.org/news/beitrag/data-sharing-principles-and-legal-interoperability-for-essential-biodiversity-variables/13f96ba8031d1c42c4519d3863e203e8/.

[40] Open Biodiversity Knowledge Management System (OBKMS): http://pro-ibiosphere.eu/getatt.php?filename=oo_4670.pdf.

[41] Research Data Alliance: Federated Identity Management Interest Group: https://rd-alliance.org/groups/federated-identity-management.html.

[42] Creative-B Roadmap: http://www.slideshare.net/dmanset/20140909creativeb-roadmap-interactive.

[43] EarthCube: https://docs.google.com/document/d/10OhZntRpizn-KaYECXtGY_tcVbanG2k R0OFJ7JZpnWw/edit#.

[44] GEO BON Information Architecture Principles: https://www.earthobservations.org/documents/cop/bi_geobon/geobon_information_architecture_principles.pdf.

[45] Walls et al. (2014): See Reference section.

[46] Traitbank: http://eol.org/info/516.

[47] Hagedorn et al. (2011): See Reference section.

[48] Canadensys—Open Licenses: http://www.canadensys.net/2012/why-we-should-publish-our-data-under-cc0.

[49] iNaturalist—Open Licenses: http://inaturalist.tumblr.com/post/138557593458/changes-to-gbif-licensing-requirements.

[50] Penev, L., Agosti, D., Georgiev, T., Catapano, T., Miller, J., Blagoderov, V., et al. (2010). Semantic tagging of and semantic enhancements to systematics papers: ZooKeys working examples. *ZooKeys, 50*, 1–16. doi:10.3897/zookeys.50.538.

[51] Penev, L., Lyal, C., Weitzman, A., Morse, D., King, D., Sautter, G., et al. (2011). XML schemas and mark-up practices of taxonomic literature. *ZooKeys, 150*, 89–116. doi:10.3897/zookeys.150.2213.

[52] Sautter, G., Agosti, D., & Böhm, K. (2007). Semi-Automated XML Markup of Biosystematics Legacy Literature with the GoldenGATE Editor. In *Proceedings of PSB 2007, Wailea, HI, USA, 2007*. http://psb.stanford.edu/psb-online/proceedings/psb07/sautter.pdf.

[53] Michener, W. K., Brunt, J. W., Helly, J. J., Kirchner, T. B., & Stafford, S. G. (1997). Nongeospatial metadata for the ecological sciences. *Ecological Applications, 7*(1), 330–342.

[54] Wieczorek, J., Bloom, D., Guralnick, R., Blum, S., Döring, M., De Giovanni, R., et al. (2012). Darwin Core: An evolving community-developed biodiversity data standard. *PLoS ONE, 7*(1), e29715. doi:10.1371/journal.pone.0029715.

[55] Wooley, J. C., Godzik, A., & Friedberg, I. (2010). A primer on metagenomics. *PLoS Computational Biology, 6*, 1000667. http://dx.doi.org/10.1371%2Fjournal.pcbi.1000667.

[56] Inspire Thematic Working Group Species Distribution. (2013). Data specification on species distribution—*Draft Technical Guidelines v.3.0rc3*.

[57] Catapano, T. (2010). TaxPub: An extension of the NLM/NCBI journal publishing DTD for taxonomic descriptions. In *Proceedings of the Journal Article Tag Suite Conference (JATS-Con) 2010 [Internet]*. Bethesda, MD: National Center for Biotechnology Information (US). http://www.ncbi.nlm.nih.gov/books/NBK47081/.

Chapter 12
Using Data for Decision-Making: From Observations to Indicators and Other Policy Tools

Matt Walpole, Melodie A. McGeoch, Philip Bubb and Neil Brummitt

Abstract Previous chapters in this handbook have focused primarily on how to develop observing systems and generate biodiversity observations. Drawing on these foundations, this chapter explores the use of such data in decision-making processes. It reflects on what data might be used for, how it is packaged, what the challenges are and what to consider in getting it right. It is intended to be a thought-provoking look at insights gained in communicating biodiversity science for policy purposes over the last several years. With a particular focus on indicators, one of the most common forms in which observations are used by decision-makers, the chapter considers the context in which indicators are used and how they are developed. It explores the realities of indicator development and use, including some of the key challenges and ways around them. It also touches on assessments and assessment processes as another important tool linking science to policy.

Keywords Indicator development · Data use · Policy impact · Assessments

M. Walpole (✉) · P. Bubb
United Nations Environment Programme World Conservation Monitoring Centre
(UNEP-WCMC), 219 Huntingdon Road, Cambridge CB3 0DL, UK
e-mail: Matt.Walpole@unep-wcmc.org

P. Bubb
e-mail: Philip.Bubb@unep-wcmc.org

M.A. McGeoch
School of Biological Sciences, Monash University, Melbourne, Victoria 3800, Australia
e-mail: melodie.mcgeoch@monash.edu

N. Brummitt
Department of Life Sciences, Natural History Museum, Cromwell Road,
London SW7 5BD, UK
e-mail: n.brummitt@nhm.ac.uk

© The Author(s) 2017
M. Walters and R.J. Scholes (eds.), *The GEO Handbook on Biodiversity Observation Networks*, DOI 10.1007/978-3-319-27288-7_12

12.1   Introduction

Previous chapters in this handbook have focused primarily on how to develop
observing systems and generate biodiversity observations. Drawing on these
foundations, this chapter explores the use of such data in decision-making processes. It reflects on what data might be used for, how it is packaged, what the
challenges are and what to consider in getting it right. It is intended to be a
thought-provoking look at insights gained from communicating biodiversity science for policy purposes over the last several years. With a particular focus on
indicators, one of the most common forms in which observations are used by
decision-makers, the chapter considers the context in which indicators are used and
how they are developed. It explores the realities of indicator development and use,
including some of the key challenges and ways around them. It also touches on
biodiversity assessments and assessment processes as another important tool linking
science to policy.

12.1.1   What Are Indicators?

Indicators are communication and decision-support tools. They tell a story to help
stimulate and guide action. They are part of a process, not an end in themselves.
A useful definition of an indicator is that used by the Biodiversity Indicators
Partnership (BIP; http://www.bipindicators.net) as '*a measure based on verifiable
data that conveys information about more than itself*' (Biodiversity Indicators
Partnership 2011). In essence, this is saying that data requires an external context in
which to be delivered before it becomes valuable as an indicator. As noted by Bubb
(2013, p. 403) '*indicators are purpose-dependent and so the interpretation or
meaning of the data depends on the issue being examined*'.

Consider some examples. One form of biodiversity observation, or measure,
may be about land cover, or the size of a patch of particular habitat. Collected
regularly over space and time, these observations can be amassed to create a
variable (or metric) about habitat or ecosystem extent. An example might be forest
extent. At this stage, this is not an indicator, it is simply data telling us about itself,
i.e., change in forest extent. The way that the data are used, alone or combined with
other ancillary information, will determine its role as an indicator. For example, the
data could be used to generate an analysis of rates of forest loss (or gain) that could
be used to track progress towards a target to reduce such as rates of loss (or one to
achieve certain levels of forest restoration). Alternatively, the data could be transformed into measures of carbon storage in forests that could be used to track
progress towards a target concerned with improving biodiversity contribution to
carbon stocks. Alternatively, in combination with information on the location of

protected areas, the data could be used to say something about how effective protected areas are at conserving forests, and to track progress towards a target concerned with improving the effectiveness of conservation interventions like protected areas.

In this example, one kind of observation (of forest cover), when collected over time, creates a change variable (forest extent) which, depending on the context, can underpin a range of different indicators. This example is not fictional—all of these targets and indicators exist in the context of the Convention on Biological Diversity (CBD). Nor is it unique; existing observations on populations, species and ecosystem services can also all be used in various ways to create multiple indicators for different purposes. More on this later.

A good indicator has to be scientifically valid (based on reliable, verifiable data and with a clear relationship between the indicator and its purpose), sensitive to change in the issue of concern within appropriate timescales and spatial scales, and produced on a sufficiently regular basis using repeatable methods to track change over time. A successful indicator is one that is actually used to support decision-making (Biodiversity Indicators Partnership 2011).

In order for an indicator to be useful to the non-specialist it usually requires a storyline, a narrative that interprets the meaning of the data on which it is based. The annually updated Aichi Passport (Chenery et al. 2013; http://www.bipindicators.net/resource/aichipassport) contains examples of a number of indicators combining time series data (usually presented as a line graph, but sometimes in other ways such as a pie chart or change map) with narrative storylines explaining what they mean in the context of the particular Target to which they relate. At a national level, the UK government's annual biodiversity indicators publication (DEFRA 2013; available from https://www.gov.uk/government/publications/biodiversity-indicators-for-the-uk) includes a traffic light scheme for illustrating which metrics (termed measures in this publication) indicate improvement, no change, or decline. The Millennium Development Goals (MDG) Annual Reports (http://www.un.org/millenniumgoals/reports.shtml) provide another example of how indicators can be communicated as a combination of data and storylines. Importantly, indicators are designed to effectively communicate information that is relevant to one or more policy objectives and to do so in a way that translates raw data into clear messages. Packaged and communicated in the right way, information on biodiversity change can have real policy impact.

12.1.2 The Policy Context for Biodiversity Indicators

Ultimately, the value of biodiversity observations and observing systems of the kind supported and promoted by GEO BON is in their use. There are a variety of policy contexts in which biodiversity indicators are required to assist in monitoring, assessing and reporting progress towards targets in plans and strategies.

At the international level the pre-eminent biodiversity context is the CBD and the collective commitments that Parties (primarily national governments) have made. The adoption in 2002 of the 2010 Biodiversity Target led to considerable effort to identify and develop indicators at a global level (Walpole et al. 2009). At the same time, Parties to the CBD were expected to report periodically on their own contributions and progress towards achieving this collective goal. In 2010 a new Strategic Plan for Biodiversity 2011–2020 was adopted by Parties to the CBD, including the twenty Aichi Targets (http://www.cbd.int/sp/). In 2014 the first national reports of progress in implementing the strategic plan, and the first global assessment of progress towards these targets was delivered (Leadley et al. 2014; Tittensor et al. 2014), with further reporting and assessment expected later in the decade.

There are a range of other global biodiversity-related conventions with strategies, goals and targets that require indicators of both national and international implementation and progress, including the Convention on International Trade in Endangered Species of Wild Fauna and Flora (CITES; www.cites.org), the Convention on the Conservation of Migratory Species of Wild Animals (CMS; http://www.cms.int/) family and the Ramsar Convention on Wetlands of International Importance (www.ramsar.org). More broadly, biodiversity and ecosystem service indicators are used in international development contexts, most notably to track progress towards the Millennium Development Goals up to 2015 (United Nations 2013; Sachs et al. 2009), and the Sustainable Development Goals subsequently (http://unstats.un.org/sdgs/). Besides these global agreements, there may be regional commitments for which governments are required to report, such as the EU Biodiversity Strategy (European Union 2011).

Nationally, biodiversity indicators are required to track progress towards national goals and targets including those defined within National Biodiversity Strategies and Action Plans (NBSAPs) (http://www.cbd.int/nbsap/; http://nbsapforum.net/), as well as for reporting against international commitments. They may be used more broadly, for public outreach and communication or for specific sectoral plans and policies, as well as for biodiversity management and threat reduction. In an innovative example, near real time, publically available satellite data on deforestation in Brazil has been used to boost law enforcement efforts that has yielded huge reductions in deforestation rates over the past decade (Secades et al. 2014). They are also likely to be useful in the context of national development planning and, increasingly, in national accounting to provide more balanced, inclusive measures of national wealth and well-being.

It is also worth noting that, for various conservation investment stakeholders, biodiversity indicators are essential to aid evaluation of the impact and success of conservation investment actions (Stephenson et al. 2015).

12.2   Developing Indicators

Developing indicators successfully involves a number of steps. A useful tool for this process is the biodiversity indicator development framework (Fig. 12.1). This covers ten steps grouped into three areas: Purpose (actions needed for selecting successful indicators), Production (steps essential to generate indicators) and Permanence (mechanisms for ensuring indicator continuity and sustainability).

Fig. 12.1 Biodiversity indicator development framework. (*Source* Biodiversity Indicators Partnership 2011)

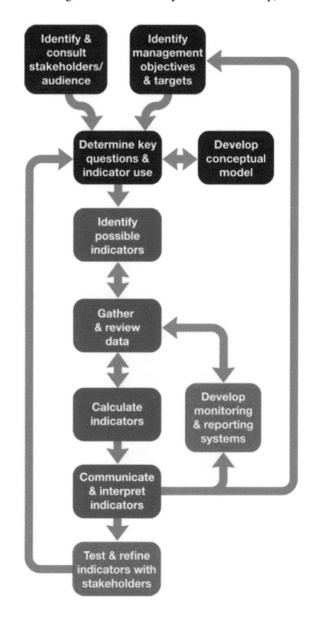

12.2.1 Starting with the Question

How are good biodiversity observations turned into good indicators? The key, in fact, is not to start with the data, but rather with the question that decision-makers need answering. Such questions can be framed in the context of explicit policies, plans, goals and targets. Are we on track to meet a particular target? Is our policy moving things in the right or wrong direction? These questions determine the kinds of indicators required and the kinds of data needed to produce them.

Once the key questions are defined, it is possible to consider which metrics would most adequately address these questions and what the most robust methods would be to deliver those metrics. It is also important to consider how they would be interpreted and what possible misinterpretation or bias might result.

12.2.2 Then Find the Data

Whilst the theory of indicator development appears straightforward, there are often significant challenges, with data availability (in particular consistent trend data with reliable baselines) being one. As an example, during the fourth round of national reporting to the CBD in 2010, Parties were encouraged to use indicators and yet few presented data or figures as part of their storylines. When surveyed, almost half of respondents indicated that they did not have, or did not know whether their country had, indicators relevant to the CBD (Bubb 2013).

This may not be the full story—many countries may have other sources of data that are not recognised or readily available. For those countries that did report national indicators, only 15 % noted that source data was primarily obtained from dedicated biodiversity monitoring systems. It was far more common for data to be sourced from monitoring systems developed for other purposes and from other sectors (such as forestry, agriculture or fisheries), or from academic research, surveys and assessments (recognising that one-off, time-bound studies are more useful in assessment processes than for indicators). Around 10 % obtained data from external, regional or global sources (Bubb 2013).

This illustrates an important point. Mobilising such existing data, which may come from a wide range of different types of organisation including universities, NGOs, government agencies and the private sector, and from a range of sectors, can be a practical first step (see Box 12.1 for an international example). A number of countries which were hitherto lacking biodiversity indicators have subsequently used this approach to develop national biodiversity indicator reports.

Box 12.1. From Ground Zero to an Indicator for Biological Invasions

When the 2010 Biodiversity Target was set calling for an indicator of trends in invasive alien species there was no obvious option at hand. At the time

there were no invasion indicators that had been developed for reporting at a
global scale (Genovesi et al. 2013). There were many sources of information,
but no collated body of data with global coverage on which species were
where, and what impact they were having, particularly not for a range of
taxonomic groups (McGeoch et al. 2012). Apart from many local case studies
and a few regional ones (notably for Europe) there were also very few data on
changes in the numbers of species threatening biodiversity over time.

The solution was to tackle the problem from three angles (McGeoch et al.
2010). First, using an operational definition that distinguished alien from
invasive species (McGeoch et al. 2012), a stratified, random subset of
countries was chosen to calculate the number of invasive species per country.
This provided a robust, representative baseline measure of invasion pressure
on countries that can now be tracked over time. Second, the well-known
IUCN Red List Index (http://www.bipindicators.net/rli/2010; Butchart 2008)
was used to illustrate trends in the extinction risk of a subset of species
threatened by invasive species. Finally, information on policy trends relevant
to invasion was used to indicate how countries were responding to the
problem.

In sum, by using a combination of systematic data collation, adaptation of
an existing indicator, and data on policy trends, an informative indicator for
biological invasion was born.

Certain metrics can be used to create multiple indicators for different purposes,
as illustrated earlier for forest extent (see also Box 12.2). This demonstrates the
value and importance of focusing on the key metrics, or Essential Biodiversity
Variables (EBVs; Pereira et al. 2013) that, when cut in certain ways and/or combined with ancillary data, can provide the most information on biodiversity change.
Indicators for many of the Aichi Targets can be derived from a relatively small set
of such metrics (Geijzendorffer et al. 2016; for an example, see Fig. 1.2).

Equally, where metrics can be standardised across scales, there is great potential
for efficiency. Although national priorities are generally to develop indicators for
specific national needs (including national goals and targets), there is a lot to be said
for exploring and enhancing harmonised indicator use as a means of (i) increasing
the availability of data for tracking progress towards broader scale, regional and
global goals and targets and (ii) unlocking the value of global data sets, such as
those in products derived from satellite remote sensing, for wider national use (Han
et al. 2014).

Box 12.2. Indicator Pragmatism: The Living Planet Index

The Living Planet Index (LPI) is a metric of aggregate change in vertebrate
population abundance over time in reference to a baseline year at which the
index is set at a value of 100 (Loh et al. 2005; Collen et al. 2009). It is built up

from individual population time series sourced from published and grey literature. The global database currently runs into thousands of such time series, and the index is calculated from a 1970 baseline to the current day.

The global LPI, like many other metrics used to underpin indicators (Walpole et al. 2009) is not perfect. The data it is built upon are patchy both taxonomically and geographically, being particularly rich in data from bird populations and temperate regions (Collen et al. 2008, 2009), although the construction of the index attempts to offset these imbalances.

Despite this, it has significant strengths as an indicator in a number of ways. First, it taps into a vast resource of existing data, and so is cost effective. Population abundance, for vertebrates at least, is one of the most commonly collected measures, both in discrete, time-bound studies and from continuous monitoring. Second, it tells a simple, easily understood message of overall, aggregate change. The LPI has achieved prominence as a tool for communicating global biodiversity change to the public via its central role in WWFs periodic Living Planet Report (WWF 2012). Third, it can be cut in various ways to answer different questions and provide indicators for different policy targets. From the global dataset it is relatively simple to draw out subset analyses, or cuts, of the LPI focusing, for example, on wetland-dependent species (of relevance to the Ramsar Convention) or migratory species (of relevance to the CMS) or harvested species (of relevance to sustainable use and human wellbeing concerns). Fourth, it can be improved with new data sources. Knowing where the data gaps are enables a focus on filling those gaps, whether through unlocking more existing data or by investing in new monitoring. Fifth, it can be applied in different settings and different scales. A national LPI built from within-country studies of population abundances of different species is simple and cost-effective to construct. Moreover, the same approach can be applied to create an index of aggregate change in habitat extent, as has recently been created for wetland extent for the Ramsar Convention (Dixon et al. 2016).

In the same way, harmonising the use of metrics across policy contexts is not only an efficient use of resources but also creates greater awareness and potential for mainstreaming biodiversity into other development sectors. Globally, metrics of forest change, fisheries, threatened species and protected area coverage have been used to provide indicators in the context of both the CBD and the Millennium Development Goals. Nationally, there are increasing efforts to incorporate 'natural capital' (including biodiversity and ecosystem services) into national accounts (King et al. 2015). The more that biodiversity data can be used in these contexts, the greater impact it will have on decision-making.

12.2.3   Trade-Offs and Compromises Between Data Availability and Policy Needs

In an ideal world, data coverage would be universal, observations would be regularly repeated and the indicators derived would be tightly linked to the targets or policy processes for which they were being used, so that even slight changes would tell a decision-maker in a timely manner what action was required to keep things on track. In reality, it is never this simple. Data are patchy in space, time and thematic/taxonomic coverage. Even where gaps can be filled, detecting meaningful (significant) change is not straightforward, ascribing causes and appropriate action less so, and sustaining consistent data collection over the long term difficult to resource. Moreover, policy targets are not always determined with suitable metrics and indicators in mind, such that their interpretation and translation for monitoring purposes can be difficult.

This does not mean, however, that we cannot develop useful indicators that influence policy and action, if we remember that indicators are at heart a communication tool, and if we are very clear about what they are communicating. Consider the 2010 Biodiversity Target and the metrics used to track progress towards it. The data were patchy and far from perfectly aligned to the needs of the headline indicators (Walpole et al. 2009), but were able to tell a compelling story around pressures, state and responses at global scale (Butchart et al. 2010) that alerted people to a need for greater action and fuelled the debate that led to a more comprehensive, explicit Strategic Plan for 2011–2020.

Part of the challenge for 2010 was the relatively late stage at which indicators began to be considered. Although the 2010 Biodiversity Target was agreed in 2002, the headline indicators were not agreed until 2006. Concerted effort to populate a suite of indicators only took place in the last few years before the target deadline, meaning that indicators had to be adapted from what was available (see Box 12.1 for an example of what was achieved). Post-2010 the need to develop indicators early in the process, in tandem with targets, has been recognised. Yet we are still largely retro-fitting indicators to targets rather than creating targets with indicators in mind. Given that few of the Targets are quantitatively specific, indicators can largely only communicate whether things are heading in the right direction or not (Tittensor et al. 2014). This is still policy relevant, however, and of great value to decision-makers in pointing to where things need particular attention.

Developing indicators can be a journey of gradual improvement. Even if existing data quality and quantity are not optimal, using what we have can be a major incentive to leverage governments, scientists and data providers to do better if it stimulates scrutiny and debate about the robustness of the data and its suitability for indicators in the context in which it is being used. This has the potential to stimulate investment and improvement in both data and indicators. Indeed, within the GEO BON community such scrutiny and investment has yielded innovate new candidate indicators, based on several of the EBVs, that make use of state of the art modelling techniques, large datasets and remote sensing (GEO BON 2015).

The Biodiversity Indicator development Framework (Fig. 12.1) includes steps and feedback loops concerned with testing and refining indicators in line with the policy questions posed of them.

12.3   Beyond the Data—Partnerships and Other Enabling Factors

Data gaps and limitations are not the only challenges for developing and using indicators, so that focusing solely on improving observing systems will not guarantee more evidence-informed decision-making. A lack of funding and human capacity particularly for data integration, analysis and reporting is a widespread constraint, hindered further by a lack of awareness, interest, and political will (Bubb 2013).

The fact that data are often derived from multiple institutions signals the importance of both a co-ordinating body and a functioning network or partnership within a country. The BIP (see Box 12.3) is a global example of the kind of partnership approach to indicator development that could be taken nationally, to bring together the best data sources and providers covering the range of indicator information needs. Most important, however, is a coordinating body, a national office or institution responsible for co-ordinating analysis and communication of biodiversity data (Bubb 2013). Many countries have government bodies for related sectors such as fisheries, forests, etc., but not often for biodiversity as a whole. Those which do, including China, South Africa, Brazil and Mexico, are able to develop and report indicators regularly as an integral part of government processes.

It is also crucial to engage decision-makers from the outset. Whilst starting with the question is key, having those asking the questions owning the process is equally important. A key player in any network will be the government statistical office which in adopting particular metrics signifies an official stamp of approval and increases the likelihood of government use, as well as credibility and uptake more widely beyond environment ministries and the biodiversity community.

Box 12.3. The Biodiversity Indicators Partnership

The CBD-mandated Biodiversity Indicators Partnership is a global initiative to promote and coordinate development and delivery of biodiversity indicators in support of the CBD, Multilateral Environmental Agreements (MEA), IPBES, national and regional governments and a range of other sectors. The Partnership brings together over forty organisations, including UN agencies, NGOs, universities and research bodies, working internationally on indicator development to provide the most comprehensive information on biodiversity and related trends worldwide. The BIP was originally established in 2007 to assist in compiling indicators to track progress towards the 2010 Biodiversity

Target. This built on earlier work under the CBD to define 'headline indicators' for the 2010 Target and of the wider academic community to explore the state of the science of biodiversity indicators and to identify promising avenues (Balmford et al. 2005 and related papers in the same journal special issue arising from a Royal Society discussion meeting on "monitoring wild nature for the 2010 target").

The BIP partners provided a range of metrics focusing on biodiversity and ecosystem service trends, pressures and threats, and responses. Some 31 time series metrics were gathered. Not all of the CBD headline indicators were populated (Walpole et al. 2009), but this still represented a large increase in available data for the Third Global Biodiversity Outlook in 2010 compared to its predecessor in 2007 prior to the formation of the BIP.

Post-2010, the BIP has reoriented to the Aichi Targets and a focus on 2020. The partnership is strengthening to include a deeper and a wider breath of data providers. As a result, the BIP was able to deliver a first indicator-based analysis of progress towards the Aichi Targets using a larger number of time series metrics than in 2010 (Tittensor et al. 2014). The partnership also serves to raise awareness of the Targets amongst the observing community, creates links to other processes and agreements requiring indicators, and provides opportunities to share global methods and metrics with national governments and indicator practitioners to help develop capacity and to harmonise across scales.

The BIP is a complementary mechanism to GEO BON. Whilst GEO BON focuses on improving biodiversity observations that can be used in policy tools such as indicators, the BIP focuses on compiling and delivering those indicators for policy users. The two are mutually supporting and closely linked, with several organisations participating in both networks. Individuals from each network are also represented in the governance structures of the other.

12.4 A Word on Assessments

Indicators can be used in various ways and in various products, including assessments. Whilst indicators tend to be thought of as relatively continuous monitoring tools, assessments are more punctuated—one-off or periodic activities intended to draw together the best available evidence with which to answer a set of specific questions. In some cases these may focus on progress towards policy targets, as is the case with the CBD's periodic Global Biodiversity Assessment, which is heavily based on indicators. In others they may be more focused on understanding past and potential future change in a key thematic or sectoral field. The Millennium Ecosystem Assessment is a good example—it amassed an evidence base to explore how and why the world's ecosystems and the benefits they provide to society have changed over time, and constructed some future scenarios of how the world might look given certain broad policy choices. The International Assessment of Agricultural Knowledge, Science and Technology for Development (IAASTD 2009) is another example of a global assessment designed to answer specific policy questions, in this case relating to reducing hunger and improving nutrition in socially and environmentally sustainable ways.

Since its inception in 2012, the Intergovernmental Platform on Biodiversity and Ecosystem Services, IPBES, has provided a platform for delivering a range of thematic, regional and global assessments related to biodiversity and ecosystem services. In 2015, IPBES initiated, in response to requests from governments and non-governmental stakeholders, a set of regional assessments of biodiversity and ecosystem services in Africa, the Americas, Asia-Pacific, and Europe and Central Asia, which will be using indicators drawing from observations. In 2016, IPBES launched a global assessment which will draw information from the regional assessments. It will be key to select observations and indicators which allow comparisons among and within regions as well as aggregation at the global level.

Whilst assessments draw on diverse sources of information, spatio-temporal biodiversity and ecosystem service metrics are an important element, not only for revealing past trends and current status, but also, where these can be modelled, for exploring plausible future scenarios (Collen and Nicholson 2014; Newbold et al. 2015). Yet, as with indicators, assessments can fail as a communication and decision-making tool for reasons unrelated to the data and observations upon which they are built.

Assessments tend to be ignored if they are not undertaken with sufficient user engagement. In that regard they are best conceived of as a process rather than purely as a product—the key messages, synthesis and technical and regional reports commonly delivered by assessments are the culmination of, not the starting point for, communication and engagement. Those assessments which have had the most significant policy traction tend to be those that have had 'client' involvement from the outset (often governments or intergovernmental bodies).

Examples:

- The 2010 Global Biodiversity Outlook (SCBD 2010), requested by Parties to the CBD, contributed to renewed, more explicit, more tangible commitments from the world's governments in the form of the Strategic Plan for Biodiversity 2011–2020 SP including the Aichi Targets.
- The UK National Ecosystem Assessment (2011), called for by the UK government, provided a significant part of the information base for England's Natural Environment White Paper, 2011, which included commitments to invest in ecosystem services and natural capital locally whilst exploring means to embed natural capital into accounting processes nationally.

Assessments with a clear audience who are shaping the questions it asks and who feel part of the process get noticed. It also helps to have policy champions in government (and preferably beyond the environment sector) who can open doors and help to 'sell' the assessment to a broader or more influential audience. Assessments that are built into (or align correctly with) planning processes also have greater impact since this ensures that their findings are delivered at the right time when they can be used in new or revised policy.

12.5 Summing up

12.5.1 Take Home Messages

This chapter has considered how biodiversity and related observations generated and curated using the kinds of methods, structures and processes promoted by GEO BON and described elsewhere in this book, may be used within policy-making processes to influence decisions that impact on biodiversity, with a particular focus on indicators. Packaged and communicated in the right way, information on biodiversity change can have real policy impact regardless of scale.

Successful examples all rely on the kinds of engagement between scientists/data providers and policymakers described in this chapter, using data to provide a service to decision-makers, with the process and delivery mechanism defined with and by those decision-makers. Indicators, and assessments, are potentially very powerful policy tools, but in all cases it is crucial to begin with the questions, not the data and to ensure policy-maker buy-into the process. When it comes to the data, a lot can be achieved by first using what is there with an eye to how it can be improved and important gaps filled. This may be by mobilising currently inaccessible existing data before investing in new observing systems, and can involve multiple partners from a range of fields. Nevertheless, however good the data, information management can be a major bottleneck to progress in delivering timely, relevant and comprehensive products; ensuring adequate co-ordination of the process and management of the data, often through a centralised body, should not be overlooked.

12.5.2 Where to Go for More Information and Support

- The BIP provides various resources via its website www.bipindicators.net, including guidance documents, indicator fact sheets and national case studies as well as the Aichi Targets Passport, an annual indicator update also available as a smart phone app.
- The CBD (www.cbd.int) and the NBSAP Forum (www.nbsapforum.net) both include resources for planning, including data and indicator use.
- IPBES and Future Earth both have working groups focusing on data, monitoring and indicators, the latter helping to define the scientific criteria for indicator development.

- NatureServe have developed a Biodiversity Indicators Dashboard (http://dashboard.natureserve.org) which showcases how global datasets can be disaggregated for national use, utilising creative visualisation methods to bring the data alive.
- GEO BON includes a cross-cutting working group on indicators, which draws representation from each of the other GEO BON working groups, as well as additional membership from relevant organisations and individuals worldwide. The group's objectives include:

  (a) Ensuring the GEO BON community of practice is aware of and able to respond to user needs, both in terms of information to support indicators and capacity to generate such information, at national, regional and global scales,
  (b) Incorporating biodiversity information and analyses from GEO BON into indicator-based policy products designed and delivered to meet user needs,
  (c) Linking GEO BON to existing initiatives that improve information delivery to policy users, such as the Biodiversity Indicators Partnership (BIP), and
  (d) Helping to communicate the value of GEO BON to end users.

References

Balmford, A., Crane, P., Dobson, A., Green, R. E., & Mace, G. M. (2005). The 2010 challenge: Data availability, information needs and extraterrestrial insights. *Philosophical Transactions of the Royal Society B: Biological Sciences, 360*, 221–228.

Biodiversity Indicators Partnership. (2011). *Guidance for national biodiversity indicator development and use* (40 pp). Cambridge, UK.: UNEP-WCMC. http://www.bipindicators.net/LinkClick.aspx?fileticket=brn%2FLxDzLio%3D&tabid=157

Bubb, P. (2013). Scaling up or down? linking global and national biodiversity indicators and reporting. In B. Collen, N. Pettorrelli, J. E. M. Baillie, & S. M. Durant (Eds.), *Biodiversity monitoring and conservation: bridging the gap between global commitment and local action* (464 pp). Oxford, UK: Wiley-Blackwell.

Butchart, S. H. M. (2008). Red List indices to measure the sustainability of species use and impacts of invasive alien species. *Bird Conservation International, 18*, 245–262.

Butchart, S. H. M., Walpole, M., Collen, B., van Strien, A., Scharlemann, J. P. W., Almond, R. E. E., et al. (2010). Global biodiversity: Indicators of recent declines. *Science, 328*, 1164–1168.

Chenery, A., Plumpton, H., Brown, C. & Walpole, M. (2013). *Aichi targets passport* (2013 ed., 90 pp). Cambridge, UK: UNEP-WCMC. Download the Aichi targets passport indicator factsheets here: http://www.bipindicators.net/LinkClick.aspx?fileticket=-YdK–Zxv4g%3d&tabid=349)

Collen, B., Loh, J., Whitmee, S., McRae, L., Amin, R., & Baillie, J. E. M. (2009). Monitoring change in vertebrate abundance: The living planet index. *Conservation Biology, 23*, 317–327.

Collen, B., & Nicholson, E. (2014). Taking the measure of change. *Science, 346*, 166–167.

Collen, B., Ram, M., Zamin, T., & McRae, L. (2008). The tropical biodiversity data gap: Addressing disparity in global monitoring. *Tropical Conservation Science, 1*, 75–88.

DEFRA. (2013). *UK biodiversity indicators in your pocket 2013: measuring progress towards halting biodiversity loss* (47 pp). London, UK: Department for Environment, Food and Rural Affairs. http://jncc.defra.gov.uk/pdf/BIYP_2013.pdf

Dixon, M. J. R., Loh, J., Davidson, N. C., Beltrame, C., Freeman, R., & Walpole, M. (2016). Tracking global change in ecosystem area: The wetland extent trends index. *Biological Conservation, 193*, 27–35.

European Union. (2011). *The EU biodiversity strategy to 2020* (28 pp). Luxembourg: Publications office of the European Union. http://ec.europa.eu/environment/nature/info/pubs/docs/brochures/2020%20Biod%20brochure%20final%20lowres.pdf

Geijzendorffer, I. R., Regan, E., Pereira, H. M., Brotons, L., Brummitt, N. A., Gavish, Y., Haase, P., et al. (2016). Bridging the gap between biodiversity data and policy reporting needs: An essential biodiversity variables perspective. *Journal of Applied Ecology, 53*, 1341–1350. doi:10.1111/1365-2664.12417

Genovesi, P., Butchart, S. H. M., McGeoch, M. A., & Roy, D. B. (2013). Monitoring trends in biological invasion, its impact and policy responses. In: B. Collen, N. Pettorelli, J. E. M. Baillie & S. M. Durant (Eds.), *Biodiversity monitoring and conservation: bridging the gap between global commitment and local action* (pp. 138–158). Cambridge, UK: Wiley.

GEO BON. (2015). *Global biodiversity change indicators* (14 pp). Leipzig: GEO BON.

Han, X., Smyth, R. L., Young, B. E., Brooks, T. M., de Lozada, A. S., Bubb, P., et al. (2014). A biodiversity indicators dashboard: Addressing challenges to monitoring progress towards the Aichi biodiversity targets using disaggregated global data. *PLoS ONE, 9*(11), e112046.

IAASTD. (2009). *Agriculture at a crossroads: Synthesis report* (104 pp). Washington, D.C., USA: Island Press.

King, S., Wilson, L., Dixon, M., Brown, C., Regan, E., Blaney, R., et al. (2015). *Experimental biodiversity accounting as a component of the system of environmental-economic accounting experimental ecosystem accounts (SEEA-EEA)*. Cambridge, UK: UNEP-WCMC.

Leadley, P. W., Krug, C. B., Alkemade, R., Pereira, H. M., Sumaila U. R., Walpole, M., et al. (2014). *Progress towards the Aichi biodiversity targets: An assessment of biodiversity trends, policy scenarios and key actions* (500 pp). Technical Series No. 78, Montreal, Canada: Secretariat of the Convention on Biological Diversity.

Loh, J., Green, R. E., Ricketts, T., Lamoreux, J., Jenkins, M., Kapos, V., et al. (2005). The living planet index: Using species population time series to track trends in biodiversity. *Philosophical Transactions of the Royal Society B: Biological Sciences, 360*, 289–295.

McGeoch, M. A., Butchart, S. H. M., Spear, D., Marais, E., Kleynhans, E., Symes, A., et al. (2010). Global indicators of biological invasion: Species numbers, biodiversity impact and policy responses. *Diversity and Distributions, 16*, 95–108.

McGeoch, M. A., Spear, D., Kleynhans, E. J., & Marais, E. (2012). Uncertainty in invasive alien species listing. *Ecological Applications, 22*, 959–971.

Newbold, T., Hudson, L. N., Hill, S. L. L., Contu, S., Lysenko, I., Senior, R. A., et al. (2015). Global effects of land use on local terrestrial biodiversity. *Nature, 520*, 45–50.

Pereira, H. M., Ferrier, S., Walters, M., Geller, G. N., Jongman, R. H. G., Scholes, R. J., et al. (2013). Essential biodiversity variables. *Science, 339*, 277–278.

Sachs, J. D., Baillie, J. E., Sutherland, W. J., Armsworth, P. R., Ash, N., Beddington, J., et al. (2009). Biodiversity conservation and the millennium development goals. *Science, 325*, 1502–1503.

Secades, C., O'Connor, B., Brown, C., & Walpole, M. (2014). *Earth observation for biodiversity monitoring: A review of current approaches and future opportunities for tracking progress towards the Aichi biodiversity targets* (183 pp). CBD Technical Series No. 72, Montreal, Canada: CBD. http://www.cbd.int/doc/publications/cbd-ts-72-en.pdf

Secretariat of the CBD. (2010). *Global biodiversity outlook 3* (94 pp). Montreal.

Stephenson, P. J., Burgess, N. D., Jungmann, L., Loh, J. O'Connor, S., Oldfield, T., et al. (2015). Overcoming the challenges to conservation monitoring: Integrating data from in-situ reporting and global data sets to measure impact and performance. *Biodiversity, 16*, 2–3.

Tittensor, D. P., Walpole, M., Hill, S. L. L., Boyce, D. G., Britten, G. L., Burgess, N. D., et al. (2014). A mid-term analysis of progress towards international biodiversity targets. *Science, 346*, 241–244.

UK National Ecosystem Assessment. (2011). *Synthesis of the key findings* (85 pp). Cambridge: UNEP-WCMC.

United Nations. (2013). *The millennium development goals report 2013* (60 pp). New York, USA: UN. http://www.un.org/millenniumgoals/pdf/report-2013/mdg-report-2013-english.pdf

Walpole, M., Almond, R. E. A., Besancon, C., Butchart, S. H. M., Campbell-Lendrum, D., Carr, G. M., et al. (2009). Tracking progress toward the 2010 biodiversity target and beyond. *Science, 325*, 1503–1504.

WWF. (2012). *Living planet report 2012: Biodiversity, biocapacity and better choices* (160 pp). Gland, Switzerland: WWF. http://www.un.org/millenniumgoals/pdf/report-2013/mdg-report-2013-english.pdf

Chapter 13
Case Studies of Capacity Building for Biodiversity Monitoring

Dirk S. Schmeller, Christos Arvanitidis, Monika Böhm,
Neil Brummitt, Eva Chatzinikolaou, Mark J. Costello, Hui Ding,
Michael J. Gill, Peter Haase, Romain Julliard, Jaime García-Moreno,
Nathalie Pettorelli, Cui Peng, Corinna Riginos, Ute Schmiedel,
John P. Simaika, Carly Waterman, Jun Wu, Haigen Xu
and Jayne Belnap

Abstract Monitoring the status and trends of species is critical to their conservation and management. However, the current state of biodiversity monitoring is insufficient to detect such for most species and habitats, other than in a few localised areas. One of the biggest obstacles to adequate monitoring is the lack of local capacity to carry out such programs. Thus, building the capacity to do such monitoring is imperative. We here highlight different biodiversity monitoring

D.S. Schmeller (✉)
Department of Conservation Biology, Helmholtz Center for Environmental Research—UFZ,
Permoserstrasse 15, 04318 Leipzig, Germany
e-mail: ds@die-schmellers.de

D.S. Schmeller
ECOLAB, Université de Toulouse, CNRS, INPT, UPS, Toulouse, France

C. Arvanitidis · E. Chatzinikolaou
Hellenic Centre for Marine Research (HCMR), Institute of Marine Biology, Biotechnology
and Aquaculture, P.O. Box 2214, 71003 Heraklion, Crete, Greece
e-mail: arvanitidis@hcmr.gr

E. Chatzinikolaou
e-mail: evachatz@hcmr.gr

M. Böhm · N. Pettorelli · C. Waterman
Institute of Zoology, Zoological Society of London, Regent's Park, London NW1 4RY, UK
e-mail: monika.bohm@ioz.ac.uk

N. Pettorelli
e-mail: nathalie.pettorelli@ioz.ac.uk

C. Waterman
e-mail: carly.waterman@zsl.org.uk

N. Brummitt
Plants Division, Life Sciences Department, Natural History Museum, Cromwell Road,
London SW7 5BD, UK
e-mail: n.brummitt@nhm.ac.uk

© The Author(s) 2017
M. Walters and R.J. Scholes (eds.), *The GEO Handbook on Biodiversity
Observation Networks*, DOI 10.1007/978-3-319-27288-7_13

M.J. Costello
Faculty of Science, Marine Science, Leigh Science Building, 160 Goat Island Rd, Leigh
0985, New Zealand
e-mail: m.costello@auckland.ac.nz

H. Ding · C. Peng · J. Wu · H. Xu
Nanjing Institute of Environmental Sciences, Ministry of Environmental Protection, Nanjing
210042, People's Republic of China
e-mail: nldinghui@qq.com

C. Peng
e-mail: cuipeng1126@163.com

J. Wu
e-mail: wujun@nies.org

H. Xu
e-mail: xhg@nies.org

M.J. Gill
Environment Canada, 91780 Alaska Highway, Whitehorse, YT Y1A 5B7, Canada
e-mail: mike.gill@polarcom.gc.ca

P. Haase · J.P. Simaika
Department of River Ecology and Conservation, Research Institute Senckenberg,
Clamecystraße 12, 63571 Gelnhausen, Germany
e-mail: peter.haase@senckenberg.de

J.P. Simaika
e-mail: john.simaika@senckenberg.de

P. Haase
Faculty of Biology, Department of River and Floodplain Ecology, University of
Duisburg-Essen, Universitätsstr. 5, D-45141 Essen, Germany

R. Julliard
Muséum National d'Histoire Naturelle, Département Ecologie et Gestion de la Biodiversité,
UMR 7204 MNHN-CNRS Centre d'Ecologie et des Sciences de la Conservation, 55 rue
Buffon, 75005 Paris, France
e-mail: julliard@mnhn.fr

J. García-Moreno
ESiLi, Het Haam 16, 6846 KW Arnhem, The Netherlands
e-mail: jaime_gm@yahoo.com

U. Schmiedel
Biocentre Klein Flottbek and Botanical Garden, University of Hamburg, Ohnhorststrasse 18,
22609 Hamburg, Germany
e-mail: ute.schmiedel@uni-hamburg.de

C. Riginos
Teton Science School, 700 Coyote Canyon Road, Jackson Hole, WY 83001, USA
e-mail: Corinna.Riginos@tetonscience.org

J. Belnap
U.S. Geological Survey, Southwest Biological Science Center, Moab, UT 84532, USA
e-mail: jayne_belnap@usgs.gov

efforts to illustrate how capacity building efforts are being conducted at different geographic scales and under a range of resource, literacy, and training constraints. Accordingly, we include examples of monitoring efforts from within countries (Kenya, France, and China), within regions (Central America and the Arctic) and larger capacity building programs including EDGE (Evolutionarily Distinct and Globally Endangered) of Existence and the National Red List Alliance.

Keywords Monitoring capacity building · Citizen science and volunteers · Key biodiversity areas · Public awareness raising

13.1 Introduction

Monitoring the status and trends of species is critical to their conservation and management. However, the current state of biodiversity monitoring is insufficient to detect such for most species and habitats, other than in a few localised areas. One of the biggest obstacles to adequate monitoring is the lack of local capacity to carry out such programs. Thus, building the capacity to do such monitoring is imperative. The capacity building needed includes finding stable lead institutions with adequate funding and staff, and the training of local personnel in the development of new programs of biodiversity monitoring where gaps currently exist and linking together existing and planned observation systems around the world (Henry et al. 2008). In addition, common technical standards among all monitoring efforts are needed, such that data from the huge variety of national monitoring programs, regional biodiversity observation networks (RBONs), and global non-governmental organisations (NGOs) can be combined into coherent data sets that allow the assessment of status and trends of biodiversity across the world (Hoffmann et al. 2014). However, it is clearly unrealistic to attempt simultaneous monitoring of all species in all places. Therefore, the first step is to identify and focus on topical priorities for a given monitoring effort (e.g., species or habitats of special concern) to determine where to focus initial programs. These efforts can then be subsequently complemented by other local, national and regional monitoring activities determined by national responsibilities (e.g., Schmeller et al. 2008a, b, 2012) or topical priorities (Henle et al. 2013).

In this chapter we highlight different biodiversity monitoring efforts to illustrate how capacity building efforts are being conducted at different geographic scales and under a range of resource, literacy, and training constraints. Accordingly, we include examples of monitoring efforts from within countries (Kenya, France, and China), within regions (Central America and the Arctic) and larger capacity building programs including EDGE (Evolutionarily Distinct and Globally Endangered) of Existence and the National Red List Alliance, which are capacity building frameworks similar in structure to Group on Earth Observations-Biodiversity Observation Network (GEO BON).

13.2 Building Monitoring Capacity at the Country Scale

13.2.1 Kenya

Northern Kenya is a region of high biodiversity conservation value. It is a hotspot of mammalian diversity in Africa and is critical habitat for several endangered and threatened large mammal species. This region is largely under tribal communal tenure, and the rich flora and fauna of the region are threatened by heavy and continuous grazing by domestic livestock, which has caused moderate to severe land degradation over large areas (Georgiadis et al. 2007).

Several NGOs, the largest of which is the Northern Rangelands Trust (NRT), have been working to promote wildlife conservation and better land stewardship in this region. In 2008, NRT and other scientists developed a simple protocol useable by community members to monitor rangeland condition, which, in turn, determines the amount and quality of forage for wildlife. These methods were based upon pastoralists' traditional knowledge and monitoring practices that were also be scientifically defensible. Through a series of conversations at which the NGOs, scientists, and community members were present, all participants agreed that the indicators currently used by pastoralists to make management decisions were insufficient, as they were largely focused on livestock condition and grass conditions affected by rainfall. All agreed that a new protocol was needed to capture information about long-term (>3 year) trends in rangeland functionality. However, these methods would need to be simple so that they could be learned and used by community members with low literacy.

With support from the United States Agency for International Development (USAID-East Africa), a team of scientific experts was assembled, as well as a diverse advisory panel, to guide this project. This team conducted focus group interviews with community members representing the Maasai, Samburu, Borana, Afar, and Karyu ethnic groups in Kenya and Ethiopia. In these interviews, researchers asked community members to describe the changes in their rangelands they had observed over the last several decades. The responses were markedly similar across diverse social and ecological contexts: increased bare ground, decreased perennial grasses, increased woody vegetation, and increased soil erosion. These observations also matched the general indicators of degradation that have been observed by scientists working in rangelands around the world.

Researchers then used this traditional knowledge to adapt an existing set of rangeland monitoring methods (originally developed and thoroughly tested in the western United States) for the East African context. Notably, the US methods were simplified by basing them on a stick 1 m long (rather than measuring tapes or other manufactured tools) and created a single graphical data sheet on which data could be collected by circling icons. The intent was that these methods could be used by people who could not read or write but could count and recognise simple icons (e.g., a picture of a shrub representing a sample point with shrub cover).

At the same time, it was agreed with NRT and other partners that a person with a higher level of education would assist with the design for data collection (e.g., selecting sites and deciding upon number of replicates), training of the pastoralists, and the analysis and interpretation of results. To guide this process, a 50+ page manual was developed to explain the core methods, as well as the key steps to designing, implementing, and describing results from a monitoring program to the communities. This guidebook included numerous photos and graphics to illustrate messages in an accessible format.

The first version of the data collection protocol and the guide book were then subject to rigorous and critical feedback from a diverse community of development and conservation practitioners, community members, and scientists through written feedback, a round-table discussion, and field testing of the methods. Following this process, the data sheets were modified to eliminate some areas of confusion. The guidebook was expanded to a 100 page document with additional case studies and appendices. In 2010, Version II of Monitoring Rangeland Health: A Guide for Pastoralist Communities and Other Land Managers in Eastern Africa was printed in full colour on plastic coated paper to provide a durable and appealing product.

Following release of the printed product, several training sessions were held for the staff of NRT and other key partner organisations. These 'train the trainer' sessions helped to further identify areas of confusion and ensure that the senior staff members were competent in the core field methods (Fig. 13.1).

Guidance was also provided to NRT senior staff as they decided upon a design for their monitoring program. They were then accompanied to the field when they introduced the new monitoring methods in the initial five communities. The NRT

Fig. 13.1 Training sessions of the Northern Rangelands Trust in Kenya. *Source* Jayne Belnap

staff trained and supervised community members to collect data, while members of the development team were present to answer any additional questions and observe issues that arose as these methods were implemented in diverse field conditions.

After one year of using these methods, senior NRT staff returned to the development team to discuss possible further simplifications to the core data collection protocol. In their experience, certain data were confusing to collect and certain other data were not necessary to answer the management questions set out by community managers. Some of the simplifications requested were easy to agree upon based on their objectives, while others were considered oversimplifications by the scientific team. Through an extended conversation, new, streamlined protocols that all parties could agree upon were developed and the data sheets modified accordingly (Fig. 13.2).

The NRT staff were very pleased with this product as a data collection tool that could be used by community members, but remained concerned that outside "experts" would be required to analyse and interpret long-term trends, decreasing the likelihood that these results would feed back into community decision-making. They therefore asked for assistance in developing a simple Access database tool— with the objective that a literate community member could enter the data into the database and generate simple graphical reports (e.g., trends in key indicators, such as perennial grass cover, over time and over sites). The development team has since been working with NRT staff and a hired database developer to create this tool.

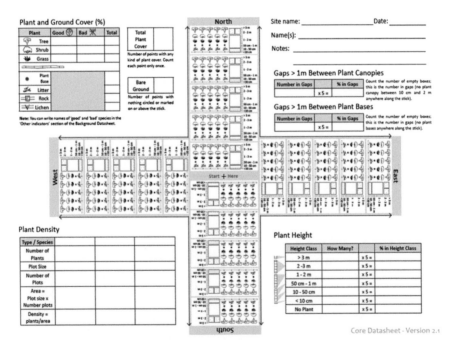

Fig. 13.2 Data collection sheet developed by the Northern Rangelands Trust in Kenya. *Source* Jayne Belnap

13.2.2 *France*

To a large extent, biodiversity monitoring in France depends on Citizen Science and thus its success depends on the efficacy of project implementation at a national scale. France has many small national NGOs (e.g., French Bird Life group Ligue de Protection des Oiseaux with about 40,000 members), and these NGOs alone were not able to launch an ambitious national monitoring initiatives. Thus, success of the national effort required a larger institution that could lead and coordinate these smaller efforts. Fortunately, the French National Museum of Natural History (MNHN) was and is able to perform this role. In addition, this research institution is under Ministries who are officially designated role to document the state of biodiversity in France. Therefore, the success of the national monitoring effort in France required both building capacity among the smaller NGOs, as well as a government institution with stable funding and staff to provide a common monitoring framework.

Integrating the individual Citizen Science-driven monitoring programs started with MNHN hosting the bird ringing project, a very successful a partnership between professional researchers and volunteers. The MNHN also launched a classical Breeding Bird Survey (BBS), again using the Citizen Science provided by the smaller NGOs [together, these formed the STOC (Le Suivi Temporel des Oiseaux Communs − Vigie Nature = Temporal Survey of Common Birds) program]. Several factors facilitated the success of these schemes: (1) coupling the BBS to the already-existing ringing efforts; (2) focusing on common birds, thus avoiding an overlap with other projects concerned with all species (and thus proportionally more rare species); (3) the concomitant emergence of biodiversity indicator based monitoring schemes, based on Mean Species Abundance, (e.g., the Breeding Bird Survey), ensuring considerable political interest in these schemes, and (4) this effort coincided with citizen science becoming fashionable in France.

With the success of the national bird monitoring effort, two new schemes were put in place to develop capacity for monitoring other taxa. The first was based on the same logic as the Breeding Bird Survey: training and motivating skilled amateurs to collect data following a protocol and a sampling design for butterflies, bats plants, and dragonflies. The second program was developed to train and coordinate efforts of the general public to monitor garden butterflies and snails, bumblebees, birds, flower-dwelling insects, and wild plants in cities. All these schemes are coordinated by the same scientific team based at the MNHN, but each also relies on a specific NGO partner, which is dedicated to the success of the (specific) participant network. The NGO partner trains participants, ensures that each new scheme capitalises directly on preceding experience to maximise the chance for joint data analysis, and coordinates all efforts with MNHN. Several characteristics have made this effort an outstanding success: (1) different monitoring schemes for different species groups were integrated from the beginning, (2) the same research group was involved in citizen science schemes for both skilled amateurs and the general public, (3) strong involvement of researchers in designing citizen science projects maximised the chances that the future database will allow sophisticated and robust statistical analyses of the large datasets and (4) development and training of personnel in the use of technology to facilitate reliable data where detection or identification is difficult (e.g., ultrasound recording for bats, photography for spiders and flower-dwelling insects). Combined, such an organisation is very cost-effective, as researchers are keen to commit themselves to such projects to ensure good quality data.

Three additional planned projects will expand the span of national biodiversity monitoring through citizen science. The first is to implement more experimental approaches in addition to simple counts. This opens scientific opportunities while keeping participants motivated to participate by offering renewed ways of looking at biodiversity. The second project expands citizen science to primary and secondary schools. Together with educational staff, students will collect 'real' data (i.e., using the same protocol as the general public) during school time (i.e., as part of the official school program). The ultimate target is for half of French children to experience citizen science at least once during their schooling. A third project is to work with farmers, encouraging them to monitor biodiversity on their farms, an approach that worked well with 400 farmers in the first year of its existence. Working with local farmer organisations was essential in this effort, but the combined launched of this project with the MNHM guaranteed its integrity and longevity.

13.2.3   China

The first ecological research station was developed in China in 1978. In 1988, the China Ecosystem Research Network (CERN) was established by the Chinese Academy of Sciences. It includes 39 research stations that include ecosystems as diverse as farmland, forest, grassland, desert, marsh, lake, ocean, and cities. At each station, the structure, function and dynamic patterns of the ecosystem, as well as abiotic measures, is recorded. In 2003, the China Forestry Ecosystem Research Network (CFERN) was established, followed by the China Wetland Ecosystem Research Network (CWERN) and the China Desert Ecosystem Research Network (CDERN). In 2005, CFERN, CWERN, and CDERN were combined to form the China National Ecosystem Research Network (CNERN). In addition to CNERN, the Ministry of Agriculture also launched the China National Grassland Resource Monitoring project (CNGRM) in 2005. CNGRM focuses on the monitoring of vegetation growth, productivity, and utilisation, as well as the effects of disaster conditions and construction projects on biodiversity. Planning, establishing, and running such a large network of sites has required stable institutions and funding to build the needed scientific capacity to organise and guide this effort, as well as to recruit and train volunteers to help with data collection. For example, in 2012, over 4500 volunteers and professionals were trained and organised to measure 8000 plots in over 450 counties in 23 Chinese provinces. Most of the field workers were technical staff of local administrations, complemented by students from agricultural universities.

While ecosystem level monitoring has been on-going for some time, species level monitoring networks were established in 2005. These have included a coastal waterbird project, covering nearly all the wetland sites along coastal areas of the East and South China Sea (Fig. 13.3), utilizing about 150 volunteers. In 2011, the Nanjing Institute of Environmental Sciences, under the umbrella of the Ministry of Environmental Protection of China, began a bird and amphibian monitoring effort. Again, this project was planned, developed, and coordinated by scientists, but the field surveys were conducted by workers from colleges, research institutes, museums, and other organisations that were recruited and trained by the scientists.

Despite these efforts, more biodiversity monitoring is needed at the national level, covering major ecosystems and indicator species, to better understand the status and trends of biodiversity in China. In 2014, the Ministry of Environmental Protection of China began developing a biodiversity monitoring network. Based on the existing organisations and frameworks, this effort will start with further development of a bird and amphibian monitoring network (Fig. 13.3). However, the end goal is a comprehensive national monitoring scheme that covers mammals, birds, reptiles, amphibians, fishes, and vascular plants, with a special focus on endangered species. Thus, similar to France, successful monitoring efforts in China

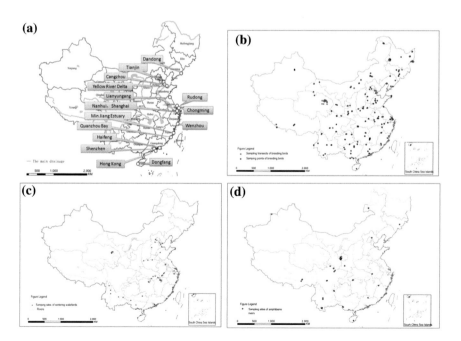

Fig. 13.3 Distribution of biodiversity monitoring sites in China for **a** the coastal wetland monitoring scheme of China, **b** the breeding bird survey of China, **c** the wintering water bird survey of China, and **d** the amphibian monitoring sites of China. *Source* Cui Peng

have relied on a stable source of funding (e.g., the government) and the scientific expertise needed to plan, develop and execute successful monitoring protocols, as well as to recruit and train volunteers for data collection.

13.3 Building Monitoring Capacity at the Regional to National Scale

13.3.1 Pan-Arctic

Arctic ecosystems and the biodiversity they support are experiencing growing pressure from various stressors (e.g., development, climate change, contaminants). However, established research and monitoring programs have remained largely uncoordinated, and therefore lack the ability to effectively monitor, understand and report on biodiversity trends at the pan-Arctic or regional scale (MA 2005). The maintenance of healthy arctic ecosystems is a global imperative, as the Arctic plays a critical role in the Earth's physical, chemical and biological balance. A coordinated and comprehensive effort for monitoring Arctic ecosystems is needed to facilitate effective and timely conservation and adaptation actions.

While all Arctic states, as well as a number of non-Arctic states and organisations, conduct monitoring of various elements of Arctic biodiversity, the lack of coordination has limited their geographic, thematic, and temporal scope and are not evenly spread across the Arctic. In particular, northern areas of Canada, Greenland and Russia have very limited biodiversity monitoring, whereas areas in northern Scandinavia, the Bering Sea, Aleutian Islands and Iceland have relatively intense, on-going biodiversity monitoring (Fig. 13.4) and in many cases, long-term datasets. Given that the area in question is 32 million km^2 (three times the size of Europe) and is comprised of largely remote and extreme ecosystems, it is not surprising that current biodiversity monitoring efforts are seen as inadequate. Indeed, recent issues regarding state finances and priorities have made it more difficult to sustain even existing efforts.

The current situation facing the Arctic demands a well-designed, scaled, pan-arctic, ecosystem-based approach that not only identifies trends in biodiversity, but also identifies underlying causes of these trends. It is critical that this information be made available, as plans for adaptation and mitigation need development, which ultimately depend on rigorous, integrated and efficient monitoring programs that have the power to detect change within a 'management' time frame.

To meet these challenges, the Conservation of Arctic Flora and Fauna (CAFF) Working Group of the Arctic Council launched the Circumpolar Biodiversity Monitoring Program (CBMP) in 2005. The CBMP is working with over 80 global partners in building the capacity to expand, integrate and enhance existing Arctic biodiversity monitoring efforts, thus facilitating more rapid detection, communication and response to significant trends and pressures. It is strategically linked to a

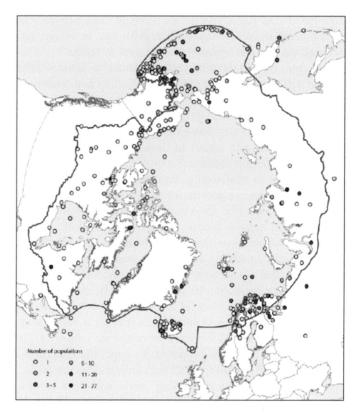

Fig. 13.4 Distribution of population time series data across the Arctic, 1951–2010. The number of populations per location is indicated by *colour*. The *red line* indicates the core area of the Circumpolar Biodiversity Monitoring Program. *Source* Arctic climate impact assessment (2005)

number of international conventions and programs including the Convention on Biological Diversity (CBD), UNEP's Biodiversity Indicators Partnership and is one of four recognised regional Biodiversity Observation Networks of the GEO BON initiative. In order to connect the diversity of biodiversity observing networks, such as scientific field stations, community-based monitoring programs, theme-based monitoring networks (e.g., caribou) operating at different scales across the Arctic, the CBMP is establishing four Expert Monitoring Groups representing major Arctic themes (Marine, Freshwater, Terrestrial and Coastal). Each group, representing a diversity of disciplines, is tasked with developing and implementing pan-arctic integrated biodiversity monitoring plans with a focus on harmonizing existing monitoring networks and methodologies, as well as rescuing and aggregating existing data to establish historical baselines. To date three (Marine, Freshwater and Terrestrial) monitoring plans have been developed and are being implemented. To facilitate effective reporting and data management, the CBMP has developed a number of headline indicators targeting CBD 2020 Targets (McRae et al. 2012) as well as a web-based data portal (Arctic Biodiversity Data Service www.abds.is; http://www.youtube.com/watch?v=ONdmmIcuqNE) that is improving metadata and the discovery, access, and interpretation of data to bridge the science-policy gap. The output from the CBMP biodiversity monitoring plans is being used to populate both the ABDS and headline indicators which, in turn, are being translated into policy-targeted reports sub-national, national and regional (e.g., annual Arctic Report Cards http://www.arctic.noaa.gov/reportcard/index.html) in scope to facilitate more timely and effective decision-making.

The CBMP's approach with regard to developing the pan-Arctic monitoring plans is to first identify the sub-national, national, and regional reporting mandates of governments relevant to biodiversity. Implementation of the monitoring plans will only be sustained if they can provide information that supports these mandates. The next steps are to (1) develop conceptual models of the ecosystems/biomes in question and through an iterative process, (2) identify the priority focal ecosystem components (FECs) and processes that should be monitored, (3) identify the biodiversity variables (e.g., attributes) and specific parameters for these FECs that should be measured, and (4) identify common methodological approaches and sampling frameworks for measuring these parameters. In most cases, these monitoring plans focus on ways to harmonise existing methodologies and data, rather than standardise them, as many monitoring networks have been using particular methodologies and data standards for many years and are unlikely to change their approach. Where new variables are proposed, an opportunity to adopt a specific methodology and standard is available and recommendations are made.

The implementation of these monitoring plans will then involve building national or thematic teams that allows for a hierarchical and efficient approach to connect to the many monitoring networks and practitioners operating at different scales across the Arctic. The resulting data will be mostly managed within existing national biodiversity data centres which the ABDS can access via the Internet. This will provide an efficient means to access the most up to date information on various aspects of Arctic biodiversity status and trends. The development of the CBMP grew from a concept team of ten in 2005 to over 80 organisations representing hundreds of scientists and local resource users around the Arctic in 2013 with a concurrent ten-fold increase in its budget over this time-span.

13.3.2 *Central America*

Key Biodiversity Areas (KBAs) are sites of global importance for conservation of biodiversity. As an important tool in conservation planning, these areas are considered critical for the persistence of one or more globally threatened species and are identified using simple standardised occurrence data (Eken et al. 2004). This approach has been modified to identify important sites for different taxonomic groups, such as important bird areas (developed by BirdLife—www.birdlife.org); plant areas (Anderson 2002), butterfly areas (van Swaay and Warren 2003), or for specific biomes (e.g., freshwater key biodiversity areas; Darwall et al. 2011). This is an international effort currently led by the International Union for the Conservation of Nature (IUCN).

This same process is also being used at the national scale. In Central America, the NGO Conservation International identified a local partner institution in each of the seven countries and built up their capacity to identify KBAs in each country. This required gathering occurrence data for species assessed as Vulnerable, Endangered, or Critically Endangered by the IUCN Red List, using as many sources as possible globally. This, in turn, provided each country with a more complete set of records for species of interest. With the help of Conservation International, each institution involved then defined the KBAs for these species, while also defining monitoring objectives. All collected information is entered into a global database—the World Biodiversity Data Base, currently managed by BirdLife International—and shared with the relevant authorities of the different countries and regional bodies (e.g., Central American Commission for the Environment and Development).

13.4   Building Monitoring Capacity at the National and Global Scales

13.4.1   *International Union for the Conservation of Nature Programs*

The IUCN Red List of Threatened Species is arguably one of the most important tools for global nature conservation. It facilitates the flow of biodiversity information from the point of data collection to policy- and decision-makers around the world, and drives research into biodiversity conservation. However, such lists at the national scale are also needed and this will require capacity building within each country. The production of National Red Lists is gaining momentum (Miller et al. 2007; Zamin et al. 2010), as they provide a valuable tool for national decision-making and priority-setting for conservation, while also aiding national reporting against global biodiversity targets such as the Aichi Biodiversity Targets (Szabo et al. 2012). Integrating global and national Red List processes provides one way in which we can dramatically increase taxonomic and geographic coverage of the IUCN Red List (Rodriguez 2008), while national assessments can benefit greatly from the expertise provided by the IUCN Red List in conducting species assessments and utilising the data for maximum conservation benefit, thus building capacity and filling gaps from global to regional scales and vice versa (see also Schmeller et al. 2014).

Previous efforts promoting National Red Lists have primarily focused on the establishment of an online resource for Regional and National Red Lists (see www. nationalredlist.org). The website provides a hub for collating National Red Lists and Action Plans from around the world and at present holds in its online library more than 140 National Red Lists and Action Plans from 33 countries. It also contains a species database with more than 85,000 national assessments for over 60,000 species. This has also sparked gap analyses in the global coverage of National Red Lists, which can help in prioritising funding for National Red List development (Zamin et al. 2010).

While this centralisation has provided a good starting point for promoting National Red Lists and Action Plans, information flows much more effectively if regional hubs are established as well, again requiring the recruitment and training of local personnel. Efforts have recently been made to formalise the development of National Red Lists (NRLs) under the auspices of the ZSL and IUCN via the newly created National Red List Alliance (NRLA). The Alliance builds on previous steps taken at international congresses, e.g., the IUCN World Conservation Congress 2012 in Jeju, South Korea, to promote National Red Lists and seek discussion with interested partner organisations. The efforts culminated in a workshop held at the Zoological Society in London in 2013 to formalise the partnership. The aim of the Alliance is to create a regional network of National Red List partners committed to supporting the development and implementation of National Red Lists, with members acting as focal points for National Red Lists. Initial regional hubs will be established in China, Brazil and South Africa, all of which have at present a strong presence in terms of National Red List development. It is hoped that the strengthened network and the joint ownership of the National Red List website will facilitate capacity building where needed as well as ensure the sharing of national-level tools and data for compilation and analysis of National Red Lists, promote the upload of additional species assessments to the National Red List website, and promote sharing of resources for training and best practice for National Red List development.

These programs show that in order to achieve a monitoring programme that is global in coverage but sufficiently resolved spatially to allow national decision making (Scholes et al. 2012), building networks with strong capacity is of utmost importance. Networks function most efficiently via key individuals or institutions, which are inter-linked with each other and which can act as regional hubs to interact with many individuals/institutions at a more local level.

13.5 The Conservation Leadership Programme and EDGE of Existence Programme

13.5.1 The Conservation Leadership Programme

An important aspect of building capacity is the training of young professionals. The Conservation Leadership Programme (CLP—www.conservationleadership programme.org) is an example of a partnership that has been training conservationists for over two decades and now has a broad network of alumni around the world. The Zoological Society of London (ZSL) has also helped establish a number of global networks for the purposes of monitoring and conservation, mostly through training conservationists in wildlife management techniques.

13.5.2 EDGE of Existence Programme

The ZSL is also actively working towards conservation of evolutionarily distinct and globally endangered (EDGE) species. As the reason most of the top 100 EDGE species are generally ignored conservation is because they occur in countries where the capacity for effective monitoring and conservation is lacking (primarily developing countries). Hence, to conserve EDGE species, ZSL has developed a grassroots capacity building program that focuses on training and supporting aspiring in-country conservation scientists to establish larger-scale conservation projects in which long-term monitoring of status and threats to EDGE species and their habitats will be a major component. This is accomplished by providing two years funding to study a priority EDGE species, attend regional training courses, study online modules in relevant topics, receive one-to-one support from a scientific advisor based at ZSL or a partner organisation, and at the end, attend a two-week conservation leadership course that includes modules on leadership and management, project planning, monitoring and evaluation, facilitation and conflict resolution, communication skills, proposal writing, and writing for publication, as well as technical one-to-one clinics to help with analysing data and writing up the results of their Fellowship projects (for details on the fellowship, see www. edgeofexistence.org/conservation/become_fellow.php). To date, EDGE has supported 41 EDGE Fellows focusing on 39 EDGE species in 26 countries since 2007, with 97 % of the Fellows still working in conservation and research. These early successes of the EDGE of Existence program suggest that targeted funding and training of key individuals can help to build lasting networks for conservation. The approach taken here is via a structured Fellowship program, although this approach can easily be adapted to target key institutions in key countries to expand the global biodiversity monitoring network.

13.6 Cross Cutting Lessons from Capacity Building Efforts

The need to monitor biodiversity is becoming increasingly apparent to scientists, the public, and policy makers around the world. However, because the highest level of biodiversity is located in developing countries where funding, biodiversity institutions and formal skills may be limited, capacity building is especially urgent in these regions. Whereas each nation and region has its specific challenges in meeting this need, there are several cross-cutting lessons that can be learned from past and current efforts. These include: Identify a well-known organisation with a stable source of adequate funding and staffing to design and implement the monitoring program. This may be a governmental organisation, a network of NGOs, or other groups. Where possible, this organisation should already have an established network throughout the region.

- Develop and maintain a simple, efficient internal organisational structure with roles of team members clearly defined;
- Include people of influence ('champions') within national governments and funding sources in the program's governance structure;
- Utilise a close partnership among practitioners, NGOs, and scientists in designing and implementing the monitoring program. Identify talented and driven individuals and key institutions to receive training or take part in the monitoring network. Include a mix of young and senior experts in the design and implementation of the program components to ensure program integrity and continuity;
- Develop a very focused and detailed implementation plan that is closely adhered to during the development of the program;
- Focus on harmonizing existing monitoring capacity and information to increase statistical power and cost-efficiencies, rather than attempting to impose new standards on existing programs or developing new monitoring programs;
- Ensure a reasonable allocation of funds for data management, analysis and reporting as well as on communications (program promotion) and fundraising;
- Ensure outputs are relevant to both decision-makers and funders while maintaining scientific integrity, the latter achieved through having engagement with not only scientists and local peoples, but also decision-makers and funders in the program design;
- Heighten public interest and concern to increase funding opportunities;
- Start small and build support for the program through the promotion of early results that showcase the value-added gains for scientists and decision-makers in coordinating and scaling existing biodiversity monitoring efforts;
- Develop a program 'brand' that positions the program as the source for credible information on biodiversity monitoring for the region and one whose endorsement is sought after by other monitoring networks;
- Practice regular communication from conception through design, implementation, testing, and refinement of the monitoring effort. This iterative process is key to building trust between the scientific development team and the intended audience of practitioners;
- Where possible, integrate efforts with an international political body that can provide a more formal mechanism and mandate for engaging scientists, local

peoples and other monitoring networks and coordinate or take advantage of the publicity associated with other events, such as the International Polar Year of 2007/08;

- Provide continuing support and mentorship to individuals and organisations involved in the monitoring effort, especially in developing countries.

References

Anderson, S. (2002). *Identifying important plant areas: A site selection manual for europe, and a basis for developing guidelines for other regions of the world.* Plantlife International: London.

Arctic climate impact assessment. (2005). *Arctic climate impact assessment.* Cambridge: Cambridge University Press.

Darwall, W. R., Holland, R. A., Smith, K. G., Allen, D., Brooks, E. G., Katarya, V., et al. (2011). Implications of bias in conservation research and investment for freshwater species. *Conservation Letters, 4,* 474–482.

Eken, G., Bennun, L., Brooks, T. M., Darwall, W., Fishpool, L. D., Foster, M., et al. (2004). Key biodiversity areas as site conservation targets. *BioScience, 54,* 1110–1118.

Georgiadis, N., Olwero, J., Ojwang, G., & Romanach, S. (2007). Savanna herbivore dynamics in a livestock-dominated landscape: I. Dependence on land use, rainfall, density, and time. *Biological Conservation, 137,* 461–472.

Henle, K., Bauch, B., Auliya, M., Külvik, M., Pe'er, G., Schmeller, D. S., et al. (2013). Priorities for biodiversity monitoring in Europe: A review of supranational policies and a novel scheme for integrative prioritization. *Ecological Indicators, 33,* 5–18.

Henry, P. Y., Lengyel, S., Nowicki, P., Julliard, R., Clobert, J., Celik, T., et al. (2008). Integrating ongoing biodiversity monitoring: Potential benefits and methods. *Biodiversity and Conservation, 17,* 3357–3382.

Hoffmann, A., Penner, J., Vohland, K., Cramer, W., Doubleday, R., Henle, K., et al. (2014). The need for a biodiversity policy support process—Building the European contribution to a global Biodiversity Observation Network (EU BON). *Nature Conservation, 6,* 49–65.

MA (Millennium Ecosystem Assessment). (2005). *Ecosystems and human well-being: Synthesis.* Washington, DC: Island Press. http://www.millenniumassessment.org/documents/document.356.aspx.pdf

McRae, L., Böhm, M., Deinet, S., Gill, M., & Collen, B. (2012). The Arctic Species Trend Index: Using vertebrate population trends to monitor the health of a rapidly changing ecosystem. *Biodiversity, 13,* 144–156.

Miller, R. M., Rodriguez, J. P., Aniskowicz-Fowler, T., Bambaradeniya, C., Boles, R., Eaton, M. A., et al. (2007). National threatened species listing based on IUCN criteria and regional guidelines: Current status and future perspectives. *Conservation Biology, 21,* 684–696.

Rodriguez, J. P. (2008). National Red Lists: The largest global market for IUCN Red List categories and criteria. *Endangered Species Research, 6*, 193–198.

Schmeller, D. S., Gruber, B., Bauch, B., Lanno, K., Budrys, E., Babij, V., et al. (2008a). Determination of national conservation responsibilities for species conservation in regions with multiple political jurisdictions. *Biodiversity and Conservation, 17*, 3607–3622.

Schmeller, D. S., Gruber, B., Budrys, E., Framsted, E., Lengyel, S., & Henle, K. (2008b). National responsibilities in European species conservation: A methodological review. *Conservation Biology, 22*, 593–601.

Schmeller, D. S., Maier, A., Bauch, B., Evans, D., & Henle, K. (2012). National responsibilities for conserving habitats—A freely scalable method. *Nature Conservation, 3*, 21–44.

Schmeller, D. S., Evans, D., Lin, Y. P., & Henle, K. (2014). The national responsibility approach to setting conservation priorities—Recommendations for its use. *Journal for Nature Conservation, 22*, 349–357.

Scholes, R. J., Walters, M., Turak, E., Saarenmaa, H., Heip, C. H., Tuama, E., et al. (2012). Building a global observing system for biodiversity. *Current Opinion in Environmental Sustainability, 4*, 139–146.

Szabo, J. K., Butchart, S. H. M., Possingham, H. P., & Garnett, S. T. (2012). Adapting global biodiversity indicators to the national scale: A Red List Index for Australian birds. *Biological Conservation, 148*, 61–68.

van Swaay, C., & Warren, M. (2003). *Prime butterfly areas in Europe: Priority sites for conservation.* Ministry of Agriculture, Nature Management and Fisheries.

Zamin, T. J., Baillie, J. E., Miller, R. M., Rodriguez, J. P., Ardid, A., & Collen, B. (2010). National red listing beyond the 2010 target. *Conservation Biology, 24*, 1012–1020.

Printed in the United States
By Bookmasters